Harvesting Timber Crops

Harvesting Timber Crops

A.E. WACKERMAN

2002
BIOTECH BOOKS
Delhi - 110 035

First Indian Edition 2002

ISBN 81-7622-077-9

Published by : **BIOTECH BOOKS**
1123/74, Tri Nagar,
DELHI – 110 035
Phone: 7109765
e-mail: biotechbooks@yahoo.co.in

Showroom : 4762-63/23, Ansari Road, Darya Ganj,
NEW DELHI - 110 035
Phone: 3245578, 3244987

Printed at : **Chawla Offset Printers**
New Delhi – 110 052

PRINTED IN INDIA

PREFACE

This text has been written to describe and discuss the principles involved in correctly harvesting timber crops, with special emphasis on the aspects related to forestry. Many phases of the actual operations concerned in harvesting are of an engineering nature, such as the construction of roads, railroads, bridges, and camp structures, and specifications for mechanical tools and machines such as power saws, tractors, and cable skidders. While these subjects are covered, detailed consideration of those items that relate particularly to engineering operations is omitted. Rather, the forestry aspects of harvesting are emphasized to assist foresters in obtaining a thorough understanding of this undertaking, which is so important in forestry.

Presentation of specific costs of various operations in the several forested regions of the United States has been purposely avoided because of the rapid fluctuations of costs and returns, even in one locality. Changing economic conditions, variable topography, ground conditions, weather, and the accessibility of the timber being harvested prevent standardizing costs. Likewise, detailed descriptions of tractors, cable skidders, power saws, trucks, railroad rolling stock, and other harvesting equipment have been omitted because of the rapidity with which new models appear on the market.

Although such material, concerning costs and descriptions of equipment, is not given in this text, it is the author's opinion that much of it should be presented to forestry students in class or as reference reading by instructors in the subject. The necessary data can be obtained from current literature in technical and trade journals and from bulletins that are published from time to time by trade associations, educational institutions, and public agencies. Catalogues issued at frequent intervals by equipment manufacturers offer an excellent and authentic source of factual data concerning current models of tools and machines used in harvesting.

All such material is readily available and it is assumed that instructors in harvesting would prefer to present such data, kept currently up to date, rather than have it included in a text, where it would soon become obsolete. The principles of harvesting remain the same, however, regardless of the changes occurring in methods of performing the various harvesting steps. Thus, all skidding by mobile power units is similar whether the actual work is performed by animals or Diesel-powered tractors.

The very helpful and valuable cooperation of W. D. Hagenstein, Fred C. Simmons, George Banzhaf, Emanuel Fritz, W. C. Hammerle, Ed. R. Linn, and Thomas J. Orr, Jr., in contributing sections on regional harvesting practices for Chap. 18 is thoroughly appreciated. These men, each widely recognized as an authority on forestry and harvesting practices in his region, have generously contributed of their time and knowledge to make this text much more useful than would have been possible otherwise. Likewise, the generous cooperation of equipment manufacturers, the American Forest Products Industries, Incorporated, and numerous individuals and agencies in making available photographs and figures is gratefully acknowledged. Without such illustrative material the usefulness of this text would be greatly impaired.

If this text will assist instructors in presenting the harvesting of timber crops more logically and more thoroughly than would otherwise be possible, and if students find it helpful in grasping the significant features of harvesting and their relationship to forestry, then the author will feel amply repaid for the effort expended in preparing this material. If it should also prove to be of use to foresters and others already engaged in forestry activities or in conducting actual harvesting operations, then he will indeed be gratified.

A. E. Wackerman

CONTENTS

INTRODUCTION

A Tribute to American Loggers

Prior to the present century, lumbering always occurred in unsettled regions far in advance of agricultural expansion and industrial concentration in rapidly growing cities of the East and Middle West, both of which required tremendous volumes of construction lumber and other timber products to build structures for a growing nation. Loggers were required to pioneer in developing transportation facilities for extracting logs from the forest and delivering them to sawmills. At first it was animal power and stream driving, and later steam skidders and forest railroads that were used to supply the ever-expanding volume of lumber needed; now it is chiefly tractors and trucks.

Life formerly was rugged in the logging camps established in remote forest areas, often inaccessible during many months of the year. Nevertheless, the logs came out of the forest as all of the mighty obstacles of nature were overcome by loggers who have never known the phrase, "it can't be done."

The tasks accomplished in felling and transporting large, bulky, and tremendously heavy logs in great volume over raw terrain gave birth to the fabulous Paul Bunyan, the heroic mythical logger of the North Woods, who with his big blue ox, Babe, was credited with mighty feats of logging. The tales related by imaginative loggers during long winter nights around bunkhouse stoves have become truly American folklore and have been perpetuated in print by noted authors. Although no one has actually seen Paul Bunyan (the closest approach being to know someone who knew him), he is a very real person in the minds of all true loggers and he still is their idol and inspiration.

Logging has changed greatly in many respects from the hearty days of the nineteenth century but it still requires men of the same caliber. Our ever-expanding nation has reached its last frontiers, and no unexploited forest areas remain in this country for the industry to move into. From a migratory industry, formerly moving from cutout areas to virgin forests farther on, the industry has, of necessity, settled down and a constantly increasing number of large and small operators are changing from the practices of exploitation to those of conservation.

Now, instead of isolated bachelor camps far in the forest, timber harvesters usually live in comfortable, well-appointed, and modern camps, in many instances with their families. Such camps are, in effect, villages, with churches, schools, and up-to-date stores. Machine shops and garages have replaced barns and stables, and sawing with power saws, skidding with tractors, and hauling by trucks are the modern expression of timber harvesting. The rough life and work of early loggers thus has been transformed and workers as skilled as any in modern factories are needed. In addition, the spirit of Paul Bunyan is essential in all whose jobs are to perform or to conduct harvesting operations in forests.

The transformation of forest harvesting from a "cut out and get out" policy to modern operations in keeping with the growing of successive timber crops on the same lands requires no less ingenuity and woodsmanship than formerly, and those who successfully harvest timber crops today have as much claim to kinship with Paul Bunyan as the rugged individuals of former years who logged under difficult living and working conditions but without the necessity, at the same time, of preparing for future cuts on the same land.

Loggers often work in inclement weather, harassed by insects, and confronted by uncertain ground conditions. Unlike factory and shop workers, whose environment, including light, temperature, and humidity, is closely controlled, loggers must be adaptable and efficient under varying conditions of topography, terrain, volume of timber per acre, transportation methods, and weather. The work, however, is healthy and invigorating to sturdy men and each day's task is variable enough to be interesting. With fully modernized camps that provide for comfortable and pleasant living, the American logger is no longer the "lumberjack" of former years.

Loggers, past and present, merit a tribute by foresters to their skills, knowledge, initiative, and indomitable spirit, which are so essential in harvesting timber crops. Without them there could be but little forestry practiced.

A. E. Wackerman

PART ONE

CONSIDERATIONS PRELIMINARY TO ACTUAL HARVESTING

Before analyzing the principles and practices involved in actually harvesting timber crops, consideration should be given to those factors bearing on present procedures and to those factors requiring attention prior to the initiation of harvesting operations proper. Part One of this text, therefore, is devoted to an analysis of the place of harvesting in the field of forestry, to a review of the historical background of timber harvesting in America and the development of present methods, to considerations involved in planning a timber harvest, and to the necessary provisions that must be made for workers engaged in conducting the various steps in the harvesting operation. In subsequent chapters those principles and procedures involved in the actual cutting and transporting phases of harvesting are considered.

PART ONE

CONSIDERATIONS PRELIMINARY TO ACTUAL HARVESTING

[illegible]

CHAPTER 1

TIMBER CROPS—THE GOAL OF FORESTRY

Forest owners who go to the trouble and expense of raising timber crops do so because they expect a return commensurate with the expenditure of money, time, and effort necessary to grow them. When this return is adequate, the owner usually is interested in growing more crops and in maintaining his forest in a productive condition. If there is no net return or if the return is so small as to be unattractive, then the forest owner loses interest in forestry and abandons the enterprise. Profitable uses for tree crops grown in managed forests, therefore, constitute the keystone of forestry.

The return obtained by forest owners usually is in money or the equivalent of money, and most private owners desire their returns in this form. Public forests, on the other hand, being the property of the people as a whole, frequently are managed to yield intangible returns that are difficult to measure in terms of money but which constitute the most important forest crops obtained. All forest yields, therefore, whether commercial products or intangible values, are in the nature of crops, when a forest is consciously managed to obtain these ends.

Forest products such as wood, resins, saps, extractives, and wood derivatives are generally the forest crops of greatest value. Forestry usually is recommended for both private and public forests on the basis that the nation will need wood and other tree products of commercial importance.

Values other than wood substance or tree products, such as the pleasing views afforded by timberlands and the effectiveness of forest cover against soil erosion, are not readily salable. In fact, it is difficult to establish ownership over many such values. All forest assets, both tangible and intangible, are highly desirable; but if there is no substantial income from commercial forest products, the other assets can be maintained only by the use of funds obtained from other productive enterprises. Hence, tree crops are important both from the standpoint of producing revenue to maintain the forest in a financially sound condition and to create and maintain the physical effects that comprise the other, less tangible values.

FORESTS MUST BE USED TO BE CONSERVED

Non-use of forests results in waste. In virgin forests the increment of the living trees is offset by losses due to the death of overmature trees.

Thus, forests constitute a resource that must be fully used to be truly conserved.

Locking up forests against use is not wise forestry any more than is denuding the land of trees, because both practices result in timber losses. When forests are destroyed in logging, all growth is lost or deferred for decades; when forests are not cut, the mature trees gradually deteriorate and finally die, which also is a loss, although a less conspicuous one. Forestry, when correctly applied, anticipates the decadence and death of trees so that they are cut while still sound, thereby placing them in the channels of trade before their usefulness is impaired. By harvesting no more wood volume from a forest than grows, a forest can be cut annually or periodically forever.

The harvesting of timber crops according to correct forestry methods is the key to most other forestry operations in that it makes possible the prompt removal of mature trees, releasing the space they occupied for the development of new trees. When there are markets for trees of small size, thinning and improvement cuttings can be made, and crowded small trees and trees of poor form for the final crop can be harvested with much benefit to the remaining stand. This constant improvement, harvesting, and regeneration process keeps a forest productive and provides income for the owner, labor for workers, and business for the community.

IMPORTANCE OF HARVESTING OPERATIONS IN FORESTRY

Harvesting timber crops may logically be considered the principal operation in forestry, in that the culmination of all forestry effort is the mature tree. It has been grown for the purpose of creating a valuable commodity but its value can be realized only when it is harvested. All silvicultural operations, for example, are directed towards the development of timber stands as rapidly as is consistent with the quality of wood substance desired, and, whenever possible, only those species that have potential commercial value are permitted to grow to maturity. Thinning, pruning, cleaning, and other cultural operations in forests are designed to hasten the growth and development and to improve the quality of marketable crops of trees, so harvesting the crop is an achievement resulting from, or made possible by, past forestry efforts.

Harvesting time in the forest, as on the farm, is the time of cash income. Thus a forest that does not have regular harvesting periods is not a forestry enterprise in the usual business sense. Continual harvesting operations day by day and year by year are much more desirable in the forestry business and its administrative matters than infrequent or periodic harvesting.

Many forestry practices are performed with the ax, so it is usually desirable and practical to perform them while harvesting operations are being conducted. Therefore, when harvesting is done annually, other forestry practices can be carried on to advantage each year. Continuous harvesting, therefore, is to be preferred over periodic harvesting, because it permits better labor arrangements, makes possible more orderly marketing of products, provides continuous income, enhances the opportunity for silviculture, and results in a forestry enterprise which fits in best with other business affairs.

Harvesting is also extremely important to a forest because if incorrectly done, or if done without regard for the future, it probably will cause serious damage to the forest or destroy its productivity for a great many years. Unwise cutting can cause much breakage and damage in felling and skidding, both to the trees being cut and to those left to grow. Unwise cutting can also result in less useful species taking over the land to the exclusion of the more valuable ones. Stands of the valuable species may also be depleted to the point where they will grow limby and rough and be greatly reduced in quality and, hence, in value.

So many desirable effects can be achieved by well-planned harvesting methods, correctly conducted, and such serious consequences can result when cutting in a forest is done without regard to the future that harvesting is one of the most important operations in forestry, entirely aside from the fact that it provides an income to carry on the business.

CORRELATION OF TIMBER CROPPING TO OTHER FOREST VALUES

A vegetative soil cover such as is provided by forests is recognized almost universally as highly desirable when the land is not to be used for agriculture or pasture. A canopy of tree crowns over the soil and the forest ground litter covering it prevent erosion and loss of the top soil as well as the silting of reservoirs and fertile bottom lands. A forest cover over the earth's surface also tends to regulate the runoff of rainfall and, hence, stream flow. Forests give cover for game and birds and, by filtering the flow of water from the surface, maintain clear streams suitable for fish. Forests are said also to ameliorate the severity of climatic extremes and, of course, they add beauty to this world in which we live, so they are greatly to be desired for their aesthetic value.

When a forest is maintained in a productive condition—well managed, rapidly growing, and correctly utilized—then all these other benefits or values that may also be considered as crops are, in the great majority of cases, adequately maintained. Forests managed primarily for the production of commercial timber crops will, without further provision, afford places for birds to nest and food for game. Such forests will also serve to protect the soil and ameliorate the climate and control rainfall runoff. Furthermore, aesthetic values will be preserved and people will be kept permanently employed harvesting the timber crops from the forest and manufacturing commercial products from the raw materials thus provided.

Occasionally other uses are more important than the timber crops that are, or could be, produced in a forest. Forests on steep slopes subject to severe erosion, forests on shifting sands that have caused stabilization of the soil, and forests that are located at critical places in the control of water runoff are much more valuable for protection purposes than for timber production, and no cutting or harvesting should be done or permitted in such forests when there is the least danger that the protection values will be lost or impaired.

Some scenic forest areas, such as are represented by certain national parks and national monuments, have beauty due to trees or forests that is far beyond the intrinsic value of the wood or tree products their forests could yield, and the forest crop then is chiefly aesthetic.

Although forest values other than the tree crops may be the most important, it is entirely possible in many cases to maintain these values and still harvest tree crops in such a way and in such amounts that the other values are not only undisturbed but frequently are enhanced by such cutting. Timber crops even in such forests, therefore, should not be neglected whenever the harvesting of wood crops can be correlated to the other requirements to obtain the greatest forest yields.

If the wood obtained from tree crops were the only valuable attribute of forests, the opportunity to practice forestry would be greatly restricted. The fact that forests do create several forms of wealth, both tangible and intangible, gives the forest owner or manager the opportunity of realizing fully upon the productive resources of his versatile forests. Multiple use of forests includes one or more of the uses discussed briefly in the following paragraphs combined, usually, with the production of timber crops.

Wild Life. The interest of many forest owners in game production does not necessarily prevent the harvesting of timber crops, so long as cutting operations are planned and conducted with due regard to maintaining the necessary game cover. Frequently owners of hunting properties would not be interested in, or concerned with, their timber if it were not for the favorable influence of certain types and conditions of forest stands on the propagation of game and on the opportunity for hunting. Trees, as they grow in size, cause heavier shade and alter the ground-cover conditions to the point where wild-life populations change in kind and amount. By harvesting a predetermined amount of the timber in a given stand, however, the conditions desired in a forest can be maintained, if present, or created, if lacking. Harold Titus[1] has shown that virgin, unutilized forests offer a poor habitat for game of all kinds and that only at the forest edge is game

[1] Titus, Harold, "Timber and Game—Twin Crops," American Forest Products Industries, Inc., 1945.

abundant. Harvesting timber crops according to sound forestry principles creates multiple edges in forests, and game animals and birds become much more abundant. Timber-harvesting operations and game management, therefore, can readily be integrated so that both the game crops and the wood crops can be harvested.

The timber-crop income obtained from game-producing, timber-growing forests usually is in dollars, while the game-crop income is in the form of personal enjoyment, unless the hunting rights are leased or sold. In the Southeastern states of South Carolina, Georgia, and Florida, there are many forest properties managed intensively for the primary crop of wild life, but which also yield valuable timber crops of pulpwood, saw logs, naval stores, poles and piles, veneer bolts, and other tree products.

Soil Protection and Stream Control. When protection of the soil against erosion, either by wind action or by the action of water, is the primary purpose of forest ownership and management, the value of such protection usually is greater than the use of the trees for wood. When soil is in such a critical position as to require forest cover to hold it in place, no cutting for harvesting purposes should be undertaken without definite knowledge that no harmful erosion will develop. If erosion in a substantial amount should develop as a result of such cutting, the cost of again stabilizing the soil would make the harvesting of the tree crops excessively expensive. Furthermore, protection forests, as soil- and water-holding forests are called, are usually on such steep slopes or in such inaccessible places that the tree crops are difficult or impossible to harvest, so when such conditions exist, the harvesting of timber crops usually is foregone.

Many of the national forests in the eastern section of the United States were purchased under the Weeks Law authorization, primarily for the protection of the watersheds of navigable streams. This does not mean that all such areas in the eastern national forests are to be considered as protection forests requiring this limitation on cutting, as noncritical areas are included where timber-harvesting operations are carried on without disturbing the more limited protection forest areas.

Recreation. The value of a forest area for recreation is fully as great when the timber crops are correctly harvested as when they are undeveloped. Certain areas of special beauty, owing to timber or other natural qualities of the location, should not be disturbed; but for the purpose of permitting the public or, in private estates, selected individuals to enjoy the outdoors by riding, driving, swimming, picnicking, and camping, the correctly utilized forest is fully as attractive and often more comfortable for use than is the forest in a so-called wild condition. Where, for particular reasons, it is desirable to have the effect of a virgin forest in certain areas, as along highways or around commercial recreational establishments, commercial cutting can be kept out of view to the casual traveler. Thus

on only a minimum area of the forest will a compromise between harvesting timber crops and recreational use be necessary.

Generally speaking, however, so long as forest crops are harvested according to forestry principles, the recreational use of the forest will not be jeopardized. In fact, by opening up roads and installing bridges and other facilities required for harvesting operations, the greater will be the opportunity for recreational use because of greater accessibility. In some sections there has been conflict of interest in using forests, chiefly public forest areas, between so-called "wild lifers," who would keep the areas in wilderness, and foresters interested in growing timber needed by the nation. Those favoring wilderness areas, however, do so more from the standpoint of keeping them as museum specimens of past forest conditions available only to a fortunate few, so actually true recreational use and harvesting do not conflict.

Grazing. Grazing and forestry are two strictly commercial land uses that supplement each other in the full use of some forests. Those who would graze cattle or other livestock in forest areas to the entire exclusion of other uses and those who would grow trees to the exclusion of grazing and other uses are more prejudiced than logical, in that both groups neglect opportunities to practice better land use by developing both businesses or by managing one and leasing the other in which they are not primarily interested.

In most forests there are quantities of grass at one stage or another in the development of the tree crops. Unutilized, this grass is not only wasted but constitutes a fire hazard in dry periods. Controlled grazing adds to the forest income and reduces fire hazards, both of which are of material advantage to the owner. Here again, however, conflicts can readily develop between those who would prevent any other use but their own, chiefly from prejudice.

Certain kinds of grazing, it is entirely true, are incompatible with forestry, such as sheep and goat grazing under almost any conditions. These animals browse even in the presence of abundant grass forage and by so doing kill back or altogether destroy young trees and even large saplings. Even cattle grazing is harmful to hardwood forests in sapling-age classes, as the cattle browse on the small trees and destroy them. When the trees have reached such size that they are beyond reach of the animals, however, limited grazing is possible. Hogs, also, are particularly destructive in longleaf pine stands, as they uproot seedlings and saplings to consume the fleshy roots, killing the trees. In certain hardwood forests, however, swine thrive on the mast, or tree seeds, and do no harm.

Generally, cattle grazing can be correlated with timber growing in pine forests better than in hardwood forests. Such combined forest use, therefore, holds more promise of profitable application in the Southern pine

forests and in the pine forests of the West than in other forest regions. In these areas natural grasses are abundant and of sufficient nutrient value to sustain cattle, and when overgrazing is prevented, the cattle do not seriously damage pine reproduction.

Grazing is an important use of Western forests, as about half of the area of Western national forests is used for grazing domestic livestock. In 1940, national forest ranges provided forage for 1,218,874 cattle, 77,011 horses, 17,749 swine, 4,958,685 sheep, and 9,616 goats.[1] The combination of grazing and forestry in the pine forests of eastern Washington has recently been recognized as having possibilities of profit, and the forest lands of at least one progressive lumber company in that region are definitely being managed to grow and harvest both forage and timber crops.[2] R. S. Campbell and H. H. Biswell, in discussing forest grazing in the South,[3] state that cattle grazing and timber growing are generally recognized as an ideal combination in the southern coastal plain. Although there are many problems involved in coordinating grazing with growing timber crops, the added income from such combined use is sufficient to justify forest owners in harvesting both their forage and their timber crops.

The crop yields of forests, therefore, are many and the most successful forestry results when forests are managed for multiple use. However, since the operations involved in harvesting the timber crops require the use of specialized equipment and are of most economic importance and have such critical effects on the forest itself, this text deals only with the harvesting of timber crops.

HARVESTING CREATES EMPLOYMENT

More labor is required, even in managed forests, in harvesting operations than in any other forestry activity. Even in European forests that receive careful cultural treatment far in excess of that given most American forests, nearly 50 per cent of all labor used in forests is required for harvesting operations. Table 1 shows, in man-days, the employment in Bavarian state forests; and logging, which involved only cutting and delivery of products to roadsides, provided by far the most employment.

Reynolds[4] has shown that labor requirements in harvesting and milling timber cut from the Crossett Experimental Forest in southeast Arkansas,

[1] "Forests and People: Report of the Chief of the Forest Service," U.S. Department of Agriculture, 1941.

[2] Arnst, Albert, Forage crops on J. Neils cutover land, *Timberman*, vol. XLVI, No. 12, 1945.

[3] "Better Management on Southern Coastal Forest Range," *Forest Serv., U.S. Dept. Agr.* AIS-17, 1945.

[4] Reynolds, R. R. Employment requirements of well-managed timberland in the lower south, *Jour. Forestry*, vol. 42, No. 12, 1944.

TABLE 1. EMPLOYMENT IN BAVARIAN STATE FORESTS, BY MAIN ACTIVITIES, 1913, 1920, AND 1929–1933, IN THOUSANDS OF MAN-DAYS

Activity	1913	1920*	1929	1930	1931	1932	1933	
							Regular	Emergency
Logging	2,410	3,955	2,200	2,012	1,538	1,520	1,674	
Planting and Cultural operations	1,187	815	1,024	720	584	240	154	510
Road construction	727	1,183	662	601	336	186	141	810
Other	263	947	517	492	451	241	171	2
Total	4,587	6,900	4,403	3,725	2,909	2,187	2,240	1,322
Total man-days per acre of forest	2.2	3.3	2.1	1.8	1.4	1.0	1.1	0.6

* A large amount of emergency work to relieve unemployment was done in 1920.
Source: Sparhawk, W. N., "Forests and Employment in Germany," *U.S. Dept. Agr., Cir.* 471, 1938.

are as shown in Table 2. From his studies he has determined that the 1,600-acre experimental forest provides 460 man-days of labor per year for harvesting overhead work and 4,319 man-days of labor per year for direct harvesting work such as cutting, skidding, loading, hauling, and milling or processing. He shows further that woods labor alone in connection with harvesting comprises about 50 per cent of the total and that 150 acres of managed forest will provide one man-year of woods labor continously.

TABLE 2. MAN-HOUR REQUIREMENTS FOR HARVESTING AND PROCESSING FOREST PRODUCTS FROM THE CROSSETT EXPERIMENTAL FOREST

Sawlogs............31.18 man-hours per 1,000 ft, International ¼-in. scale, of logs made into finished lumber

Pulpwood...........21.28 man-hours per standard cord of wood made into kraft pulp or paper

Acidwood...........19.84 man-hours per standard cord of wood manufactured into charcoal and recoverable chemicals

In the United States as a whole, the lumber industry employs, on the average, about 650,000 regular woods workers.[1] In addition, part-time employment is provided for some 2,500,000 farmers who cut timber from their farms for their own use or for industrial consumption. In the sawmills and woodworking factories, including pulp and paper mills, another 650,000 people are employed. The United States census for 1935 shows that 20 industries are directly dependent upon forests for raw materials and that the forest-industry group ranks fourth among the 16 important manufacturing industries in number of wage earners employed.

[1] "Living and Forest Lands," *U.S. Dept. Agr. Misc. Pub.* 388, 1940.

CONTINUING NEED FOR TIMBER CROPS

Wood is one of the basic raw materials needed by modern civilization This fact was amply demonstrated during the Second World War, when forest products of practically every kind were classed as critically urgent materials. Methods of utiiizing wood have been enlarged with the introduction of new processes, such as some comparatively recent pulping methods, and ways of dissolving wood pulp for rayon articles. The plastic field is looking to wood substance as a source of cellulose, and the use of wood in this and similar processes is rapidly growing.

Even the use of lumber, one of the earliest of forest products, has been maintained in tremendous volume in the United States and the world as a whole. The annual consumption of lumber in the United States averages between 25 and 30 billion board feet. Only in periods of economic depression, when construction is deferred and industrial activity is low, has the annual consumption of lumber in this country fallen below these levels. Wood is still the most important construction material and there is no reason to believe that wood will be replaced in modern times in appreciable amounts by any other building substance.

The rapid growth of the pulp and paper industry in the United States has greatly increased the use of pulpwood, which has now become an important product of the forest in most sections of the country. In 1899 only 1,600,000 cords of pulpwood were harvested in the United States, but now the annual cut is around 17,000,000 cords and it is still increasing. There is no reason to believe that the use of wood for pulping purposes will ever decline except for periods of short duration. Based on trends in the recent past, in fact, it is likely that pulpwood will become an even more important timber crop.

Veneer and plywood manufacturing, a major industry in recent years, also seems destined to expand still further and furnish markets for important quantities of wood from American forests. The development of waterproof bonding glues has made possible a much wider use of plywood than formerly, causing an increasing demand for veneer and plywood products. The growth of the plywood industry is well illustrated by the production record for the Douglas fir segment of the industry. In 1925 only 153 million square feet were produced, but the output steadily increased to 1,800 million square feet in 1945.

Tree crops furnish other well-established industries with basic raw materials. Poles and piles are almost entirely of wood and there is no indication that they will be supplanted by other materials. Current needs for poles and piles from American forests average around 2 million pieces, or 33 million lineal feet, annually. The railroads require quantities of wooden crossties, as no substitute has ever been found equal to a well-made wooden tie that has been treated with preservative compounds. Railroads and industrial users require in the neighborhood of 48 million hewed and sawed crossties and nearly 140 million feet of sawed switch ties per year to maintain railroad tracks and for new construction.

Table 3 shows the current annual cut of wood from American forests in cubic feet for various commodities and indicates the relative importance of each.

TABLE 3. TIMBER REMOVED FROM COMMERCIAL FORESTS OF THE UNITED STATES IN 1944, IN MILLIONS OF CUBIC FEET

All timber*

Item	Total	Softwoods	Hardwoods
Lumber	6,711	4,778	1,933
Fuel wood	2,203	719	1,484
Pulp wood	1,306	1,163	143
Hewed ties	363	153	210
Fence posts	220	59	161
Veneer logs	392	165	227
Mine timbers (round)	225	40	185
Cooperage	174	40	134
Shingles	72	72	—
Other	516	159	357
Total	12,182	7,348	4,834

* Includes saw-timber and cordwood trees. The volumes, necessarily shown in cubic feet, include the tops (cordwood size and larger) of the softwood saw-timber trees, and the tops and limbs of the hardwood saw-timber trees. Bark is not included.

Source: "Gaging the Timber Resource of the United States," U.S. Dept. Agr. Forest Service, 1946.

Tree crops, therefore, are assured of adequate markets, since the growing and harvesting of timber crops is the foundation of a large and permanent segment of American industry. Harvesting, because it is the forestry operation that places forest products in the channels of trade, brings income to forest owners, governs other forestry activities, furnishes large-scale healthy employment, and requires highly specialized methods and equipment, is therefore one of the most important operations in forestry.

CHAPTER 2

THE DEVELOPMENT OF FOREST INDUSTRIES IN AMERICA

Wood has been used by mankind since the dawn of civilization. From the days of primitive construction and the crude implements of antiquity to today's modern buildings, vehicles, and articles of commerce and trade, wood always has been important, and undoubtedly will always be a prominent item in our civilized needs.

Use of wood from the forests is known since legendary times. Charcoal burning is perhaps the oldest use of wood by man, aside from the gathering of faggots and forest debris for wood fires and rough lean-tos. Perhaps of considerable magnitude in the aggregate, these early uses for timber nevertheless were not large in the sense that an organized industry harvested timber on an important scale.

Timber-cutting operations assumed prominence when the needs of mills or other processing plants for particular timber products grew to proportions large enough to require special efforts to supply them. Charcoal burners usually worked singly or in small groups and only intermittently. Their operations did not have the complexion of an industry. But when sawmills were built, organized timber-harvesting operations were necessary, and specialized equipment and methods were devised.

The use of wood constantly increased as civilization progressed and, in addition to the large amount of wood used at present for fuel and shelter, wood is, as has been noted, an important raw material for many industries. Pulp mills require millions of cords of wood each year for the manufacture of paper and fiberboard to supply the needs of our nation and others dependent upon us. Distillation of wood yields various products such as turpentine, rosin, alcohol, charcoal, and acetic acid. Particular woods, such as chestnut, are treated to extract tannins or dyes. Ties, posts, poles, and piles are made directly from the trees and usually are treated with preservatives to ensure them of lasting qualities. Mine timbers of various types also constitute an important part of the volume of wood used today. Particularly choice portions of many trees are cut into veneer bolts and then rotary "peeled," sliced, or sawed to produce thin sheets of wood to be glued together into strong and attractive plywood. Some wood is ground into flour and used in linoleum and plastics, an important new wood product, and some is shredded to make excelsior.

Lumber is, however, still the most important purpose for which trees are cut, with respect to the volume of wood used each year. As a construction, crating, and fabricating material, the woods native to the United States have no peer, and sawmills dot the entire nation, wherever trees are growing. Timber harvesting, therefore, is constantly in progress to supply sawmills with logs, and other wood-using plants with their wood raw materials.

HISTORY OF LOGGING

Lumber manufacturing has had a long and colorful career as a pioneering and typically American industry. Starting in the Northeastern and Southern colonies, lumbering grew in magnitude as the country expanded and as export trade developed. From these points of origin the industry migrated into the Central states and then into the Lake states. Beginning in 1899, the South led in lumber production, until 1926, when the West assumed the lead and has maintained it ever since.

The first lumber manufacturing was done by hand-powered pitsawing, such as still is practiced in some primitive countries. This kind of operation requires two men, one above and the other below in a pit, to pull the long saw lengthwise through a log, sawing off one board at a time by a slow and laborious process. Naturally, no large supply of logs was needed at such mills, and only a few wagons and teams were necessary at most. Somewhat later, power mills of the Mulay type, similar to a hand-pitsawing setup, but with the saw blade actuated by geared wind or water power, came into general use. Later several saws were installed to form a gang saw, cutting several boards at once. Such mills could cut more logs per day and the problem of keeping logs at the mills began to assume importance.

There are records of European mills using wind or water power as early as 1322. In 1663 a power mill was installed in London, causing riots by the pit sawyers, who feared the expansion of such power mills would drive them from their jobs. Water-power mills in the United States naturally were along water courses; and since the timber being used floated, it was natural that logging first developed along the streams so that the logs could be floated to the sawmills.

During the early part of the Nineteenth century, steam power was developed and used in sawmills in the United States. Steam power created much greater productive capacity for sawmills and by 1869, according to the Ninth Census, steam power was used in about 50 per cent of the sawmills, the remainder being driven by water power. Steam power rapidly replaced water power and by 1904 the Thirteenth Census reported that 90 per cent of the sawmills were driven by steam power.

The daily and annual production of mills was rapidly multiplied as the application of steam power to heavier and larger machinery was perfected and as circular saws and, later, band saws were introduced. By the latter part of the Nineteenth century the sawmill had about reached its present state of perfection, except for certain recently introduced refinements. Hence, mass production in the lumber industry was achieved about 1890, while the industry was at its peak in the Lake states. Consequently logging became a mass-production operation, to keep mills cutting from 40,000 to 80,000 ft of lumber a day supplied with logs.

The shift in lumber production from one region to another is well illustrated by the records of leading lumber-producing states for past years. Between 1830 and 1840, for instance, Maine led the nation in the volume of lumber produced. From 1840 to 1850, New York was the leading lumber-producing state; and from 1850 to 1860, Pennsylvania was the most important state in lumber production. The leading states since then, listed in order of production at 10-year intervals in Table 4, show later shifts in regional lumber production.

TABLE 4. LEADING LUMBER PRODUCING STATES BY DECADES

Year	States
1869	Michigan, Pennsylvania, New York, Wisconsin, Indiana
1879	Michigan, Pennsylvania, Wisconsin, New York, Indiana
1889	Michigan, Wisconsin, Pennsylvania, Minnesota, Washington
1899	Wisconsin, Michigan, Minnesota, Pennsylvania, Arkansas
1909	Washington, Louisiana, Mississippi, North Carolina, Arkansas
1919	Washington, Louisiana, Oregon, Mississippi, Alabama
1929	Washington, Oregon, Mississippi, Louisiana, Alabama
1939	Oregon, Washington, California–Nevada, Alabama, Mississippi
1946	Oregon, Washington, California-Nevada, Alabama, North Carolina

The relative importance of the forest regions of the United States in lumber production since 1850 is shown in Table 5. Both the Northeast and the Lake states regions, although formerly important, now produce only a small amount of lumber as compared to the total. The South, however, although it lost the leading position to the West in 1926, is a close second. In 1946 the West produced 14.4 billion feet of lumber, or 40 per cent of the total; and the South 13.8 billion feet, or 38 per cent.

TABLE 5. PERCENTAGE OF LUMBER PRODUCED IN SECTIONS OF THE UNITED STATES

	1850	1870	1890	1899	1909	1919	1929	1939
Northeastern	54.8	37.8	19.8	16.3	11.7	7.5	3.3	3.9
Central	18.6	20.0	13.1	16.1	12.3	8.7	6.4	4.9
Southern	13.6	9.4	20.3	31.7	44.9	46.6	41.9	39.7
Lake states	6.3	24.4	34.6	24.9	12.3	7.8	4.8	3.1
Western	5.9	4.9	9.6	9.9	18.4	29.2	43.4	48.2
Other	0.8	3.5	2.6	1.1	0.4	0.2	0.2	0.2
Total U.S.	100.0	100.0	100.0	100.0	100.0	100.0	100.0	100.0

TABLE 6. LUMBER PRODUCTION BY YEARS SINCE 1909

Year	South, per cent	West, per cent	Other, per cent	Total, M bd ft
1909	44.8	18.4	36.8	44,509,761
1910	43.6	22.0	34.4	40,018,282
1911	43.1	22.7	34.2	37,003,207
1912	46.3	21.7	32.0	39,158,414
1913	47.7	23.7	28.6	38,387,009
1914	47.6	22.5	29.9	37,346,023
1915	48.6	22.0	29.4	37,011,656
1916	49.3	24.3	26.4	39,807,251
1917	47.9	27.8	24.3	35,831,239
1918	43.2	31.3	25.5	31,890,494
1919	46.6	29.1	24.3	34,552,076
1920	42.5	35.7	21.8	33,798,800
1921	50.2	30.2	19.6	26,960,864
1922	45.5	38.2	16.3	31,568,888
1923	44.3	39.1	16.6	37,165,505
1924	45.2	37.9	16.9	35,930,986
1925	44.7	39.6	15.7	38,338,641
1926	42.2	42.9	14.9	36,935,930
1927	41.9	43.8	14.3	34,532,420
1928	40.9	45.2	13.9	34,142,123
1929	41.9	43.4	14.7	36,886,032
1930	39.0	46.6	14.4	26,051,473
1931	36.2	50.7	13.1	16,522,643
1932	38.8	50.2	11.0	10,151,232
1933	41.3	48.9	9.8	13,961,134
1934	38.2	47.3	14.5	15,493,639
1935	40.0	46.6	13.4	19,538,731
1936	39.1	48.1	12.8	24,354,884
1937	39.6	47.8	12.6	25,996,857
1938	43.2	45.2	11.6	21,646,271
1939	39.7	48.2	12.1	24,975,474
Average.....	43.3	36.8	19.9	
Total.......				960,467,939

Source: *South. Lumberman*, May 15, 1941.

Table 6 shows how well the production of lumber from the South has been maintained as compared to other regions.

The building activity of a fast-growing nation caused the rapid expansion of lumber production, forcing the industry to tap new regions to obtain an adequate supply of logs. Thus, the industry has been a pioneering one, forging its way into the vast undeveloped forest areas of, first, the Northeast and later the Lake states, building mill towns and logging camps. Settlement and agriculture did not precede lumbering but followed it, to a limited extent, with unfortunate results in many localities. Numbers of settlers attempting to clear and cultivate cutover lands were left stranded as the industry declined and mills ceased operations, and as the communities dependent upon them disintegrated.

The movement of the lumber industry into the South differed from previous migrations in that the South was already settled and populated, largely by farmers and residents of rural communities. The inland forests of the South, however, were practically intact, except where farms and plantations had been developed on the better soils. Logging was conducted in much the same way that it had been in the Lake states except that an already available local labor supply made it unnecessary to import labor for mill and woods work except for key positions. Logging camps soon became family camps as farmers and other established residents found industrial employment more attractive than small-scale farming, but did not want to abandon their families.

Because lumbering did not decline seriously in the South, it became the basis of economic development and has assisted in building up the region. Many important Southern communities have their origins in sawmilling and logging although today other industrial activities may have become more prominent.

The tremendous volume of excellent timber in the Western forests attracted the industry to that section on a large scale as soon as transcontinental railroads were completed, which afforded outlets to Eastern markets for lumber. The lumber industry early became and still is of great economic importance throughout the forested areas of the West. Because of heavy demands for lumber and readily accessible forests of large-sized high-quality species, the lumber industry of the West soon perfected large-scale production methods both in the mills and in the woods.

Although lumbering preceded agricultural and other developments, except mining, it has not resulted in serious land-use problems in cutover areas of the West, as was the case in the Lake states. Recognizing that the West was the last forest frontier in America, a large percentage of the industry in that region has retained ownership of its cutover forest lands and is taking the necessary steps to grow timber and place its operations on a sustained yield. The lumber industry of the West, therefore, is well established as a permanent and beneficial factor in the economic life of that important region.

The migration of lumber manufacturing is now largely over, as there are no unexploited forest regions in the United States and because lumbermen have learned that forests can be grown if the principles of forestry are followed in harvesting and in caring for growing crops of trees. Many sawmills both in the South and in the West and some in the Northeast and in the Lake states are now permanent because plans were adopted in time to correlate timber growing with cutting, and timber exhaustion has been prevented.

The lumber industry thus has settled down and many lumbermen, instead of moving after cutting out, now consciously plan to grow their log requirements in managed forests. Farmers and other individual and non-industrial forest owners also have learned that they can grow and market timber crops profitably, and by properly treating their timberlands, they add to the permanently available supply of saw timber and other tree products, thus assisting in the stabilization of lumber manufacturing and other wood-using industries.

The Northeast Region. The first sawmills of record in the United States were located in New York state near the site of Albany in 1623, and at Berwick, Maine, in 1631. By 1633 there were mills in Massachusetts and others in New York state. The first European-type sawmill, using a gang saw, was built in Berwick, Maine, in 1650, but the first circular saw, forerunner of the modern type, was not used in this country until 1820 or shortly thereafter.

Sawmilling during this early period of development in a new country was coincidental with settlement. Lumbering was a localized business, each mill serving its immediate territory and perhaps some adjacent and readily accessible population centers. Logging, therefore, was still a small-time operation in so far as any one mill was concerned. But it was a growing enterprise and by 1682 the number of mills in Maine alone was 24, and logging operations consequently were expanded sufficiently to supply them with logs.

The Northeast early assumed the lead in lumber production because the extensive forests of large white pine, characteristic of the region, yielded logs that floated readily and produced excellent construction lumber. In Pennsylvania the Swedes and Dutch had built sawmills before the time of William Penn, *i.e.*, before 1681, and by 1786 there were 16 mills near Lancaster, Pa., and others elsewhere in the state. At the time of the American Revolution there were 52 mills along the Hudson River in New York state.

Many of the mills of these early times were combination grist mills and sawmills, run by water power. As they increased in size and as logging on a larger scale became necessary, logging camps were operated up the streams from the mills, cutting timber and hauling logs during the winter to the frozen lakes and rivers. The accumulated logs were then floated to the mills during the spring freshets in sufficient quantities to supply the mills during the summer.

The Lake States Region. Although the first mills were established about 1832, it was not until after the Civil War that the Lake states (Michigan, Minnesota, and Wisconsin) became the center of lumber production. The Lake states region assumed leadership in lumber production soon after 1870 and held it until it shifted to the South in 1899.

Exhaustion of timber near streams in the Northeast and Central states and the favorable transportation facilities offered by the Erie Canal for lumber shipments east turned the attention of lumbermen westward to the Lake states. In this region white pine also was the predominant timber tree, as it was in the Northeast. Sawmills were built in numbers along the shores of Lake Huron and Lake Michigan and the inland streams were extensively used for floating tremendous quantities of logs from the interior forests to lake-shore mill sites. Large steam-powered mills were common by then, and the production of lumber in individual mills had been increased many times over by the development and perfection of modern manufacturing machines such as the carriage, circular saw, edger, trimmer, and other features of modern sawmills.

Logging on a large scale was done in the winter with teams and sleds for hauling, as the solidly frozen earth provided excellent ground conditions and snow- and ice-covered hauling roads provided a minimum of resistance to the loaded sleds. All winter long the logs were hauled and banked on the frozen rivers and lakes or in rollways on the riverbanks and lake shores. During the spring freshets, when streams were full, the logs were floated to their destinations. When several mills put their logs in the same river, each company registered its brand with the state authorities and it was cut or stamped into each log. Thus, the legal ownership of each log was shown by its brand and the owner's rights to it were protected by law. The sorting of logs in the drives was frequently done by associations of owners to protect impartially the interest of each member.[1]

Logging operations were definitely seasonal at this time, due to the climate of the Lake states. Cutting and hauling of timber were done in the winter; stream driving in the spring; and sawmilling in the summer, frequently with the same crews. During the fall months, preparations were made for the winter cutting and hauling by locating and building camps, roads, banking grounds, splash dams, and similar improvements.

The only way to ship lumber from the mills to Eastern markets was by boat during the open, or ice-free, season on the Great Lakes. This contributed to the seasonal nature of early Lake states logging and milling, as it was preferable for the mills to saw only when the lumber could be loaded out as rapidly as it was cut and air-dried.

When railroads were built into the Lake states region in 1860, many logging operations were moved to sites away from streams where timber

[1] "Michigan Log Marks," *Mem. Bul.*, 4, Mich. State Col. and Agr. Expt. Sta., 1942.

had become scarce; hence, the great expanse of Lake states timber not tributary to streams became accessible by means of logging railroads. Inland mills were located on the railroads as they provided transportation outlets for lumber.

The generally level topography of the Lake states, with heavy stands of timber per acre, permitted the construction of standard-gauge logging railroads at a reasonable cost and, by using special log cars and either rod or geared locomotives, no unsurmountable difficulties were met in this new type of logging. Coincident with the development of railroad logging, steam skidders capable of moving logs uphill as well as down on a quantity-production basis were invented. Skidders at that time were steam-power units with cables, drums, and tongs or chokers, used to pull the logs to the railroad track either on the ground or raised on overhead cables, and sometimes also equipped to load them on cars. The first power skidder was designed by Horace Butler, of Ludington, Mich., in 1883, according to L. H. Gault of the A. Leschen & Sons Rope Co.

During the reign of the lumber industry in the Lake states, sawmills were perfected to about their present state, and logging became a specialized and important part of the industry's operations, since the large mills required large amounts of logs, and well-organized woods operations became imperative. The band saw used in modern large mills was invented in England in the early 1800's and was introduced into the United States in 1869. It came into general use in the Lake states from 1880 to 1890.

The principal contributions of Lake states logging to the lumber industry were railroad logging, power skidding, and power log loading, which eliminated the industry's dependence upon streams for transportation and permitted year-round logging and milling.

The Southern Region. Some sections of the South were prominent in early lumber manufacturing, especially along the Atlantic and the east Gulf coasts. George Washington was actively engaged at one time in logging a 40,000-acre tract of timber, in which he had a personal interest, in the Dismal Swamp of Virginia and North Carolina. This swamp has been logged for many years and still is yielding valuable timber. Mount Vernon was built of Southern pine lumber in 1743. Before 1770, however, there was no extensive development of lumber manufacturing in the South and, of course, logging or woods operations were scattered, small, and relatively unimportant.

Prior to the Civil War period only a limited amount of lumber was shipped north from Southern lumber mills. That which was shipped went to Baltimore and Philadephia by water from the eastern shores of Maryland, tidewater Virginia, and eastern North Carolina.

As early as 1803 a steam sawmill began operations in New Orleans, but it was destroyed by workers who were afraid it would decrease the need for labor. Along the Mississippi River and its tributaries in Louisiana, Mississippi, and Arkansas, lumber was whipsawed by slave labor until about 1825, when this method was gradually replaced by steam-power mills using sash saws. The R. F. Learned & Son mill, cutting cypress, poplar, and willow "under the hill" at Natchez, was originated at that time and is still in operation. In 1875, markets north of the Ohio River made their first important demands on the Southern pine region for pine lumber.

During the period 1870–1900 there was a rapid migration of the industry from the Lake states, as the mills there cut out, to the unexploited forests of the South. This expanding demand upon Southern forests for lumber to be shipped North was due to the growing scarcity of cheap Lake states timber resulting from the exhaustion of readily accessible timber and the still-continuing upbuilding of the Middle West and Prairie states, which created a heavy demand for large amounts of construction lumber.

One of the first species of timber cut in the South was cypress, well known for its durability and strength properties and much in demand for construction purposes. Specialized logging methods were necessary to get cypress timber from the swamps and wet river bottoms to the mills. At first, choice cypress trees were deadened by girdling them with an ax a year or two in advance of logging, which was done during the dry summer seasons. The deadening operation caused the trees to dry out enough to float readily, and they were left on the dry ground after cutting to be floated out during high-water periods.

Later pullboats, in reality floating power skidders, were perfected which permitted year-round production of tree-length cypress logs in quantities sufficient for large mills. Most of the early cypress logging was done without regard to timber ownership as but little was known about the titles to overflow lands along rivers and swamps. Cypress is still an important Southern lumber species, being obtained in limited volume now from river and coastal swamps or incidental to hardwood logging along streams and rivers from North Carolina to Texas.

Some of the earliest Southern logging was in Florida, where timber was squared by hand hewing and then floated down creeks and rivers to tidewater and exported. Some of the first products of forests along the Gulf Coast of Florida were ship timbers and planking. It is reported that Navarez, in 1528, built a boat in Florida and sailed it to Europe. Several such sailings started from near Apalachicola. As long as wooden ships sailed the seas, limb sections from live oak trees were much in demand for bracing ship members. In 1826 President John Quincy Adams established a navy yard on Pensacola Bay because of the live oak trees found to be so prevalent there. Long before this time these semitropical waters were the hangout of pirates and other riffraff of the sea, who drew upon the forests for timbers to repair ships.

A number of mills in south Alabama and west Florida, some of which are still running, started in 1870–1880. Logs were transported in ditches and some old ditches are still in evidence along natural stream courses, which were straightened and made navigable for large longleaf pine "floater" logs. Floater logs were those cut from the largest trees, which were mostly heartwood and hence light enough to float. The logs were rolled by ox teams or carried by carts to the ditch banks, where they were put in the water, dogged together end to end, and pulled to the millpond by ox teams; a slow but economical method adapted to the locality, the timber, and the time. Some of the early mills that are still operating and which at first were logged by means of ditches now use the most modern timber-harvesting methods and are adapting their harvesting operations to a sustained yield. Their early and subsequent cutting of only the largest trees being, quite by accident, good forestry.

The Western Regions. The first settlements in the Pacific Northwest obtained their lumber needs from the large, straight-grained cedar trees, which could be readily split or rived into boards. Shortly after 1828 the first sawmill was constructed at Fort Vancouver, on the Columbia River in the state of Washington. Much of the lumber manufactured from it went into the construction of Walla Walla, Wash. It was built by Dr. John McLaughlin of the Hudson's Bay Company and it was the first sawmill west of the Mississippi River.[1]

Migration to the Oregon Territory was greatly stimulated by interest aroused over the controversy with Great Britain, about this time, when the American slogan was "fifty-four forty or fight," referring to the latitude many felt should have been the north boundary of the United States. The growing population naturally needed lumber in increasing amounts.

In 1844 on the then future site of Astoria, Ore., the first Oregon mill was built by Hunt, and five years later there were 30 sawmills in Oregon alone.[2] In 1847 at Olympia, Wash., a water-power sawmill was built, and in 1849 the first cargo of forest products from Puget Sound was shipped to San Francisco aboard the brig "Orbit" to meet the needs created by the gold rush to California. The gold rush of 1849 and the boom developments that followed in its wake were a great stimulus to the lumber industry of Puget Sound, as lumber could readily be shipped by water to California and other coast ports.

[1] Cornwall, George F., New sidelights on Hudson's Bay sawmill operations, *Timberman*, November, 1942.

[2] Vinnedge, Robert W., "The Pacific Northwest Lumber Industry," *Yale Univ. School of Forestry, Lumber Indus. Ser.*, 1923.

In 1852 the first steam sawmill was erected on Puget Sound by Henry Yester, and in the same year a Swede, Nicholas Denlin, built a water-power mill at the same site. Captain Talbot, formerly of the lumber industry in Maine, started a mill at Port Gamble, Puget Sound, about that time and it has developed into one of the large modern companies still important in the industry.

Until transcontinental railroads were opened for traffic in 1880, the lumber industry of the Pacific Northwest was limited to the West for its domestic markets. Export shipments to the Far East, the west coast of South America, and Pacific islands, also offered additional outlets for lumber. However, the opening up of rail connections between the Far West and the Middle West and Eastern regions at once greatly enlarged the opportunities for marketing Western lumber, which naturally resulted in stepping up production to higher levels.

In 1880 Washington and Oregon had 265 lumber manufacturing plants with a combined annual production capacity of 500 million feet. Western lumber, lath, and shingles flowed into the Middle West between 1880 and 1890, and the 1890 Census showed a lumber capacity three times as great as 10 years previously. To stimulate a still greater movement of lumber from the West to the East, the railroads put lower freight rates on lumber into effect between 1890 and 1900; *i.e.*, 40 cents per 100 pounds to the Twin cities, 50 cents to Chicago, and 70 cents to New York.

Soon after the turn of the century, many Southern lumbermen feared for the future of their industry because of the rapid removal of virgin timber in the South and the almost unlimited Western timber supply with greatly enlarged markets. Much of the lumber industry capital in the South, therefore, was taken West and invested in timber and mills. Likewise many Lake states companies whose timber had lasted until this period also transferred their operations to the West. Independent investors also bought Western timber, confident that within a very few years all the lumber used in the United States would have to come from Western forests because of the erroneous "timber famine" theory held by many prominent foresters and lumbermen of that time.

The rapid development of second-growth timber in the South, the declining per capita consumption of lumber, and other factors combined to cause much smaller needs for Western timber than were anticipated. Many timber investors had to start milling in order to obtain some income with which to pay taxes and interest on borrowed funds invested in timber. Periods of so-called "dumping" of Western timber were largely due to these causes. Naturally such conditions greatly influenced the kind of logging that was done. Conservative cutting had no attraction for timber owners whose ownership burdens had become so great as to require sacrificing timber on an already flooded market.

Washington was the leading state in lumber production from 1909 to 1937 (except for 1914, when Louisiana ranked first). Since then Oregon has been in first place. The peak year for lumber production in the West was 1929, with a combined cut of Douglas fir, ponderosa pine, redwood, and other species of nearly 16 billion feet.

California has long been an important lumber-producing state because of its redwood and sugar pine forests, both containing trees unusually large in size and noted for the specific technical qualities of their woods. Beginning in the middle 1860's redwood was cut and milled along the Mendocino coast of California for the San Francisco market. Steam logging began about 1882, being introduced by the Caspar Lumber Company, prominent at that time. About 1890 the production of sugar pine became important.

The small forest area of Nevada adjoins California and the production of lumber in California and Nevada is usually combined for statistical purposes. Nevada, a relatively nontimbered state, nevertheless had important sawmilling and, consequently, logging operations in the early days, beginning as far back as 1853. The mining industry created a demand for lumber that took most of that produced. Fuel and charcoal wood also were cut, and because of this heavy drain on Nevada's limited forests, they soon were severely depleted. The peak of lumber activity occurred between 1875 and 1880, with an annual production of about 25 million feet.[1]

The forest areas of Idaho, eastern Washington and Oregon, and western Montana are known as the Inland Empire to the lumber industry. This region is important particularly in the production of Idaho white pine. In 1840 at Lapwai, in north Idaho, the first mill of record in that state was built to supply the lumber needs of the growing mining industry. The first mill in Montana was built near Stevensville in the late 1840's. These mills were combined grist and sawmills, powered by water wheels.[2] The only substantial outlets for lumber produced in the Inland Empire are by rail, which limits the general markets for the Inland Empire lumber to its own region, the Middle West, and the East. Specialty grades of white pine, of course, find wide distribution.

The first sawmill in the Southwest was built at Flagstaff, Ariz., in 1881, primarily to supply ties for the Santa Fe Railroad. Another mill was built shortly after that at Williams, Ariz., for the same purpose. From these beginnings the industry, although limited in extent, has become important to the region and supplies considerable lumber to Texas and interior southern California. Now there are from 70 to 80 mills in New Mexico and 30 to 35 in Arizona, only 8 of which are large.

[1] Anonymous, Queen of them all was Virginia City, *Timberman;* June, 1941.

[2] Anderson, I. V., and E. F. Rapraeger, "Highlights of the Lumber Industry," North. Rocky Mountain Forest and Range Experiment Station, 1940.

HISTORY OF PULPWOOD PRODUCTION

The use of wood in the manufacture of pulp and paper is of much more recent origin than the use of wood in its natural form, as for lumber, implement handles, staves, ties, and similar products. Until 1840 there was no satisfactory known method of pulping wood. In that year, however, Keller, a German, developed a roughened stone grinder for disintegrating wood into a pulpy mass. From this beginning the pulping of wood has grown to be the important industry it now is.

The first pulp mill in the United States was built by William Rittenhouse at Germantown, Pa., in 1690, and the original mill structure is still standing. This mill used linen rags, not wood, in the manufacture of pulp for paper. The expanding use of paper, however, created a shortage of rags for pulp and forced pulp makers to search for other raw materials that would yield fibers in large quantities. Due to this need, wood was tried and found suitable for pulping, after many refinements in methods were perfected.

Grinding was the first method used for pulping wood and it is widely used even today in the Lake states, the Northeast, and in eastern Canada. Although Keller started grinding wood for pulp in 1840, it was 1867 before the first ground wood pulp was made in the United States, at Curtisville, Mass. Woods used at that time were chiefly spruce and balsam.

In 1852, in England, a chemical method, the soda process, of separating the wood fibers from the cementing materials was discovered, perfected, and developed for practical use, and in 1863 two soda mills were built in the United States. This made a much wider range of wood species, particularly hardwoods, suitable for pulping and naturally it enlarged materially the production of pulpwood from the forests. It also made possible the expansion of the industry into other forest regions.

In 1867 the sulfite method of pulping wood by chemical action was discovered. It was perfected commercially in 1874 in Sweden. This made possible the making of stronger pulps from spruce, balsam, hemlock, and aspen, since the fibers are not torn as in the grinding process.

The sulfate, or kraft, method of chemically pulping wood was discovered in 1883 by C. H. Dahl in Danzig and was almost immediately used in Sweden, Finland, and Norway. In 1907 it was introduced into Quebec, Canada, and soon thereafter into the United States. This process is particularly well adapted to pine species, so pine forests soon became an important source of pulpwood.

The early development of the pulp and paper industry was in New England and New York. Pulping at first was limited to the groundwood process and later to the sulfite chemical method, and both used spruce and balsam, although some hardwoods in New York and Pennsylvania were processed by early soda mills. Furthermore, the water power available from rapidly flowing streams, excellent winter logging conditions, and stream driving all combined to foster the industry in the Northeast.

From the Northeast the industry expanded into the Lake states of Minnesota, Michigan, and Wisconsin. Excellent forests of balsam and spruce were to be found in this region also. Aside from the early pioneering developments in the South, such as at Orange, Tex., Moss Point, Miss., and Roanoke Rapids, N. C., the industry turned westward to the Pacific coast region. Here, also, suitable species of spruce, fir, and hemlock were present and no pioneering of methods was necessary in developing the Western pulp industry, as these species are closely related to those used in the Northeast and the Lake states. In the South, on the other hand, before mills could be built, the process to be used had to be perfected, so it is the most recent forest region in which pulp mills have become of major importance.

The extensive area of Southern pine forests early attracted the attention of the pulp and paper industry as a source of raw material for sulfate mills. The first attempt to make paper from Southern pines was before the origin of the sulfate process, at Pensacola, Fla., by Smith and Thomas in 1903.[1] Their attempt, however, never was perfected to the point of being practical for commercial production. A soda mill was built soon afterwards but it had only indifferent success and was moved to Orange, Tex., in 1911. The process used was changed from soda to sulfate and from then on the plant was successfully operated for several years by E. H. Mayo, who was responsible for perfecting the application of the sulfate method to Southern pines in the manufacture of kraft paper. The production of this mill was 20 tons a day. From this beginning sulfate pulp production has increased to the present volume of 12,000 tons daily at Southern mills, utilizing Southern pines and a small percentage of hardwoods.

Once demonstrated to be feasible and practical, the growth of the kraft pulp and paper industry in the South was phenomenal. In 1913 a mill designed to use Southern pines was constructed at Moss Point, Miss. It was largely financed in England. Mr. Mayo, the successful operator of the Orange, Tex., plant was employed by this company and he soon became manager. This mill operated continuously and in 1928 was sold to the International Paper Company.

In 1906 the Roanoke Rapids Paper Company built a mill at Roanoke Rapids, N. C., and manufactured a butcher paper, with a capacity of 30 tons a day. It used pine and the sulfate method in making this product. The Bogalusa Paper Company in 1917 built a pulp and paper mill in conjunction with the large sawmill of the Great Southern Lumber Company, in Bogalusa, La. The sawmill, one of the largest ever in operation in the South, has now ended its operation but the pulp mill still runs and it is expected to be permanently supplied with wood from the local vicinity,

[1] Allen, J. H., History of making pulp and paper in the South, *Cypress Knee*, University of Georgia, School of Forestry.

largely obtained from company lands, including extensive areas of pine plantations that have already grown to merchantable size.

From these original beginnings, the kraft pulp and paper industry in the South has become of important proportions in practically every section of the Southern pine region from Virginia to Texas. Close to 20,000 cords of wood are required daily by these mills, but with the rapid advances in forestry by public agencies and private forest owners, no dearth of timber nor serious threats of a timber famine have developed.

TABLE 7. ANNUAL PULPWOOD PRODUCTION IN THE UNITED STATES*

Year	Cords
1870	2,000
1880	40,000
1899	1,617,100
1910	3,146,500
1920	5,014,500
1930	6,099,200
1936	7,506,100
1941	15,400,000
1944	14,819,000
1946	17,818,000

* Source: *Pulp and Paper Industry*, vol. 21, No. 5, May, 1947.

Wherever the industry has built, it has remained as a permanent fixture. Even in New England, the region of its first development, a great many mills still are actively in production. The pulp industry is essentially a permanent one in that the mills are costly installations as compared to sawmills. Its history differs from that of the lumber industry in that it did not migrate but expanded.

TABLE 8. PULPWOOD CONSUMPTION BY REGIONS IN CORDS OF 128 CU FT

Region	1921		1931		1941	
	Cords	Per cent	Cords	Per cent	Cords	Per cent
New England	1,345,709	29.5	1,170,439	17.4	2,277,738	14.5
Middle Atlantic	1,107,654	24.3	875,985	13.0	1,295,935	8.3
Lake states	1,218,274	26.7	1,405,443	20.9	2,081,400	13.3
Pacific coast	342,560	7.5	1,345,754	20.0	3,391,762	21.6
Southern	542,982	12.0	1,925,145	28.7	6,630,665	42.3

Indicative of the fact that the pulpwood industry has not appreciably declined in any one region but has simply expanded into others, increasing pulp production and, therefore, total wood requirements is the comparative data by regions given in Table 8 on the regional growth of the pulp and paper industry in the United States.

HISTORY OF OTHER FOREST PRODUCTS

Specialized products of forests such as poles, piles, ties, veneer bolts, cooperage stock, naval stores, acid wood, and so on have been needed in varying amounts for many years. One of the oldest products is cooperage bolts, used in the manufacture of staves for casks and barrels. This product requires particular woods such as white oak, frequently from a particular region. Thus, the Ozark region of Arkansas and Missouri is famed over the world for its white oak for wine and whiskey staves. Chemical wood, as it is called, has been produced for a long time but its source is limited to the heavy hardwoods such as beech, birch, maple, and oak. Its production is important locally in the vicinity of the comparatively few plants that destructively distill wood to obtain charcoal, acetic acid, and methyl alcohol.

Veneering is an old art but recently it has assumed great importance with the perfection of stronger glues having waterproof qualities. Plywood, made up of alternate layers of wood veneer sheets of thicknesses from 1/8 to 1/28 of an inch, is used both for construction and for decorative purposes in the form of panels, as well as in furniture manufacturing. This industry, drawing directly upon the forest, has greatly expanded recently and gives promise of still greater expansion, since it affords a means of getting the most clear surface from choice sections of trees. These sections are either sawed, sliced, or cut on a rotary lathe in such a way as to give volume or quality products, depending upon the uses to which they are to be put.

Ties, both hewed and sawed, have been a prominent product of the forest since railroads became a method of transportation. Many efforts have been made to substitute materials other than wood for railroad ties but none has proven the equal of wooden ties. Hewed ties are a direct product of the forest, being hand-made in the woods. Sawed ties, on the other hand, are sawmill or tie-mill products and are extracted from the woods in the form of standard-length logs or tie logs.

Poles have been cut from forests since the telephone and telegraph were put into commercial use. The need for poles of much larger sizes for the cross-country transmission of electricity multiplied tremendously during the development of modern communications and power transmission. Today the pole requirements of power and telephone and telegraph companies constitute a major item of harvested forest products. The strict requirements with respect to straightness, strength, and freedom from decay restrict the use of trees for these purposes to only the best of particular species. Thus, the value of trees suitable for poles is usually sufficiently higher than that of other products the same trees could yield, so that pole production is attractive to forest owners.

Piles are products similar to poles, in many respects, from a harvesting standpoint. Strong, straight, and sound trees are required, as the piles produced from them are used in the construction of bridges, wharves, and similar structures, and as foundation supports for large buildings.

In the past and even at present most of the wood substance harvested from American forests is used in the form in which it grew and existed in the tree: as lumber, furniture, implement handles, and in innumerable other ways. However, the use of wood substance as a raw material to be converted into other forms is making marked headway and may, in time, be the dominant use. The pulp industry is already strongly established and still expanding, the pulp being used not only for paper but for rayon and plastics as well. The destructive-distillation process also converts wood to other forms, but no large increase in this industry seems imminent. Even veneer and plywood, especially when impregnated with resin compounds and compressed, lose resemblance to wood as such. Regardless of the end or intermediate use of the wood substance, the primary product is wood; and as it grows and matures in forests, it must be harvested before it can be processed.

Gum naval stores, obtained by distilling the gum of longleaf and slash pine trees, are manufactured by one of the oldest American forest industries. It was developed in the days of sailing ships, which required pitch and tar for calking seams and treating ropes. The gum obtained by cupping the trees is a product of the living tree, as are maple sugar and true rubber latex. The industry has highly specialized methods of operating, quite different from harvesting wood crops obtained from the trees themselves, so methods of harvesting naval stores are not treated in this text.

In such utilization as tannin extracts and wood naval stores, the wood substance itself is not used, except as a by-product, but the substances naturally absorbed or deposited in it constitute the primary product. Hemlock, oak, and other barks from forest trees have long been used in tanning leather, so this use is not new. Dead chestnut, blight-killed, has been used as a source of tanning for years, but this nonreplaceable species will soon be exhausted.

The historical background of many of these products is obscure and, being of less magnitude compared to the major products of logs and pulpwood in the sense that they are harvested in relatively small amounts, it is not important to trace the development of these products in detail. Until recent years most of them, except naval stores, were produced by hand laborers working individually or in small groups, and organized operations, as in the case of logging and pulpwood cutting, were rare.

CHAPTER 3

THE ORIGINS OF MODERN TIMBER-HARVESTING METHODS

Logging activity, or timber-harvesting operations, were carried on, of course, in the forests tributary to the manufacturing units of the lumber- and timber-products industries. In fact, the timber supply and the available methods of harvesting and delivering it to the mills determined to a large extent the location and the rate of development of the forest-products industries. Varying forest and topographic conditions in the forested regions of the United States and constant improvements in mechanization directed towards harvesting more timber within a shorter length of time resulted in vastly different methods of harvesting timber in the several stages of the development of the forest industries.

In the Northeast, winter cutting and team hauling with sleighs and spring-freshet stream driving typified logging operations in the early days, and with improvements this method still is used, along with strictly modern mechanized methods. In the Lake states, also, seasonal logging with stream driving predominated at first because other transportation was not available. The expansion of railroads into the Middle West, however, made possible the development of year-round railroad logging and power skidding with both overhead and ground cables. This type of logging was perfected in the Lake states, later widely used in the South, and then intensified in the West to cope with the heavy stands of large-sized timber in that region.

Southern logging began with the use of primitive ox teams and wagons. This method still was common even during the peak of large-scale logging for getting logs to forest railroads, where they could be loaded and hauled to the mills. Most operations in the shortleaf-loblolly pine forests were animal operations because of comparatively light stands per acre and because so-called short logs, 12, 14, and 16 ft in length, were needed at the mills and logs of this size could readily be handled by teams. Teams skidded the logs to railroad spurs, where power loaders loaded them on log cars for the rail haul to the mill. In longleaf pine forests, however, numerous steam-powered cable skidders were used to handle tree length logs because squared longleaf pine timbers, the longer the better, were greatly desired for export to Europe. Also, virgin longleaf pine grew in extensive, pure, even-aged stands with volumes up to 30,000 ft per acre, making them ideal for power skidding. Most longleaf operations, therefore, were of the power-skidder type, although animals were used to a limited extent.

Where animals were used for skidding, logging railroads with spurs at frequent intervals were necessary because of the limited distance for which animals are efficient for skidding. Within recent years, however, cable skidding has practically been discontinued, except in swamps and a few mountain locations that are inaccessible with other equipment. Likewise, most railroad logging operations have been abandoned, as trucks and tractors are now widely used because of the facility with which trucks can traverse forest roads and trails and because they are economical for hauling from scattered small tracts of timber to which railroads cannot be built. Tractors are so much faster than animals and can handle such large loads and can be moved so readily that they have displaced teams for skidding on most large-scale operations.

Western loggers always were confronted with the problem of handling very large trees and logs. At first so-called "hand logging" was done by crews of men who had only a few horses or oxen and no machinery except ropes and windlasses. Only timber felled close to water on sloping ground could be handled in this manner. The procedure was to fell and buck the timber and then, by prying, rolling, or pulling with a rope windlass operated by animal power, get the logs into the water so they could be floated to a mill or to a log market. Ox teams with up to 10 or 12 animals were also used for skidding during this period.

With the development of large-sized mills, more efficient logging methods were introduced into the West, such as big-wheel carts with animals, and later steam donkeys, skidders, loaders, and railroads.

Tractors and trucks are now widely used throughout the West for skidding and hauling. The large volume of timber required by the mills and the large size of the trees, combined with the heavy stands per acre, favored the rapid development and perfection of extremely heavy and powerful equipment for skidding, yarding, and transportation of a tremendous volume of timber per day. Since tractors designed for logging have proven their worth and adaptability under many different operating conditions, they in turn are supplanting much of the large, powerful stationary yarding equipment because fixed expenditures are much smaller and because the mobility of trucks and tractors makes them adaptable to partial cutting.[1] With yarding systems using overhead cables or other forms of cable skidding, partial cutting is practically impossible, as any trees not cut for logs are destroyed by lashing cables and new stands must start from seed. Even with power-cable skidders, however, areas can be left uncut to reseed adjacent cut areas.

[1] Van Tassel, Alfred J., "Mechanization in the Lumber Industry," Report M-5, Work Projects Administration, National Research Project, 1940.

Much study and thought has been given to the problems of developing economical logging methods that are also adaptable to harvesting tree crops under a practical forest-management plan. Continued improvement in logging technique, therefore, is to be expected along lines that will make it practical to remove specified amounts of timber per acre in accordance with a forest-management plan without undue damage to trees left to grow, and thus make possible the practice of good forestry with economical harvesting.

The changing complexion of logging has been closely related to variations in the kind of timber being utilized. During the periods when virgin forests were being cut, heavy skidding and loading machines and railroads, or other specialized transportation facilities such as flumes, or improvements to drivable streams were necessary. Such installations required heavy investments for transportation, and cutting usually removed all merchantable trees to obtain as large a volume of timber as possible, and thereby spread the cost of logging improvements and equipment over as much volume as possible, to keep the cost per thousand feet at a minimum. Breakage of uncut trees in felling and skidding and the almost inevitable forest fires that followed logging resulted in the destruction or loss of most of the trees left after cutting. There was little or no expectation that there would ever be another cutting on the same lands, so no particular care was exercised.

INFLUENCE OF FORESTRY ON LOGGING

During the migration and development of the wood-utilizing industries, a distinct trend towards harvesting, aside from the purely developmental phases of logging mechanics, has been obvious to close observers of the timber-harvesting industries. Unending forests made any thoughts of timber scarcity seem absurd during the earlier decades of logging. As soon as the timber at one locality was cut, it was only necessary to move on a few miles to a new location. The conception of forestry, or timber growing, did not prevail upon conscious American thought, except to a limited extent, until around the turn of the twentieth century, when President Theodore Roosevelt popularized forestry by creating a number of national forests, and by laying the foundations for the present U.S. Forest Service. Even then, forestry was thought to be something idealistic and only possible for the government, not for commercial operators or individual forest owners.

Logging, therefore, continued to be conducted with the sole thought and purpose of cutting and getting to the mill all usable timber on the cutting areas at the lowest cost. No thought whatever was given to a second cut from the same land. Nevertheless, when the general movement of lumbering from the Northeast to the Central states and then to the Lake states occurred, some units of the industry remained behind, to cut scattered stands still available or to log species passed up in the first cutting that later became merchantable. In time, second-growth stands appeared and grew large enough for cutting and hence the beginnings of more or less continuous operations were started. In all of the regions where lumbering at one time or another was predominant, at least some lumber, and considerable amounts of other timber products, are still being harvested.

In the South especially, due to the rapid growth of its timber species and their early maturity, the basis of continual timber harvesting was rapidly established. During the Civil War period many cotton fields were abandoned and were never reclaimed for agriculture. They promptly reseeded to pine and, growing rapidly, reached merchantable sawtimber size about 1920, within the memory of local citizens. Such stands became known as "old-field" pine and they were visible proof that timber could be grown. This demonstration of timber growth attracted widespread attention and stimulated interest in forestry.

Before the South had declined too far from its peak as a lumber-producing region, those still in the business witnessed the distressed condition of lumber manufacturers in the West due to overproduction, costly mountain logging, and the great distance from markets. The log supply of Southern mills was being augmented by timber from many small, formerly isolated tracts inaccessible for large-scale logging operations, including old-field timber coming on the market as road systems developed and trucks were perfected. To stay in business required perpetuating timber supplies. This could be accomplished by reducing mill capacity in some cases or shifting some pine production to hardwoods not previously merchantable, and, above all, by applying the rudiments of good forestry to cutting operations. The unattractive Western situation during the period of dumping, and the advantages of maintaining going Southern mills, further stimulated Southern operators to permanence. At present a considerable and increasing number of Southern lumber companies have perpetuated their operations through the adoption of planned harvesting schedules and correct forestry practices.

So long as there was more timber to move to, the incentive to work out permanent operations was not great. But with declining virgin timber supplies, it became obvious that if existing American timber supplies were used up, the only places where lumbering could continue to pioneer would be in Siberia or the Amazon, not very attractive lumbering locations for American businessmen. Many landowning lumber companies whose timber had not been too closely cut originally or who had logged with teams and had accidentally practiced good forestry by selecting only the then most valuable species or sizes of trees, were able to reorient their business affairs and make their operations permanent.

Many Western companies, likewise realizing that they were on the last remaining timber frontier in America, also have adopted cutting policies that have perpetuated their operations. This has become possible largely because of the perfection of tractors that are adaptable to removing only a portion of the stand at one time. Much Western timber is so large and the terrain so rugged that animal operations are impractical. Without powerful tractors, therefore, power-cable skidders would be necessary, and as they are so destructive, harvesting timber for permanence is seldom possible with such equipment.

Now in all regions many private owners have taken steps to perpetuate their operations by means of fire protection, integrated utilization, partial cutting, and by relating the volume cut to the volume grown. The gradual trend away from simply cutting all trees that had value at the mill towards a harvesting type of cutting operation that gives due consideration to provisions for subsequent cuts has materialized in all forest regions. This intensification and expansion of timber growing in the United States offers much promise to foresters interested in careers with private industry.

In all places except the West this harvesting type of timber-cutting operation did not develop until after most of the virgin stands had been cut. Since virgin stands are generally stagnant, growing but little, and with trees constantly dying due to age, insect attack, lightning, and windfall, there is but little or no net growth, and frequently losses exceed the growth. Even-aged virgin stands, therefore, often must be cut clean, as even the smaller trees are as old as the largest, being suppressed and weakened by their more vigorous neighbors and not worth leaving to grow. The next stands usually must come from seedlings, when virgin timber in even-aged stands is logged. When the virgin timber is mixed with younger trees in uneven-aged stands, as often occurs, the virgin trees can be removed by partial or selective cutting and the smaller, younger trees will continue to grow when not destroyed by logging. Second-growth even-aged stands also can be cut in such a way that rapid growth can be maintained as the smaller trees recover vigor.

The pulp industry, having followed the lumber industry in practically every important region, has profited by many of the experiences of the lumber industry. One of the lessons it learned by observation is the wisdom of owning forests and of growing timber to assure adequate raw materials. Furthermore, the large investment, usually from 4 to 6 million dollars for an average plant, is too great to depreciate over a short span of life and the investment is fixed in place, it being exceedingly difficult or impossible to relocate a pulp or paper mill. Trees are mature for pulpwood while still small and young. Frequently stands are cut in which trees average only from 7 to 8 in. in diameter. Timber crops can, of course, be grown much more quickly for a pulp mill than they can for a sawmill, which makes forestry more attractive than when longer waiting periods are involved. Individual owners and farmers likewise can visualize timber growing as a practical undertaking when small-sized trees are merchantable, and are more readily interested in growing timber. Many mills incur considerable expense in encouraging farmers and others to grow timber by giving away seedlings, providing forestry advice, and advertising.

MODERN CONDITIONS FAVOR THE HARVESTING METHOD

Economic conditions in the United States and existing transportation facilities make possible, in most localities, the maintenance of forest properties for the continuous production of timber crops by the practice of forestry. Early logging preceded settlement and the construction of public transportation facilities; hence, roads, railroads, or waterways primarily for extracting logs and other tree products needed to be installed by the mills. These costly operations discouraged forestry.

Now permanent transportation facilities have been provided for most of the country, including common-carrier railroads, hard-surface highways, and secondary roads, and even waterways, and forest owners often find it possible, by using modern trucks and tractors, to transport the tree products from the stump to mill or market at an economical cost and without the necessity of installing costly improvements such as railroads and spurs. Thus, harvesting the maximum volume of timber per acre over a short period of time to keep the cost per unit of production to a minimum is no longer necessary. Great improvement in tractors and motor trucks has been an extremely important factor in making forestry possible because such equipment can operate economically on light cuts per acre and can be relocated readily without expense, thus facilitating the type of cutting that forestry methods require.

While lumber consumption has declined per capita and in total amount in the United States as a result of stabilization of the population, the use of wood has actually increased in a variety of other ways, such as for poles and piles, pulp and paper, rayon and plastics, veneer, and specialized products.

Utilization of timber for saw logs alone involves considerable waste. Studies have shown that practically one-half of the wood substance in a tree is wasted when it is cut and sawn for lumber, since the small portions of the trees are not suitable for boards. In addition to the top section left in the woods, the sawdust and most of the slabs and edgings have only fuel value at most plants.

When there is a lack of markets for wood in trees smaller than saw-log size, the practice of proper thinning operations is handicapped because no income is obtained to offset the cost of the thinning. However, when there are markets for pulpwood, or similar products, the tops and small trees cut in forest improvement operations can be profitably utilized. It has been shown that about ½ cord of pulpwood can be obtained from tops per 1,000 ft of saw logs cut from Southern pine trees.[1]

Use of wood for various purposes makes it possible to utilize more fully the variety of different species that ordinarily are present in the forest. During recent years there has been a rapid trend towards the greater use of all species. More different species are cut for ties than formerly and most species, many formerly nonutilized, now are in demand for pulpwood. The expansion of the veneer and plywood industry has brought about better utilization of species formerly left in the woods, such as black and tupelo gums and sap gum.

All of the refinements of lumber manufacturing, such as improved methods of drying and treating to make wood durable, assist materially in creating a demand for more different species and for better utilization of the timber cut. Many small articles are now made at sawmills, such as stepladder parts and legs and braces for card tables, and, by gluing small pieces together, ironing boards and boxes can be built up of lumber discards that formerly were burned.

The important factors that contributed to the development of harvesting operations for permanence, in place of cutting operations to liquidate, are as follows:

1. The rapid growth of timber became widely recognized.
2. The lack of undeveloped forest areas in America.
3. The development of many miles of public roads and highways suitable for truck haul to mills of logs and lumber.
4. The development of flexible yet powerful logging machinery such as trucks, tractors, power saws, and loaders, making cable skidders unnecessary.
5. Diversified utilization making it possible to utilize for products other than lumber much more of the wood grown in a forest than formerly.
6. Improved lumber-manufacturing methods, kiln drying, and the perfection of lumber dips to control sap stain and insect attacks on freshly sawed lumber.
7. Timber growing by farmers and other individual owners who sell forest products or stumpage.

FORESTRY PRACTICE POSSIBLE AND DESIRABLE

Marked changes in uses for wood and the maintenance of wood-using mills in well-developed areas have brought manufacturing establishments dependent upon forests more closely in touch with other businesses. Sawmills, pulp mills, and other wood-conversion plants are not now so remotely located as formerly; they are in the midst of other business concerns and find that many of their interests are common.

This condition has brought its problems as well as its advantages. Timber cutting is now exposed to public view along the many highways

[1] Anonymous, Pulpwood from logging tops, *South. Pulp and Paper Jour.*, May, 1940.

and roads that traverse most of the forest areas. Wood-using companies, sawmills and pulp mills primarily, must now give heed to public opinion because if it becomes aroused over the belief, justly or unjustly, that the forest is being destroyed, laws may be passed aimed at remedying the situation. Formerly, when most timber cutting occurred in localities remote from public view, the need for public understanding of timber harvesting was not so necessary.

It is generally true that forest owners are no longer confronted with the practical necessity of clean cutting. Increased accessibility of forests, modern road development, and portable timber-handling machinery have now made possible what before was unthought of, *i.e.*, permanent timber-cutting operations. Furthermore, there are sound business reasons why forest owners should adopt practical forestry and harvesting operations instead of clean cutting. One reason is that from now on the forest industries must in the majority of instances provide for future timber supplies on their own lands, since the opportunity to move on to uncut areas has greatly diminished or disappeared altogether, except at extremely high cost.

Grown timber produced under controlled conditions can be made much more adaptable to man's needs than untended forest-grown trees, in that the density of the grain can be controlled by regulating the density of the stand through thinning; knots can be limited in size and number by the same means and by pruning; and crooked and otherwise less desirable trees can be eliminated from the stand and used for other products while still small. Man-grown trees can be produced not only more rapidly than those grown by nature but they also can be of better form; *i.e.*, straight, sound, and free of excessive knots.

FORESTERS SHOULD KNOW THE BEST HARVESTING METHODS

A forester may never be called upon to fell a tree or buck logs or do many of the other jobs, skilled or unskilled, concerned with harvesting timber crops, yet it is highly important that technically trained foresters know the fundamentals of good harvesting operations. In well-managed forests the other fields of forestry, such as silviculture, regulation of cut, and management, must be integrated with harvesting. A knowledge of how much volume it is practicable and possible to remove from the forest and how products can best be transported from stump to market at a favorable cost is necessary before sound operating plans can be drawn for a forest.

Silvicultural considerations require that some stands be cut differently than others, even in the same forest, if the desirable species are to be grown naturally. Therefore, successful reproduction following cutting operations, in keeping with the silvicultural requirements of a forest, depends largely upon the harvesting skill of the forester.

Aside from the necessity of integrating harvesting with the practice of other forestry measures, there are additional important reasons why foresters who may have the supervision or control of harvesting and selling operations in a forest, government or private, should know when such operations are economically and skillfully performed. One of the important responsibilities of foresters is to write timber sale specifications for cutting timber so the purchaser of stumpage will not cause unnecessary damage or waste of timber in cutting the trees he has bought. Contracts covering stumpage sales by the U.S. Forest Service from national forests are comprehensive in this respect and can only be drawn by foresters who have a working knowledge of good harvesting methods. Similar sales of timber from privately owned forests call for a knowledge of harvesting that will protect the forest owner and still not work unnecessary hardships on the buyer.

Inefficient logging or other timber-cutting operations obviously result in wasted values, and inefficient operators selling on the same market with more efficient operators naturally can get no more than market price for their product, and frequently less. A greater production cost with the same or a smaller selling price means that inefficient operators cannot pay the same price for stumpage as can the more efficient operators. This is important to the owner of the timber, so he or his forester should be able to recognize good and poor practices in cutting timber and in organizing harvesting operations, to avoid dealing with inefficient operators or those who waste values.

A working knowledge of what methods might be employed in harvesting operations and the economical range of each is of real importance in locating permanent transportation systems. Roads installed at the wrong places for harvesting operations can add up to several dollars per 1,000 ft in logging costs that might have been avoided. Roads also should be located with respect to other needs, such as fire protection and future plans for forest development. Only when a forester is thoroughly competent in the field of forest harvesting can he weigh the various considerations involved and make wise decisions as to road locations.

Lack of knowledge concerning harvesting principles and practices can result in unwise timber sales by forest managers. For example, pole-quality trees may be sold for saw logs and saw-log trees sold for pulpwood; or veneer bolts may be sold at an attractive price but only the choicest parts of the trees used, the rest being left to rot even though of saw-log quality, resulting in less income than could have been obtained. Also, saw logs may be sold and the pulpwood in the tops left to decay.

In forests yielding several products, integrated harvesting can often be conducted so that all merchantable volume is removed at the same time, whereas not enough volume would be left after one operation to justify its salvage. A forester capable of judging the volume of timber a competent operator would have to remove to do economical harvesting would so organize the operation that all merchantable products would be removed at the same time, perhaps with the same equipment, and thereby provide for an economical harvest of all the timber crop in the forest.

In some cases it may prove more economical, if no competent operators are available, for forest managers to conduct their own harvesting operations. Obviously, when such is the case, the forester in charge must be able to plan and conduct the harvesting efficiently. Otherwise he would be sacrificing forest income to his incompetence.

For the reasons given, therefore, it is essential that a forester be able to judge the competence of operators who bid for timber and also be able to draw up practical yet correct specifications under which timber crops will be harvested from his forest. When a forester has the responsibility of obtaining the maximum income from a forest property under his supervision, he also should be capable of conducting a harvesting operation himself. The study of harvesting principles and practices in forests, therefore, is an essential part of a forester's education and training.

CHAPTER 4

PLANNING THE TIMBER HARVEST

Forests grow under a wide range of topographic and climatic conditions—from sea-level swamps to high mountain slopes and from the tropics almost to the Arctic Circle. Harvesting timber crops, therefore, is performed under very variable conditions depending upon the location of the timber to be cut, its size, species, kind of products, the distance they are to be transported to mill or market, and the available transportation facilities.

INFORMATION NEEDED ABOUT THE HARVESTING AREA

Before a harvesting operation of any kind is planned, full information about the forest area in which it is to be conducted must be obtained. Without such knowledge many costly mistakes would be made with respect to the location of camps, roads or other transportation installations, and, very probably, the choice of equipment and methods adopted for the various operations involved in harvesting.

Prior to the acquisition of forest land, some kind of a cruise or timber estimate usually is made to determine the volume of merchantable timber in it so that the purchaser can judge whether or not it is worth the price asked. Hence, some data obtained when timbered tracts were acquired usually are available for analysis in planning a harvesting operation. However, acquisition cruises, unless recent and detailed enough to furnish the data needed in laying out a harvesting plan, are not of much value.

Many changes occur in the size and condition of timber within even a few years' time, due to growth, fire damage, and cutting that has occurred subsequent to its acquisition, so earlier estimates and cruises often are no longer accurate. Also, as time passes, species formerly considered valueless come into use and hence increase the merchantable volume to be harvested. Since acquisition cruises are made primarily to determine whether the volume of merchantable timber is sufficient to justify the purchase price, information needed in correctly locating transportation systems and camps is generally inadequate. Hence, it is usually necessary to cruise or at least carefully inspect forest areas before harvesting operations begin, to supply the data needed for sound planning.

The timber in a forested tract of land varies in size and kind from place to place, except in rather rare instances when it occurs uniformly dense and of the same approximate size over extensive areas. Virgin timber in undeveloped regions is more uniform as to stand per acre and size than second-growth timber; but the variation, even in virgin forest areas, is marked enough to justify a preharvesting cruise. Even were the timber known to be uniform, variations in topography need to be known for the correct location of transportation routes and to determine the transportation method to be used.

Location of Area to Be Harvested. When timber is to be harvested from a new tract, the first information needed is its exact location and extent with respect to known points such as rivers, towns, railroads, or other base features. If a mill is to be constructed for the manufacture of the timber to be removed, it should be so located that the forest products harvested in the forest can be delivered at the most economical cost, with due consideration being given, of course, to facilities for shipping the manufactured products to market.

The most accurate and most useful way in which this information can be expressed is in the form of a map, drawn to suitable scale. Such a map should show the boundaries of the tract in question, also rivers, lakes, roads, railroads, towns, and other pertinent data. The map showing this information ordinarily will be drawn to smaller scale than the map showing the distribution of timber and topographic features within the tract itself.

When harvesting operations have already been conducted on portions of the tract and main transportation routes have already been established, it is not necessary to equal the information required for new operations in previously undeveloped tracts. However, if detailed maps of the areas to be harvested have not been prepared, they should be made, as in the case of entirely new locations.

The survey made of a tract prior to initiating harvesting operations frequently is of a reconnaissance nature and only intensive enough to indicate in a general way the principal features of forest conditions and topography. This is sufficient in many cases to permit the choice of the main transportation method and the location of routes. More detailed maps and descriptions of the actual areas to be harvested may be deferred until shortly before the operation reaches each location, as changes in cutting specifications, species utilization, or newly developed equipment may make such maps obsolete before the harvesting operation reaches the areas so surveyed and, hence, duplication of the work is required.

Distribution of Timber. Obviously it is necessary to know where timber to be harvested is located so that roads or other means of access to it can be planned. This information can be obtained only by making an inspection on the ground and recording the amount and kind of timber in each subdivision of the area.

The forester or timber cruiser making such an examination usually subdivides the area into blocks or units of suitable size and reports on the amount and kind of timber in each. When the area has been surveyed by the government according to the rectangular system of townships 6 miles square, sections 1 mile square, and "forties" of ¼ mile square, it is customary to use these units.

When other surveying systems with no regularity have been used, such as "metes and bounds," it may be necessary to establish regular units of some arbitrary but suitable size, such as 1-mile squares of the same size as sections. This is necessary if the variations in volume of timber by species and size are to be localized instead of averaged over large areas. Since harvesting requires expensive transportation construction, it is not enough to know that the timber, for example, in an area of 2,500 acres, averages 7,000 ft per acre. Actually the timber in the area may range from 1,000 ft per acre to 15,000 ft per acre in stands varying in area from 10 to 200 acres.

A well-planned harvesting road or railroad system is so laid out that the roads tap the heavily timbered areas to hold subsequent skidding distances to a minimum. The same volume of timber will move out over roads or railroads without proper planning, but the heaviest stands might be farthest from the roads and much more skidding would have to be done to get it to roadside or loading point.

The actual data about the timber throughout an area to be harvested are customarily obtained by one or more survey or timber-cruising crews, which vary from one to three or four men per crew, depending upon how much information is recorded. Usually the crew consists of a compass man and a cruiser. The compass man runs a series of courses by hand or staff compass that cross the area being investigated at regular intervals. Prior to making the actual cruise, if no other controls are available, it is customary for the cruisers to install base lines. The base lines are run through the area at appropriate intervals and are clearly marked by blazing trees or otherwise making conspicuous marks. The cruise lines, then, tie in along the base lines, which prevents the compass man from getting too far off his course. Base lines usually are from 1 to 2 miles apart.

The timber within a strip along the cruise lines, or in plots taken at intervals along the lines, is tallied and the volume computed by appropriate tree-volume tables to determine the average stand per acre. The tallies are kept separate for each of the area units, and often also by types and stand-classes, which are mapped by the cruiser. Thus the stands of timber in an area to be harvested are mapped so the location of each is known, as well as the amount, kinds, and sizes of timber it contains.

Less intensive methods are often used, in which cases the cruiser, who may have a compass man or may run his own compass lines, visually inspects each unit area and makes appropriate notes concerning the timber. Such notes may record his estimate of the total volume of merchantable timber within each forty or other area unit, or the estimate may indicate the volume separately by species classes, such as pines and hardwoods, and sometimes also by size classes, showing the volume of timber within specified diameter limits.

Topography. Since the location of transportation routes and the choice of operating equipment to be used in harvesting timber crops must be suitable to the conditions that will be encountered, a knowledge of the terrain is fully as important as information about the timber itself. The cost of constructing roads and railroads varies greatly with the steepness of slopes and soil condition, whether rocky, firm, or soft. Likewise, certain types of equipment used in skidding, for instance, are not suitable for all topographic conditions, so in planning an operation any conditions that would preclude the use of any type of machine or require a particular type must be known in advance, if costly mistakes are to be avoided.

Those making preharvesting cruises or forest surveys, therefore, show on their maps the locations of all streams by size; slopes, either gentle or steep; flat areas; swamps or other soft ground; rocky areas; and existing roads. Contours are usually shown, but where topography is not steep it is often only necessary to indicate it by showing the direction of slope, commonly done by small arrows on the map. When the forester or cruiser making the examination is experienced in road location, he often indicates on his map the most suitable transportation routes.

The more rugged the topography, the more costly it is to build transportation facilities, so the topographic information is more detailed and more accurately located in mountain areas, for instance, than in level sections. Often a transportation route is limited to only one location, as in the valley between high ridges, while in others, such as the coastal-plain regions of the Atlantic and Gulf states, the planner has the opportunity to choose from several possible road or railroad routes. When topography does not require a specific route, the one selected should place the most timber nearest to it, with due consideration being given to the cost of construction. Whenever possible, stream crossings are avoided because of the cost of installing bridges and fills.

General Considerations. When large tracts are being opened up for harvesting, more information is needed than when small, scattered tracts are to be cut. Large tracts usually do not have well-developed road systems, and as larger operations usually are planned for larger areas, more improvement work is done to make them accessible. Scattered small tracts ordinarily are intermingled among farms and suitable road systems are close by. Hence, when tracts up to about 100 acres or so are to be harvested, the crew foreman can walk over the area and choose road locations on the ground, as such woods roads only need to reach some public highway and the small volume that will be hauled over them does not justify costly construction. Such areas, also, usually contain old roads that can be improved at slight cost.

Also, when regular logging or pulpwood-cutting operations are proceeding systematically over a large forest area, the superintendent or foreman often spends a day now and then riding through it to become familiar with the area to be harvested next. In such cases, no detailed mapping is necessary, although it would be desirable to have such mental information recorded in map form for another to use should the superintendent leave or be away from work for an appreciable period of time.

When specialized products such as poles or high-grade veneer bolts are to be cut from an area, someone "looks" the tract to determine whether or not such products can be produced in sufficient volume to justify an operation. The foreman of the crew that will harvest them then goes over the tract and chooses routes to the areas of most concentrated volume and plans his operation from personal inspection. In some cases the cruise parties, in systematically sampling areas to be harvested, tally specialized products so a close estimate of their volume can be obtained.

Sites for camps are also chosen from information obtained from preharvesting surveys. Camps are placed where they can be economically supplied, as well as from the standpoint of accessibility to cutting areas. The same route often serves both purposes but in the case of some northern operations the routes for sleigh-hauling logs or pulpwood to rivers or lakes for stream driving are in opposite directions to those used to bring supplies to the camps. Also, in such cases it is desirable to choose routes with favorable downgrades in the direction of loaded travel, and a favorable grade for product hauling would be adverse for "toting" supplies to camp. Hence, different routes are sometimes chosen, at least for part of the distance.

The use of aerial photographs in planning harvesting operations is increasing. Such photographs show in accurate detail many features of both stand density and topography. With recent developments in the use of sterioscope viewing, it is possible for a trained man to interpret many features of a forested area simply by carefully analyzing aerial photographs. They are principally valuable in connection with ground surveys or cruises, as field parties with such photographs, or maps prepared from them, can plan their work to much better advantage. The location of large bare areas and nontimbered swamps, for instance, can be known in advance; and once their edges are reached by men running strips on the ground, they can offset or turn back and avoid traversing areas in which detailed information is not needed. Without advance information concerning such features of an area, field parties would have to traverse them to determine their extent, even though they contained no timber.

In forests under management, practically all of the information needed in planning harvesting operations concerning the timber and topography is already available. Data concerning the location of stands by area, age, and volume per acre are needed in determining growth and organizing forest operations, and are a matter of record in the forester's office. Good maps are essential for forests under management and these show all topographic features as well as stands of timber for which detailed information is on file. Hence, no special investigations need to be made of scheduled harvesting areas. The harvesting operation itself, however, needs to be carefully planned and gauged as to the amount and kind of products to be removed and the kind of cutting called for under the management plan.

TYPES OF OPERATIONS POSSIBLE

The diversity of conditions under which timber is harvested is reflected in the many different ways in which operations are conducted. Since transportation of timber products from the stump where they are cut to the utilization point requires most of the effort and incurs most of the expense in harvesting, the movement of tree products, starting with the first step at the stump to the final delivery at the mill or consuming point, is a very important phase of harvesting. Much thought and effort have been given to developing economical ways of handling and hauling logs and other timber products, and most of the methods now used are the result of long experience with actual operations. New equipment and new methods are justified only when they result in more economical or less damaging operations.

The harvesting operation consists of a number of distinct steps or phases, each of which can be analyzed separately, although all must be integrated closely with the others for efficiency.

The principal steps in harvesting timber crops are:

1. Selection of trees for cutting to conform to the specifications for various products and to the plan of forest management.
2. Cutting, which includes felling the trees, the removal of limbs, and bucking them to desired lengths.
3. Bunching, or the short-distance assembly of products for skidding.
4. Skidding, or the movement of products comparatively short distances to loading points.
5. Loading on the vehicles used in hauling.
6. The haul from the forest to the mill or market.

Each of the above operations can be performed in many different ways. The hauling of product from the forest to the mill may be by any one of the following methods or by almost any combination, provided that the method is adapted to conditions: railroad, truck, barge, raft, stream driving, cargo vessel or flume. Loading must be adapted to both hauling and skidding and may be by animal power, by various types of heavy-duty loaders, by hand labor, or, in some cases, by the use of gravity. Skidding may be by animal power, by tractors, by overhead or ground-lead cable systems, by gravity, or, in rare instances, by hand. Bunching may be by hand, animal power, or mechanical power.

Cutting involves severing the tree to be felled near the ground and then cutting the bole of the tree into the desired products, such as logs, pulpwood, poles, piles, ties, veneer bolts, fuel wood, or mine timbers. This may be done by one-, two-, or three-man crews using handsaws or power saws of several types, or a combination of methods, such as felling with power saws and bucking with handsaws. Some small products are felled and bucked with the ax, but only in minor amount.

Other phases of harvesting likewise are affected by conditions of climate, topography, and location. In localities where labor is lacking or not available in sufficient amount or skill, provision must be made for housing and feeding the harvesting crew. When local labor is available and suitable, then no such facilities need be provided or they may be of only an emergency type.

The harvesting operation also may vary from one producing a single product, such as pine saw logs, to one yielding a variety of products. Saw logs, pulpwood, veneer and handle bolts, ties, poles, and other products may be produced from the same area when there is a ready market for such items and the timber is of such species and quality that they can be cut from it. The various products may be removed at the same time or they may be removed by successive operations, depending upon which is the most advantageous.

The wide range of operating methods available for harvesting timber makes it possible under almost any conditions found in nature. Certain factors, however, operate to limit the choice of methods for a given forest. In planning a harvesting operation, due consideration therefore must be given to those items, concerning the timber itself, the topography, the climate, and the location, that have a bearing on the choice of operating method.

FACTORS THAT INFLUENCE CHOICE OF OPERATING METHODS

Size of Product. Since transportation of forest products within and from the forest constitutes most of the effort and expense in harvesting operations, those conditions which affect the movement of the products have most weight in determining the kind of operation to be conducted. Thus, large products such as logs require the application of power, either animal or mechanical, even for bunching. Small products, on the other hand, such as pulpwood cut to 4- or even 8-ft lengths from small trees, may be and frequently are bunched and loaded by hand labor. The choice of power unit, when hand methods cannot be used, depends to a considerable extent on the size of product. Either animal or mechanical power may be used if the products are not too large. The large trees of the West preclude the use of animal power, so some form of mechanical power usually is necessary, either tractors or cable skidders.

Daily and Annual Output Required. Another factor affecting the type of operation in harvesting is the daily and annual output to be removed from the forest. Large sawmills requiring many thousands of feet of logs a day must obtain a steady supply of logs from large-scale harvesting operations. Under such circumstances a standard-gauge logging railroad using steam or Diesel locomotives is possible and often is the most economical method because it will be used every working day of the year and for a number of years. Thus the large investment and operating cost of a logging railroad can be charged against the large volume of product that will be harvested and the cost per 1,000 ft of logs or other units will be within reasonable limits.

On the other hand, a small daily output could not bear the cost of a logging railroad, and some other form of transportation must be used, such as motor trucks. Although trucks are used on many large operations, the fleet must be large enough to supply the mill with its log requirements. The large operator, therefore, has a choice between trucks and logging railroad, but the small operator is limited to trucks.

Amount Harvested per Acre. One important requirement for railroad logging is that the timber being harvested be concentrated so the mileage of railroad is not out of proportion to the amount harvested per mile of track. Logging railroads are most common when saw logs are being harvested to supply a large mill with virgin timber, as the heavy stand per acre provides sufficient volume per mile of track to keep the cost per unit low. Large pulp mills, such as those in the Southern states, although they use up to 1,500 or 1,800 cords of wood a day, cannot operate private railroads to haul their pulpwood to the mill because the wood is produced over a wide area from a large number of scattered cutting operations that remove only a comparatively small amount of wood per acre. Many large sawmills also obtain logs from scattered small operations and, when this is the case, trucks rather than railroads must be used, as it would be exceedingly costly to build a railroad to each small tract and operate log trains for transporting the logs from the woods to the mill.

Labor. Many timber-harvesting operations must be conducted in undeveloped regions where labor is scarce or unavailable, and consequently provision must be made to house and feed the workers. In other sections forests are intermingled with farms and settled communities and woods workers can go to and from their jobs in the woods much as they would in a city or town, using their own cars or other transportation facilities. Most of the forest areas of the South, for example, are in well-developed localities and small harvesting operations can be conducted without providing accommodations for the men. When such are provided they usually are for temporary occupancy during the work week, the men driving their cars to and from their farms or homes over the week ends.

In most parts of the forested areas of the Western and Northern states and Canada, on the other hand, no adequate roads are available for travel and farms and villages are few, so labor for woods work is scarce. Conducting even small-scale harvesting operations under such circumstances obviously requires that camps be built and preparations be made for feeding the workers, who are recruited elsewhere.

Many logging camps are, to all intents and purposes, small towns with all modern facilities and services supplied by the company conducting the operation. Even where labor is plentiful, it frequently is necessary to provide quarters for skilled workers not available from the local labor supply. Some such camps are portable while others are permanent, depending upon how often the harvesting location is changed and whether or not the plan of forest management provides for a sustained yield with continuous harvesting operations within a localized area. The more frequent the change of camp location, of course, the more readily portable must be the camp buildings.

Duration of Operation. The length of time during which harvesting is to be carried on influences the kind of operation possible. Short-term operations of only a few weeks or months cannot be conducted with heavy improvement expenses, even though the daily or annual volume is large. Railroads used in harvesting saw timber require not only heavy daily volume, but also a long-time operation to permit writing off the cost over a large number of units. To build a railroad for only a few years' use would make the operation unduly costly, as much of the cost is incurred in preparing right-of-ways and building tracks, and these improvements cannot be salvaged and moved without incurring additional expense. Short-term operations, therefore, must be organized so only a minimum of fixed expense is incurred, and the cost of railroads, private truck roads, flumes, and other costly installations is prohibitive.

One important advantage in harvesting according to a plan for permanence, or sustained yield, is that roads, bridges, and other fixed-cost improvements will be used continuously, or at least will be used again and again in the future, which means greatly reduced depreciation rates in line with the usable life of the improvement rather than an arbitrary depreciation figure based on the amount of timber to be handled over or with it. Many a camp, railroad, truck road, and other installation has ceased to have value only because no more timber was available to be hauled over or handled with it, although it was physically usable for many more years.

Under permanent use, depreciation of equipment and installations can be charged off against timber harvested on the basis of actual useful life rather than economic life based only on the volume of timber to be handled during one harvesting period.

Integrated Use. Some harvesting operations are conducted for only one kind of product, such as saw logs, pulpwood, or poles. When this is the case, the type of operation planned need be based only on consideration of the problems involved in cutting and removing the one product. Other operations involve cutting and removing two or more kinds of products, such as saw logs and pulpwood, or saw logs, veneer bolts, and poles, and consideration must be given to handling them all.

When several kinds of products are to be harvested, much closer utilization of the timber generally is possible and greater financial returns are obtained because each tree and each portion of each tree can be made into the product for which it is best suited and most valuable. Different products ordinarily are not cut and handled at the same time or by the same crews. Instead, one operation follows the other, but because the harvesting operations are coordinated, each can use many of the facilities constructed or brought to the location, such as roads, loaders, and skidders. This reduces the harvesting cost of each product and returns the highest financial yield. In fact, in many cases it would not be possible to engage in harvesting one product, such as poles or veneer bolts, if other products could not be removed in bulk to carry the major portion of the cost of developing transportation installations and perhaps even camps and similar fixed improvements.

The type of harvesting operation that can be planned for a given forest, therefore, depends to a considerable degree on whether or not several or only one product is to be removed. When the operation is to be integrated, plans for handling a wide variety of products must be made. When only one product is to be cut, only that one need be considered.

Facilities Already Available. The way in which a harvesting operation can best and most economically be conducted is often influenced by existing installations, improvements, or equipment owned by the operator or available to him. A change from railroad logging to tractor and truck logging or from team logging to trucks and tractors means that the equipment and supplies already on hand must be sold or discarded. When this involves too much loss it is usually advisable to use the already available equipment and adapt it to the job if this can be done without excessive cost.

This is one reason why obsolete instead of modern methods and equipment frequently are used. To discard a logging railroad would require the construction of truck roads, unless public roads and highways are favorably situated. The discarding of a skidder or locomotive to replace it with some other type of equipment or method could mean the writing off of a considerable loss, as well as the expenditure of additional funds for the new equipment or improvements.

When the operating advantages of new equipment are sufficiently great, the obsolete equipment should be discarded and replaced with new. Frequently old methods and old equipment are used because they have been fully depreciated and only operating costs are chargeable against the timber being harvested. This apparent saving is not always real, however, and it is not uncommon to find inefficiency in an operation due to reluctance to discard obsolete and worn-out equipment, even though much greater efficiency and substantially lower cost would result from so doing.

Legal Limitations. In some localities, legal limitations prevent the use of certain harvesting methods. State highway-trucking regulations usually require that the load weight be below specified amounts, and where logs are very large and large loads must be trucked, it may be illegal to transport them over state highways. In such cases private roads or railroads must be built, if this is possible, or smaller loads must be hauled.

Stream driving is not allowed in navigable streams as it would obstruct and menace navigation and waterway developments, and so it is not always possible to transport by stream driving even though waterways are present. Generally, however, legal limitations affecting the harvesting of timber are not too prevalent or too severe and not a great deal of difficulty results from them.

ALL PHASES OF HARVESTING MUST BALANCE

The most economical harvesting operation is one in which all equipment and all labor is used to optimum capacity. Each stage of harvesting, therefore, must be in balance with other stages. In an operation where the product being cut is not stored between one stage and another, as occurs in some instances, cutting must be no faster than bunching, and bunching no faster than skidding, and skidding must be kept geared to loading. Also, hauling must be at a rate sufficient to keep the mill in logs, pulpwood, or other products, as the case may be. Likewise, receipt of wood at a manufacturing establishment cannot exceed production except for short periods to accumulate a supply against bad weather and interrupted harvesting. Careful selection of harvesting method and type of equipment, therefore, is necessary for a well-conducted operation.

Loading is one of the key operations in harvesting and the type of loading equipment used frequently determines what equipment or methods will be best adapted for other operations. Loading is the link between skidding and hauling, and particular attention must be given to this operation to avoid an unbalanced operation. Thus, if a certain type of loading equipment can, with the size of timber and kind of roads found on a given operation, load out 10 truckloads of saw logs or pulpwood per day, it would be uneconomical to use it on an operation otherwise limited to 6 truckloads a day. Similarly, if the requirement was 14 truckloads a day, it would be inefficient use of loading equipment and men to operate two such loaders. The correct balance, in such an instance, would be obtained between the various stages of harvesting by the use of some other kind of loader or loaders.

In the following chapters the stages in harvesting and the various ways of accomplishing each under different conditions and kinds of equipment are discussed. Harvesting, being an outdoor operation, must be conducted with due regard to weather conditions and a terrain that varies as harvesting progresses over an area. No standard method of harvesting can be established for all operations, as can be done for production activities housed in factories and carried on under controlled conditions. Seldom, if ever, are any two harvesting operations alike in all respects, so equipment must be adapted to each and used in the most productive and economical way for best results.

Although the application of harvesting equipment and the use of harvesting procedures vary from operation to operation, the principles of harvesting remain the same. Particular attention, therefore, will be devoted in the following chapters to harvesting principles, as they hold true even when methods change markedly. One who is familiar with the principles involved in harvesting can more readily judge how to adapt suitable equipment and methods to any given or proposed harvesting operation.

The steps involved in a harvesting operation begin with the selection of trees to be cut and end with transportation of the products to mill or market, but in planning an operation, consideration of these steps is in reverse order. Methods of felling and bucking or loading, for instance, cannot be selected until the transportation method has been determined. Likewise, skidding methods must conform to the transportation method and also to the loading method.

The first decision to be made in planning a harvesting operation, therefore, is the transportation method. It may be by any one or more methods, by land or water, or by a combination of methods, but until the method or methods of transporting products from the forest to the mill or market has been decided upon, methods of conducting harvesting steps cannot be determined. Of course, the method of felling and bucking trees may not be entirely dependent upon the method of transportation, but it is influenced greatly by it. For instance, with stream driving it would not be practicable to cut trees full length, as they would jam in the streams too easily. Furthermore, from the standpoint of an operating company, the accessibility of the tract is dependent chiefly upon the methods by which the products can be transported to the mill. Therefore, the first concern of one planning a harvesting operation is the selection of transportation method.

Once this has been decided, it is possible to select the methods to be used in loading, yarding or skidding, bunching, and finally felling and bucking. The methods and equipment must not only give a balanced operation but they must be adapted to the size of timber to be harvested and the volume to be removed per acre.

After a harvesting operation is in progress, the sequence of operations begins with the selection of trees for cutting and because they may be analyzed more clearly in this sequence, the operations are considered in that order in this text.

CHAPTER 5

WOODS LABOR

Forests commonly are not in settled communities and workers engaged in harvesting timber crops cannot, except rarely, commute daily between their homes and the location of their work in the woods. Even in the South, where the forest area does not constitute a wilderness and is traversed by highways and roads and intermingled with farms, it is too time-consuming and too expensive for the men to go back and forth daily between their homes and their work in the woods, except to a minor degree.

Some provision, therefore, for housing and feeding labor is essential for almost any harvesting operation other than one conducted by a farmer in his wood lot, or by small groups of woods workers taken each day from small towns to work in the nearby forest, as is the case with many pulpwood-cutting and small sawmill operations in the South and East. This is seldom possible in the West and northern forest regions, where there are few farmers and other local residents living near cutting locations.

Camp facilities supplied to labor on harvesting operations vary greatly, depending upon the size of the operations and the divergent customs and needs of different locations. When only a few workers are engaged, the facilities often are primitive, with but few comforts. Camps may be of the family type, the married men living in private quarters with their wives and children, with separate quarters for single men; or they may be strictly bachelor camps. Whether or not a camp is of the family type depends upon local custom and the policy of the company running the operation.

Most camps are portable and are moved from time to time to keep the crews within easy reach of the woods operations, such as cutting, skidding, loading, and road and railroad construction. When the entire area to be harvested can be reached from one camp, a permanent type is preferred. Often a permanent headquarters camp can be used with smaller portable camps operating throughout the area. The time required to get men to the job from camp, rather than distance, is the main consideration in deciding whether a permanent or temporary camp should be built. When travel time to and from work exceeds 1 hr, it is usually better to use portable camps.

Camps are more than lodging and eating places for the men; they also are headquarters for control of harvesting operations and for the maintenance and repair of mechanical equipment of all kinds used in the woods. A camp store is maintained to sell practically all items needed by the men and, in family camps, by their families. In the latter, the camp store handles grocery and meat items and, in fact, is a general store. In bachelor camps most grocery items are not available, as all hands normally obtain their meals at the boarding establishment.

Garages and repair shops are maintained at the larger camps to keep tractors, trucks, power saws, and even locomotives, power loaders, and skidders in good operating condition. This means that the garages and repair shops must be well equipped with heavy tools, heat, and power. Garages and shops are seldom portable, as they must be large enough to handle large trucks and tractors conveniently. When camp is moved they are either abandoned or demolished to be rebuilt at the new location with salvaged materials.

An office is maintained at all camps of any size for keeping pay-roll, employment, and numerous other records required by state and Federal labor laws. Also, the camp office keeps cost records on production and operating costs of trucks and other equipment, as well as accounts for the camp commissary. Inventories of all equipment, parts, supplies, fuel, tools, and all similar materials must be kept for cost-accounting purposes.

Other important features of camps include a water supply, electricity, heat, and power installations. Wells are ordinarily drilled for pure water, although surface water may be used if suitable. Water is needed not only for cooking and sanitary purposes but for boilers generating steam for the camp and for locomotives when they are used. The essential pumps and tanks are, of course, a part of water developments. Fuel may be wood, coal, or oil; the latter is used either for internal-combustion motors or for burning. Coal or wood, of course, is burned for heating and the generation of steam used to develop power for the repair shops, generation of electricity, and other purposes. When oil is burned it merely supplants coal or wood as a heat-producing fuel, but when used in Diesel motors it generates power only, either for direct application in the shops or for conversion into electric power through one or more generators. Larger camps also have recreation rooms, libraries, and first-aid stations with competent medical and nursing attendants.

Camps of the modern type, like those maintained on the larger Western and Southern operations, are occupied the year around, as there is no idle season such as is the case on many Northern operations. Such camps are kept well painted, clean inside and out, with a degree of landscaping and well-kept lawns. Being portable and subject to moving every few years, there can be no fixed permanent improvements at the camps other than those that are justified for the comparatively short period of years they will be at one location.

Harvesting operations conducted along lakes, rivers, and swamps very often are provided with floating camp facilities. Barges or houseboats are equipped for sleeping and eating, and instead of being moved by train or truck, such camps are towed from place to place by tugs or river boats. Under such logging conditions, better and more mobile camps can be maintained as houseboat camps than they could be as land camps. As there are not a great many waterside operations, floating camps are not numerous.

TYPES OF CAMP HOUSING

Comfortable sleeping accommodations are necessary for men doing hard physical work every day in the woods, and practically all large-scale modern operations now provide the best possible quarters for their employees. Camp houses vary, of course, with the locality, local customs, and the method of conducting the harvesting operations.

Individual Units. Most Western and Southern camps provide individual units or camp homes that are portable. This is the case because good transportation facilities are available, either logging railroads or truck roads over which the camp buildings can be moved. Individual units provide sleeping accommodations for from two to four men in bachelor quarters or for one family when family quarters are provided.

When the individual units are used, they must be small enough to be transported on log cars or trucks. For family accommodations two or more single units are often placed together and combined into one camp house. In some cases, with railroad operations, the units are mounted on railroad trucks and parked on spurs when the camps remain in one place only a few months to a year at a time. In similar fashion, spur tracks are laid out at a camp location and the units not mounted on railroad trucks are placed on log cars and spotted along the spur tracks. They can then be jacked up and the log cars removed to be used in hauling logs. When the camp is to be moved again, a log car is spotted beneath each house and the blocks removed, permitting it to settle on the car, after which it can readily be moved to the next location without disturbing the contents and with a minimum of preparation. Camp houses that are moved by trucks and trailers also are often placed high enough off the ground so a truck trailer can be backed under them; the house is lowered off the blocks until it rests on the trailer bed. This eliminates skidding them upon trucks or cars, which upsets the contents.

Bachelor quarters at some camps in the Southern states are old railroad coaches that have been refitted for sleeping dormitories by removing the seats and installing cots or bunks, either single or double-decked. Camp house units that are moved intact on log cars or trucks are provided with heavy skids beneath so they can be pulled onto the vehicle moving them

FIG. 1. Portable logging-camp buildings on an Alabama truck-logging operation. *A*. The camp office and commissary. *B*. Sleeping quarters for six to eight men. *C*. The superintendent's quarters. These units are moved by skidding them up on log trucks and trailers. Electric power and light are provided.

FIG. 2. Loading and transporting portable camp houses on logging railroad cars in Oregon. Units too far from the track for the loader to reach are skidded to it on the ground with tractors. Note the two stringers beneath each house to provide the necessary rigidity for handling. Such units may also be moved on log trucks and trailers. (*Photographs by Paul Hosmer, Bend, Ore.*)

or picked up bodily by a loading crane and placed on the car or truck, except when the car or truck can be spotted under them.

The camp superintendent or foreman is provided with personal quarters and the crew foremen in bachelor camps often occupy separate quarters. In the Southern states, when both white and negro workers are housed in the same camp, separate houses in separate sections of the camp are provided for each race, as it is never possible or desirable for them to occupy the same houses and other personal camp facilities, nor for their houses to be intermingled. Likewise where two or more nationalities are represented in a camp crew, it is often desirable to arrange accommodations so that each group will be together.

When individual units are provided, housing two to four men, space is usually at a premium but there is room to read and relax during leisure hours in the evening and during holidays. In addition a central recreational building often is provided and equipped with current magazines, books, games, and often motion-picture projecting equipment.

Washrooms are either included in each unit or are provided in conveniently located separate structures. In modern camps these facilities are the equivalent of those available in towns and cities, with septic-tank sewage disposal. Electric lights are nearly always provided, as portable electric light plants are available in all sizes and, in the larger camps, electric power is needed in the shops and for refrigerating food.

Large Temporary Units. Many camps in the northern portions of the United States and in eastern Canada are large bunkhouses of the traditional lumberjack type, although greatly improved in sanitary and comfort features over the old-time camps. This type is used when the principal operations are in winter with sleigh hauling, as transportation routes are not as well developed and it would be difficult to move individual units to new locations, particularly during the summer months. Therefore fabricated camp buildings of a temporary type are constructed and when operations have been completed from that camp, they are abandoned or the lumber and fixtures salvaged and re-used at the next location. Such camps are planned for at least 3 years' use, so that enough timber products will be handled to write off the camp cost at a reasonable rate per unit. The bunkhouse type of camp building is more economical when it is to be used for only a few years.

Since the introduction of portable panel-type buildings, such as were used in many Civilian Conservation Corps camps, this type has been used to a limited extent. The panels comprising walls, roofs, partitions, and floors are bolted together and can be knocked down and reassembled without too much difficulty.

Bunkhouses that will sleep up to 40 men are not uncommon. Toilet and bath facilities are provided, usually in separate buildings, and a recreation building also may be built separately, or enough room may be left at one end of the bunkhouse for relaxation and reading. Either single or double-decked bunks are installed barracks style and the men all sleep in the same large room. The disadvantage of knockdown buildings is that the contents must be removed and transported separately. In portable units the contents, such as household furnishings, can be left inside, although they should be packed or fastened down solidly.

Portable-unit buildings are used to a limited extent in eastern Canada. They are constructed with heavy runners beneath and can be moved over prepared iced or snow roads by pulling with a tractor. The runners slide over the surface in the same manner as sled runners.

Floating Camps. Housing facilities for crews driving logs or pulpwood down streams are often of the houseboat type so that they can be kept close to the crews working the drive as it progresses downstream. The living accommodations are ordinarily comparable to those provided in

FIG. 3. A stream-driving camp in Idaho. The camp is floated down behind the drive as close as practicable to it and may be towed upstream for use the following year or dismantled at the lower extremity of the drive and another floating camp built upstream for similar use the following year.

bunkhouses, except that the limited space in a floating camp does not permit accommodations for as many men as in stationary structures.

Since the floating camps for stream-driving crews are used only for a few weeks each year, the duration of the average drive, they are not so comfortably appointed as land camps. Adequate provision is made for comfortable sleeping and for drying clothes, as the men frequently get wet keeping logs from stranding and preventing log jams. Floating camps used on waterways along the north Pacific coast and rivers in the Rocky Mountains and Southern sections are in use practically all year and can be readily floated from location to location. Hence, such camp houseboats are sturdily built and equipped for continuous use.

Small-crew Camps. Some woods work is done by small crews of from two to six men who provide their own shelter and do their own cooking. These crews may be working considerably in advance of actual harvesting operations and therefore are unable to stay at the main camp. Or, when the operation is seasonal, they may be working during the summer, when the camp is unoccupied and inconvenient to reach when the ground is not frozen.

Often such work is contracted by individuals who organize a small crew to work under their direction. The work may consist of cutting roads, building bridges, constructing main camps, or recovering logs stranded during previous drives, in northern areas. Small contractors also occasionally produce pulpwood and other products requiring a minimum of animal or mechanical power, and occupy small temporary camps of this kind. The shelters built are often crude lean-tos built from round logs cut from small trees and roofed and made weatherproof with a roll or two of tar paper. Tents are also used, as they can be packed in by the men or taken in with canoes or wagons.

Camps of this type are located where pure water from springs or clear streams can be obtained for drinking and cooking. The men bathe in lakes or streams and cook outdoors or under a tent fly, and in general live much the same as hunters or fishermen on a protracted camping trip. Occasional trips may be made by one of the men by canoe, on foot, or in a wagon to obtain food and supplies from some central camp or settlement.

In the Eastern mountain regions and throughout the South, small crude camps are also occasionally used to house a few men engaged in producing logs, pulpwood, acid wood, ties, and similar small products obtained by hand labor. It is not usually necessary, however, for the men to remain continuously in camps, as passable roads are usually nearby and the men can return to their homes over week ends since there ordinarily is a car or two in a crew, or they can ride the trucks hauling the products from the operation.

The camp houses may be portable and moved from place to place on trucks, but few if any other comforts or sanitary facilities are installed. In some cases, there are even no windows and only a doorway with a gunny sack hanging in it to keep out most of the rain. Camps of this crude type are usually used by negro crews, who are accustomed to the few comforts of primitive living, and who are willing to live under such conditions for a few days at a time.

FEEDING FACILITIES

Men doing heavy woods work require nourishing meals and logging camps are noted for the good quality and large quantity of food served. Of first importance, therefore, in all camps housing woods workers is adequate provision for cooking and serving meals. A well-lighted and clean dining hall or mess hall with adjoining kitchen is one of the most important buildings in camp.

Mess halls are large enough, preferably, to serve the entire crew at one sitting, so the buildings in larger camps usually are too large to be moved intact. In camps where the buildings are on railroad trucks it is possible to have one or two mess halls that can be moved in the same manner as railroad-car bunkhouses. However, since special efforts are generally made to have attractive mess halls that can be well lighted and kept clean, it is generally preferable to erect temporary mess halls and kitchens or use the type that can be dismantled in panels and reassembled when moved.

FIG. 4. View of a Western logging-camp dining hall at mealtime. Loggers require nourishing food in adequate amounts such as is shown here.

Separate living quarters are provided for cooks and their helpers, as they work different hours than other employees. In Southern camps where both white and negro workers are employed, separate dining rooms are provided for each.

The equipment in mess halls consists of tables and either benches or chairs. Either one or more long tables or a larger number of small tables may be used. No other fixtures other than a heating stove or two and space for hanging hats and coats are needed. Food is usually on the table when the men are called and serving dishes refilled as necessary during the meal. Because of the large number of dishes to be washed and other necessary work involved in feeding large groups of men, the custom of no talking at the table and no after-meal smoking at the table has long been in effect. Also, the man who is late for a meal usually finds that he has missed it.

TABLE 9. TOTAL FOOD CONSUMPTION PER MAN IN LOGGING CAMPS OF ANACONDA COPPER MINING COMPANY, WOODWORTH, MONT., IN 1942*

Item	Pounds per Man-day	Item	Pounds per Man-day
Flour	0.6351	Sugar	0.1954
Meats and fish	1.6410	Tea and coffee	0.0901
Canned milk	0.0763	Eggs	0.2942
Fresh milk	1.1699	Cereals	0.0805
Canned fruit	0.3033	Syrup and molasses	0.1653
Fresh Fruit	0.4476	Spices and condiments	0.0906
Dry fruit	0.0438	Lard	0.0880
Dry beans and peas	0.0371	Cheese	0.0517
Fresh vegetables	1.2622	Extracts	0.0012
Canned Vegetables	0.1770	Total	7.0150
Butter	0.1447		

* Number of meals served, 163,437; average number of men, 149.
Source: *Timberman*, May, 1943.

The kitchen always adjoins the mess hall and usually is a part of the same building for convenience and efficiency. It is equipped with the usual kitchen essentials such as stove, refrigerator, food storage space, sink, dish shelves, and hot-water tank. Kitchens should not be crowded and should be large enough for the cook and his helpers to work in comfort. In large camps feeding a hundred or more men, a refrigerating room of the walk-in type is necessary to keep an adequate stock of fresh meats and other perishable items on hand.

Camp morale is largely dependent upon the quality and amount of food served as well as suitable living quarters. Kitchen and mess-hall cleanli-

TABLE 10. AVERAGE DAILY FOOD CONSUMPTION PER MAN IN WEYERHAUSER TIMBER COMPANY'S LOGGING DIVISION IN PACIFIC NORTHWEST

Item	Pounds
Fresh meats	0.9
Smoked meats	0.5
Potatoes	1.0
Flour	0.8
Fresh milk	2.0
Eggs	0.20
Butter	0.16
Canned fruits	0.5
Fresh fruits	0.5
Canned vegetables	0.33
Fresh vegetables	1.0
Sugar	0.5
Miscellaneous	0.11
Total	8.50

Source: *West Coast Lumberman*, July, 1938.

ness is also essential in the interest of the men's health, so competent camp cooks are highly desirable. The variety and quantity of food served in many Western camps is indicated by the food lists shown in Tables 9 and 10. Meals are usually served in mess halls morning and evening, the men taking prepared lunches for noon. In cold weather it is the practice in some camps to take hot lunches to the crews at noon.

THE CAMP STORE

In family camps a store or commissary is always provided to supply the needs of the men and their families. Groceries, meat, clothing, toiletries, household utensils, and similar items are carried, in keeping with the customary wants in different regions. It is usually possible for the men and their families to make occasional trips to shopping centers, so only the everyday needs of the camp residents are ordinarily carried.

Bachelor camps do not stock grocery or meat items, since the men have their meals in the camp mess hall. However, they do need work clothing, personal toiletries, tobacco, and similar items, and it is necessary that a camp store or commissary be maintained to supply the men with these articles. In northern bachelor camps occupied during the winter, when camps are practically isolated for weeks, it is essential that each camp operate a store, or "wanigan," as it is called. The articles carried are known as "van" in eastern Canada and the northeastern portion of the United States. Table 11 lists the van sold per man per day over a period of 158 calendar days from September to February, 1940-41, at a camp in Quebec, Canada.

TABLE 11. CAMP-STORE SALES PER DAY PER MAN AT WINTER CAMP IN QUEBEC, CANADA, 1940–1941

Item	Sales per Man per Day
Tobacco and smoking supplies	$0.067
Clothing	0.067
Footwear	0.021
Chocolate bars and candy	0.001
Postage stamps	0.005
Patent medicine	0.011
Miscellaneous	0.008
Total	$0.180

Source: Salkeld, W. B., The camp store, *Woodland Rev.*, vol. 12, No. 9, August, 1941.

CAMP SHOPS, GARAGES, AND BARNS

The extensive use of trucks and tractors in harvesting makes it essential that facilities be maintained at camps for the maintenance and repair of this equipment. Too much lost time and expense would be incurred, when considerable numbers of trucks and tractors are used, to have them maintained and repaired in commercial garages or shops. Hence, modern logging camps include well-equipped repair shops staffed with competent mechanics for the upkeep of mechanized equipment.

The buildings used for this purpose should be large enough to handle the average repair load without crowding and should be well lighted and, in cold weather, heated. It is seldom possible, even in portable camps, to construct shops so they can be moved intact. They are either of temporary construction or of the knockdown panel type that can be taken apart and reassembled. Space should be provided for both regular maintenance work such as greasing, tire changes, and motor tuning and for complete overhaul jobs.

Storage space is also needed for spare parts and supplies, and gasoline- and oil-storage tanks must be provided. Suitable pumps for filling trucks and tractors with gasoline or Diesel fuel are also a part of the normal equipment.

It is always desirable to have garage space for storing trucks and tractors overnight and over week ends instead of leaving them out in the weather. In moderate climates this may consist of open sheds but in cold climates during freezing weather it is a practical necessity to have enclosed and heated garages for storing trucks and tractors when not in use. Delay in starting motors in the mornings during sub-freezing weather means wasted man-hours and the extra wear that results from running them cold before lubrication is fully effective is also costly, so the expense of providing warm storage garages is justified.

When horses or mules are used in harvesting, barns and hay- and feed-storage sheds must be provided. The kinds of barns needed vary with the climate. In the South very simple shelters against rain are all that is necessary. In the North, of course, barns must be warm and they are tightly built either of framed construction or of logs, and covered with weatherproof paper. Hay and feed are stored either in one portion of the barn or in adjacent buildings where it can be fed conveniently. Room must also be provided for storing the harness and caring for it. Also, a box stall or some other arrangement should be provided where injured or sick animals can be isolated and treated. A lot of suitable size should be fenced adjoining the barn where the animals can exercise and remain outside during warm weather. Equipment necessary for shoeing the animals and caring for their feet must be provided and veterinary supplies should also be stocked for the treatment of minor ailments.

Railroad shops are located at most camps where logging railroads are used. When there is one main or headquarters camp with smaller portable camps operating in connection with it, the shops, of course, are located at the more permanent camp. Repair and maintenance of locomotives, log cars, loaders, skidders, track-laying machines, and any other rail equipment is customarily done at the camp shops or, in some cases, at shops located at the mill.

Shops equipped to overhaul and maintain heavy-duty railroad rolling stock are in buildings too large to be portable. Heavy hoists, turning lathes, and other railroad-shop equipment suitable for taking locomotives and other logging machines apart and putting them together are necessary in railroad shops.

OTHER CAMP BUILDINGS

Other camp buildings, varying with the kind of camp, its location, and size, are needed to round out the typical harvesting camp. They include most but not necessarily all of the following.

Office. A camp office force keeps pay-roll records, orders supplies, keeps inventories, and handles other routine accounting and record-keeping duties incidental to the conduct of an important department of a manufacturing company, or of an independent timber-harvesting enterprise. Timekeepers turn in daily records of men working and pay rolls are drawn up. After appropriate deductions are made for state and Federal taxes and charges for purchases, the men are paid off either by check or in cash.

Accurate cost records are kept at many camps to determine the most economical type of equipment or method of operating. This requires daily or weekly reports on fuel consumption, labor used, repairs, and other costs chargeable against particular machines or operations. These records must be checked and analyzed and the unit costs determined in terms of work accomplished or volume of product handled.

Records of areas cut, timber volume removed, and preliminary operations such as road or railroad construction, bridge building, surveying, and other activities are also made a matter of record in the camp office. When timber-cutting rights are owned by the operating company for limited periods of time, a careful record of them must be kept to be sure each is cut before the time allowed expires. Commissary charges against the men for meals and store purchases are kept in the camp office so that appropriate deductions can be made against their pay. Inventories are kept of machine parts, stocks of various kinds, and commissary and cookhouse supplies. All of this record keeping and cost accounting requires an adequately equipped office and clerical and accounting personnel.

Office buildings can be made of the portable type, as a comparatively small office can handle the accounting work for a large operation. Smaller camps may have only space with a desk where the timekeeper, commissary clerk, and foreman can each keep his own records. Small camp or large, however, an office of some kind is needed for even the most limited records.

First-aid Station. Provision should always be made for caring for men who are injured on the job or who become sick in camp. In small camps a spare room or cot may be all that is necessary but in larger camps a separate building should be, and usually is, provided for first-aid treatment. Seriously injured or sick men are taken to company hospitals or to company headquarters where they can receive suitable care.

Accidents unfortunately are not uncommon in harvesting operations and first aid, promptly administered, saves later complications, so suitable space for giving such treatment by a competent nurse or medical attendant is important. Also, it is desirable to isolate sick men, even though they are not ill enough to be hospitalized, to prevent the spread of contagious diseases and to give them better care, rest, and quiet.

The first-aid station, therefore, should be large enough to contain from one to three or four beds, depending upon the size of the crew and the illness-experience ratio. A nurse or medical attendant should be in charge to treat those requiring attention and the building should, of course, be clean, light, and airy and well equipped with instruments, medicines, and first-aid dressings.

Filing Shop. Saws must be sharpened frequently to keep them in good working condition, and in most well-conducted operations this is done by a skilled saw filer. Many foremen require that crosscut saws be sharpened daily, others only when they get too dull to function properly. Power saws are now widely used and their correct sharpening requires the use of a special power grinder.

Saw filing is close work, and skill on the part of the filers is essential, not only in setting teeth and pointing them so they will best fit the kind of timber being cut, but also so they will suit the individual preferences of the sawyers. The filing house, therefore, should be well lighted and so designed and equipped that the filer can do his work without discomfort. A great deal of space is not needed in a good filing shop, even for large camps when the filer has a helper or two, so the building can readily be a portable unit.

Laundry. Laundry installations are often provided in the washrooms for use by the men themselves. In large modern camps, however, a central laundry often is provided for washing the camp bed linen and, when they prefer, the men can have their personal laundry done for a nominal charge. When a central laundry is provided, it is usually in a separate building or included in one of the shop buildings.

Storage Sheds. Sheds are provided, varying in number and kind with the needs of each camp for storing materials and supplies used on the operation. They may be either of portable or temporary construction and each one is designed to store the kinds of articles for which it was built.

CAMP LOCATION AND LAYOUT

Camps should be located at points from which the largest volume of timber can be harvested and so keep to a minimum the camp moving cost per unit of product. Locations are chosen from the preharvesting cruise or survey with due regard to transportation of products from the forest to the mill and from the mill to the camp. Supplying the camp with food, fuel, and other supplies is particularly important when products are not

FIG. 5. View of a modern logging camp in Washington. All the smaller buildings are portable as units or by sections.

hauled to the mill but to streams and lakes, as is frequently the case with winter logging in the North.

Whenever possible it is desirable to locate a camp along a main highway so ready access can be had by car to nearby towns, thus eliminating or reducing many of the services that would be necessary were the camp more isolated. The camp site should be reasonably level and well drained and large enough to accommodate comfortably all the buildings needed. A supply of pure water in adequate volume is essential and this may be obtained from springs or from deep wells. The longer a camp is to remain at the location, the more important it is to choose a suitable site.

KIND OF LABOR REQUIRED

During the period of large-scale logging in the Lake states and, before that, in the Northeast, loggers were largely transient workers, spending the winters in snowbound camps remotely located deep in the forests, and leaving in the spring, when logging and river driving ceased, for summer idleness or odd jobs. They were called "lumberjacks," a term that persists even today, although the type of worker has changed greatly. They frequently

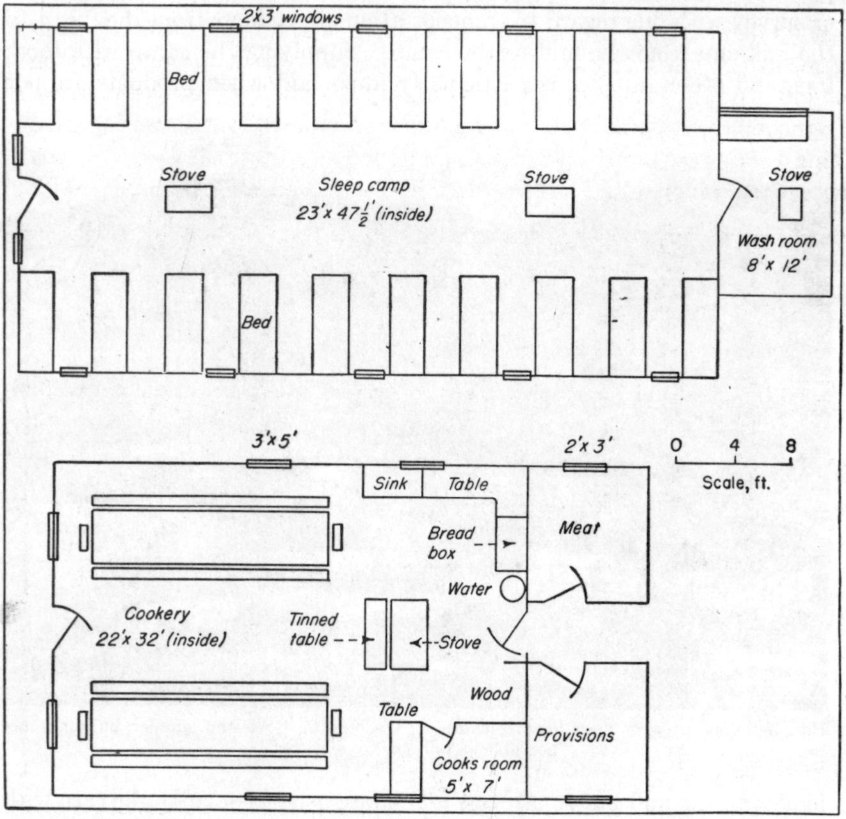

FIG. 6*A*. Floor plans of a 40-man bunkhouse with double-deck single beds and a cookery, used in eastern Canada. This building is of nonportable construction. (*From Woodlands Section, Candian Pulp and Paper Association, Index No.* 548 [B-3].)

changed jobs even during the winter, walking from one camp to another and working at their trades, such as teamster, faller, or swamper.

The camps were composed of rough but snug bunkhouses, barns, mess halls, wanigans, blacksmith shops, saw-filing shops, and storerooms. Up to forty or even more men slept in the bunkhouses in double-decked built-in bunks along each side wall. Life was rough and work was hard from early morning to dark. Meals were generous in quantity and, in the main, good, since lumberjacks avoided camps where the "grub" was scanty or too poor in quality. Paul Bunyan was created in the minds of the loggers during this period and tales of his fabulous exploits with Babe, his blue ox, were bandied about the camps for years, until now he has become a part of American folklore.

With improved labor conditions and the increase in harvesting operations in settled regions, the typical logging camp and rough bachelor crews have practically disappeared except in some northern sections. Labor's demands for better working conditions and recognition by management that better work results from satisfied workers willing to stay on the job instead of the migratory type have resulted in the development of the modern camp where every reasonable effort is made to provide good living conditions for the workmen. Permanence of employment during the entire year resulting from continuous operations and the higher skills

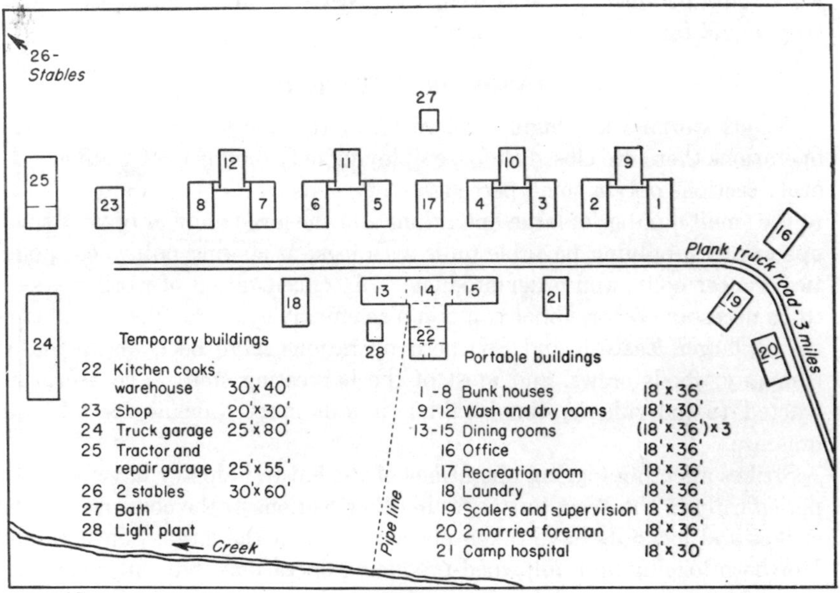

FIG. 6*B*. Ground plan of a 200-man camp with both temporary and portable buildings, used in eastern Canada. (*From Pulp and Paper Magazine of Canada, October*, 1943.)

required on mechanized operations also have contributed to the improvement in camps housing workmen engaged in harvesting timber crops.

Skilled workers now are required on all large-scale harvesting operations. Even felling and bucking have become mechanized and more than a strong back is needed. This does not mean, however, that hand fallers are not skilled workers. Correct judgment as to the best place to fell a tree, where to place the undercut, and how best to buck the fallen tree into logs or other products requires skill and experience, as will be noted later, in addition to physical strength. However, when labor was plentiful there was an abundant supply of men capable of doing these jobs well enough, so they could not be too particular about their living conditions. Power felling and bucking, however, require all these skills and, in addition, a mechanical aptitude.

Mechanical ability is also needed by truck drivers, repair men, skidder operators, riggers, locomotive crews, loader men, and by practically every other workman around a mechanized operation, except flunky jobs. Even small operations are now highly mechanized because power saws, trucks, tractors, and various types of small power loaders are used even when the crew output is comparatively small. Logging and other harvesting operations, therefore, must compete for labor with many other mechanized activities, which means that those who are attracted to woods work must be able to live as well, and receive as much compensation, as they would for equivalent work elsewhere.

LABOR ORGANIZATIONS

Woods workers are highly unionized in the Western states and most operations there are closed shops, employing only members of a union. In other sections only a small percentage of woods workers are unionized due to the small number of large operations and the large number of very small operations supplying portable mills with logs, or cutting pulpwood, poles, ties, veneer bolts, and other products. A concentration of workers seems to be necessary before labor can organize effectively. Furthermore, many of the larger Eastern and Southern operations have been disbanded as company woods crews, and most of the harvesting steps have been contracted to individuals, thus making it still more difficult for labor to organize.

Strikes affecting logging operations of the entire industry have occurred periodically in the West but never in other sections of the country. Local strikes and lockouts have taken place at some of the larger Southern and Northern logging and pulpwood-producing operations, but only one or at most a few operations have been affected at any one time. The lumber and pulp industry of the West is highly concentrated, so its woods operations are more susceptible to industry-wide strikes and labor pressure than in other sections of the country.

Union representatives and executives of the individual companies conducting logging operations negotiate with respect to wages, hours of work, and working conditions. Usually the contracts with labor organizations are drawn up for one year and must be renegotiated annually. When disagreement occurs during the contract period as to whether or not the contract is being adhered to, protests are made, and even strikes may result. If a contract expires before another has been signed, due to failure of labor and management to reach an agreement, labor may use the strike weapon to force acceptance of its demands.

Under the authority of the National Labor Relations Board, employee elections can be held to determine, for any operation, which of two or more competing unions will be the bargaining agency for the workers. The Board conducts the elections to assure impartiality and the union obtaining the majority of votes is officially responsible for negotiating with management concerning wage rates and working conditions. Occasionally, strikes result from jurisdictional disputes between two unions, with management agreeable to dealing with either or both, in so far as it is concerned.

LABOR LEGISLATION

A number of Federal and state laws concerning labor and industry apply to woods workers. Federal laws apply equally in all states, while state laws, of course, apply only within the particular state.

Child-labor Law. The federal statute prohibits the employment in industry of any person younger than sixteen years. Harvesting operations are so strenuous that persons below sixteen years have seldom been employed, but in some places boys of school age formerly were employed as helpers, water boys, or in other capacities where physical strength was not required. Under this act, however, only men over the minimum age can be employed.

Social Security. Federal social-security laws apply to workers engaged in harvesting timber, and appropriate deductions from wages together with employer contributions must be transmitted to the Federal Social Security Board to be credited to each employee. Deductions and contributions are to cover two phases of the program, old-age benefits and unemployment insurance.

Workers under social security who become unemployed are entitled to weekly unemployment payments, varying with the state. Likewise, when workers reach the age of sixty-five years and retire from active regular work, they receive monthly payments in accordance with the amounts they have individually contributed. The purpose of this act, of course, is obvious. It is designed to prevent distressed conditions both when workers are unemployed and also after they have passed the productive stage.

A deduction of 1 per cent from pay rolls is made for social-security purposes, and an equal amount is contributed by the employer. Deductions and employer contributions started at 1 per cent, and gradually are to be increased to 3 per cent. Pay-roll deductions and employer contributions are made on all wages and salaries up to $3,000 per year, so persons receiving more than that are taxed only on the $3,000 maximum recognized under the law. Accurate records must be kept for all employees and their earnings, and each must have, or promptly obtain, a social-security number, which identifies his record in Washington. Clerical work, therefore, has increased greatly owing to the Social Security Act, and many employers have chosen to be relieved of these responsibilities whenever they could do so by abandoning company operations and contracting with individuals to perform the various phases of harvesting.

Wage and Hour Law. This is another Federal statute important in woods operations. It is known as the Fair Labor Standards Act and provides for a minimum hourly wage and for a maximum number of hours per week, unless overtime is paid at the rate of 1½ times the regular wage. The minimum wage per hour specified in the law at first was 30 cents, later became 35 cents, and is now 40 cents. When this act was passed in 1938 the wages of Western and most Northern woods workers were considerably above these amounts. In the South, on the other hand, wages were considerably lower than in other sections and in many cases below the minimum. Southern operators, therefore, vigorously protested passage of this act but were successful only in delaying it for a year or two.

After passage of the wage-hour law, wages were adjusted where necessary to conform to the standards prescribed for workers on an hourly basis of employment. Piece-work wage rates also were adjusted so the worker earned at least the minimum wage. Considerable confusion developed over this feature as some piece workers, such as pulpwood cutters in the South, often worked only part of a day although spending all day on the job.

Limiting hours of work, unless overtime is paid, also required considerable readjustment of work schedules. Many employees of lumber, pulp, and other forest-products companies had been working longer hours because of the necessity of getting log trains and trucks to the mill with logs. To pay the required overtime rate of 1½ times the regular rate to very many workers would mean considerably increased costs, so log-train and truck schedules had to be revised. In many instances two crews had to be provided where one had served before, and this resulted in such workers receiving lower total wages than prior to the act.

Travel time is a matter that has not even yet been decided definitely. Since harvesting operations, in many cases, are conducted on forest properties owned by the company and company trucks or trains take the men to work and back, the Wage-Hour Division has recently maintained that travel time is working time under the law. If this interpretation should stand, operators would be severely handicapped by added cost. The point that workers riding a company train, truck, or bus to work should be allowed working time for such travel while individuals riding to and from work in their own cars, street cars, busses, or even walking should not have such allowance seems finely drawn. It is opposed by operators and may have to be settled in court.

The wage-hour law also requires that records be kept of the wages paid and working time of each employee, which has still further increased the amount of clerical time and expense required of employers.

Workman's Compensation. Laws are in force in most states requiring payments by industrial employers to workmen who are injured or become sick due to causes connected with their work. Some laws specify fixed amounts for particular disabilities such as loss of life, limb, or sight, and weekly benefits during periods when an injured or sick employee is unable to work. Others require employers to carry insurance to cover their liabilities under the law or to post a bond adequate to insure settlement of claims for which they may be held liable.

Prior to the adoption of workmen's-compensation laws, employees injured on their jobs or becoming sick from conditions attending their work either negotiated settlements with their employers or instituted legal action. This resulted in some cases of injustice and much uncertainty, so most employers and employees recognize the advantages of adequate workmen's-compensation laws, which protect both groups.

Woods work is generally considered by insurance companies as hazardous employment and rates are proportionately higher for insurance covering harvesting operations than for many other kinds of work. Safety measures have been widely adopted, however, and the normal hazards of woods work have been greatly reduced.

LABOR TRAINING

Many progressive companies maintaining large camps or a large number of small operations have instituted labor-training programs. This training is designed to equip the men better for their jobs and to decrease accidents. Several different methods for training men are followed. Classes may be held at regular intervals when the men can readily assemble, as in camps, or individual instruction on the job may be given concerning such subjects as correct use of the ax and saw, driving teams, tractor operation, and similar work. Foremen are sometimes given special courses in labor handling as well as job planning. The Woodlands Section of the Canadian Pulp and Paper Association has given both individual and class instruction in a number of subjects and has issued manuals of instruction on correct methods of doing woods work, use and care of tools, personal safety, care of equipment, and other subjects that will result in greater labor efficiency. This benefits the employer by greater labor output per man and it benefits labor by increasing the earning capacity of the workers.

Accidents are always costly both to the employer and the employee. They result in lost time as well as individual suffering, so it is in the interest of all concerned to prevent, in so far as is possible, avoidable accidents due to carelessness. Associations of wood-using industries frequently issue posters encouraging care on the job and individual companies often maintain bulletin boards showing the number of days since an accident has occurred, to stimulate interest in personal safety. The "hard hats" used in shipyards, which have prevented many serious head injuries due to falling objects, have recently been introduced in Western logging operations and are proving effective there also. Compensation-insurance rates are based on the accident experience of an industry, so in the interest of obtaining cheaper rates and disregarding other benefits it is advantageous for employers to conduct their operations safely, provide equipment with safety devices wherever possible, and train their employees to work carefully.

PERMANENT CAMPS POSSIBLE IN MANAGED FORESTS

Working and living conditions in forests can be made unusually attractive when forest management for permanent operations is practiced. No longer can the forest-owning industries depend upon labor that is willing to work and live crudely in rough camps. When forest harvesting was conducted in undeveloped areas far ahead of settled communities, it was necessary to recruit crews that would live in bachelor camps for several months, remote from civilized contacts.

Improved living standards demanded by labor generally and the development of the areas where timber is being harvested, particularly throughout most of the East and South, have changed working standards. Now woods labor needs and is generally receiving adequate housing and recreational facilities. The day of the migrating woods worker is largely past. The so-called "lumberjack" is fast passing out of existence. Only in inaccessible regions is it still necessary to conduct forest-harvesting operations with seasonal logging camps for men only. When this is necessary, the quarters provided are, in most instances, clean and reasonably attractive, with facilities for recreation and cleanliness.

Still further opportunities exist, however, for stabilizing the woods labor supply and providing a high living standard when permanent operations are made feasible on a forest tract by the practice of correct forest management. Outdoor work in the woods is attractive to most sturdy men. It is healthy work and each day's job varies enough to prevent monotony. No two trees or logs are alike and every acre offers its own peculiar problems of harvesting. The deadly sameness of production-line factory labor and office work is lacking.

The only drawbacks to woods work have been the hard physical labor required and unsatisfactory living conditions. Low wage rates have prevailed also in some regions and have discouraged the better type of labor from accepting woods work. These handicaps have been or can be largely eliminated. The use of laborsaving machines, ranging from power saws to power loaders, in harvesting products formerly handled by hand has done much to eliminate the need for labor with a "strong back and a weak mind." Intelligence along with reasonable strength is required of most woods workers now. Wages have been adjusted upward to the point where willing workers can obtain a satisfactory income from woods work, so the low-wage feature of the past is no longer an obstacle.

Under past operating conditions and even today, whenever camps are required, it is difficult to provide satisfactorily for family life for woods workers. Even when family quarters are provided they lack the aspects of home life, and the desire of most American men and women for homes does not make such living conditions too attractive. With the stabilization of harvesting, as in forests managed for permanence, it is not only possible but advantageous to establish permanent living quarters for forest workers and their families.

Although the location of the harvesting operation changes constantly in a managed forest, it is never too far for the men to go to and from work daily from a centrally located point. Public highways and roads now transect the forested areas in many parts of the country, particularly in the South, East, and portions of the Lake states, and even portions of the West, so that by means of busses or trucks, woods crews can be economically carried back and forth to the harvesting location each day. Only one or two men need remain at the location at night as watchmen or barn tenders.

Under such operating conditions, permanent centrally located camp villages can be built to provide home living facilities for woods workers. Family houses can be built close enough together to provide for community electricity and water, as well as for companionship and certain recreational installations, schools, and other desirable features of village life. Even though such a camp were some miles from town, so many people own automobiles that there would be no isolation from other communities. The radio and daily mail service would, of course, keep the residents informed of current local and national affairs.

Each house could be allotted enough land so the worker and his family could have a vegetable garden and fruit trees, raise chickens, and, if some so desired, have a cow and perhaps hogs. With the shorter working week now generally in effect, industrious workers would have adequate time to look after their gardens and livestock, and a suitable outlet for the energy of growing children would be provided.

Woods workers provided with such an opportunity to enjoy the best type of family existence would be more inclined to remain on the job than to seek other employment, and costly labor turnover would be reduced. A high standard of living could be enjoyed without a large cash income. Hence, wages could be held to reasonable levels. When shutdowns occurred, there would be no immediate hardship, as each family would have most of its subsistence already provided. Indolent workers who failed to take advantage of the opportunities afforded by such forest villages for woods workers would have to be replaced by others who would do so; and after a period of recruiting and sifting, a congenial, happy, and reasonably prosperous community would be established and the labor problem would largely be solved. Bachelor quarters could be provided for a few single men, but such camps should be predominantly family camps or preferably villages.

It should not be more costly to establish such a community village than to maintain the common type of portable cramped quarters for woods workers. The initial expense would be more but should not be excessive. Reasonable rent could be charged for the houses and facilities provided, but it should be only enough to cover the actual cost of such service and to maintain the houses and depreciate the investment in them. As the buildings are permanent structures, these charges should be nominal. Much saving in labor turnover would result and the crews could be better trained in their work, resulting in greater output at less cost. Also, such workers would be interested in the permanence of their jobs and hence in the permanence of the forest. They would respect cutting regulations and fire protection much better than if they were only working in the woods until they could get better jobs elsewhere. Labor discontent ferments on unstable and undesirable living conditions and there should be a minimum of such trouble in a competently managed permanent forest village.

Garages and shops would be located at these villages for the repair and maintenance of trucks, tractors, and other mechanical equipment and tools. Having such facilities permanently situated should result in much better care of tools and equipment with consequent longer useful life and lower costs.

The opportunity for permanent villages for forest workers is, of course, dependent upon so managing forest areas that permanent employment can be offered in harvesting the timber crops grown. The widespread adoption of the principle of growing timber crops by large primary wood-using industries, such as lumber and pulp companies, has made such a labor policy possible in many instances, and its adoption will greatly expand in the future as immature stands on large forest holdings come into merchantable size and are managed for permanence.

Correct cutting and other harvesting practices are difficult to obtain when labor is not permanent. Despite all the regulations that foresters can prescribe for harvesting operations, unskilled, careless, or indifferent workers can do much unnecessary damage to the forest in harvesting it. It is very much in the interest of good forest management, therefore, to establish a permanent and skilled force of workers to conduct harvesting operations, living in or near the forest that will provide them with a lifetime of work.

PART TWO

PREPARATION OF TREES FOR REMOVAL FROM THE FOREST

Among the most important operations in harvesting timber crops are those having to do with the preparation of trees for removal from the forest, because how timber is cut influences to a considerable extent the volume and quality of products obtained. This, in turn, has an important effect on the value of the forest crop. In some instances the finished products are made in the forest; in others only the first steps in their manufacture occur before the products are transported to mills or factories. Consideration is given here, therefore, to the selection of trees for cutting; felling; bucking; hewing, riving, or peeling operations that take place in the forest; and to the tools customarily used for these operations.

CHAPTER 6

SELECTION OF TREES TO CUT

The actual harvesting operation begins with the selection of trees to be cut, which may be either in stands fully mature and ready for the final harvest or in immature stands in need of thinning or other improvement operations requiring the cutting of trees that have commercial value and can be sold. The term "harvesting" implies the cutting of some product that can be used to supply the needs of people.

Whenever any trees are cut in a forest and sold or used, then timber harvesting is taking place. Cutting operations in young unmerchantable stands or the cutting of unmerchantable trees in older stands that does not yield usable timber products of merchantable value in excess of the cost of cutting are not harvesting operations. Such cuttings are noncommercial silvicultural operations.

FACTORS AFFECTING SELECTION OF TREES FOR HARVESTING

Kind of Operation. The selection of trees to cut in harvesting depends on the kind of operation that is to be conducted and the forest policy of the owner. Industrial harvesting operations are conducted primarily to obtain certain products, such as saw logs or pulpwood, and the trees are selected on the basis of their suitability for the kind of products wanted. Harvesting operations in forests owned and managed as a forestry enterprise not connected with any particular industry, on the other hand, are conducted to obtain the variety of products that will yield the greatest income. In such cases the markets available for different products influence the selection of trees for cutting. When a market exists for poles or veneer bolts, for example, trees are selected that will make these products, since usually they are more valuable. Similarly trees of saw-timber quality are selected for cutting into saw logs instead of pulpwood, as saw logs normally bring greater returns than pulpwood.

Harvesting Costs. Every harvesting operation is subject to a particular set of costs which, when related to the selling price of the product, limits the size of the trees that can be cut and manufactured at a profit. Businesslike harvesting requires that no tree products be cut that will cost more to manufacture than they can be sold for, and also that all trees with any net value be cut, unless they are specifically reserved from cutting for seeding or for future cutting. Instructions, in one form or another, are therefore issued to the cutting crews on harvesting operations specifying the smallest tree to cut and the minimum top utilization. If this were not done the saw crews would cut only the trees that were easiest to fell and buck and they would, in many cases, leave some profitable trees standing because of inconveniences that would reduce their output.

Great variation exists in minimum cutting diameters for various products in different sections of the country, and even between different operations cutting the same products in the same region. The minimum cutting diameter, on a purely economic basis, in some Western forests is often as large as 18 to 20 in., while for pine in the Southern forest region on small-mill operations it frequently is as small as 8 in. The reason for this difference is in the different harvesting and manufacturing costs due to accessibility and distance to markets, as well as in the selling price of the finished products.

The variation in minimum cutting size for saw timber between different operations in the same region is due to the fact that large mills are not designed or equipped for the efficient handling of small logs, the carriage, saw, and other mill equipment being designed for large logs. Portable mills, on the other hand, can cut much smaller logs because small-mill equipment is designed to handle small logs and produce narrow lumber. In small mills, large logs are frequently so unwieldy to handle that they are not wanted, whereas they would be ideal for a larger mill.

Species. The minimum cutting size also varies between different species; the more valuable species being cut to smaller sizes than the less valuable species. Thus, western white pine is cut to a 15-in. minimum diameter on some operations and at the same time and place white fir is not cut at all or only the largest and choicest trees are removed, because of the wide variation in selling price between the two associated species.

Pine in the South is cut to smaller sizes for saw logs than are most hardwood species because small pine logs can be sawed into two-by-fours and other structural lumber items, whereas hardwoods usually are sawed into 1-in. lumber for industrial use and a considerable amount of clear surface is required, such as for furniture, and such lumber does not come from small hardwood logs.

Product. There is a great variation in the sizes of trees cut for various products. Saw logs are sometimes cut to small diameters but naturally not to such small sizes as when trees are cut for pulpwood. Likewise, trees intended for other products are cut only to the smallest sizes that will produce them. Veneer bolts must be large enough and clear enough to peel, slice, or saw into valuable veneer. Poles and piles must be carefully chosen, not only on the basis of stump diameter but also with respect to the shape and condition of the entire tree. A maximum size as well as a minimum size must be considered for such products as poles, piles, posts, and ties, because their specifications require that they fall within a range of sizes.

Management Policy. Forests owned by operating companies that are liquidating their timber assets and are not planning on cutting the area again are generally cut to remove all merchantable timber for pulpwood, saw logs, or other products. This does not mean that the area is to be devastated, as it may reproduce well and grow another crop of trees if enough undersized trees or an adequate number of seed trees are left standing in a healthy condition.

Forests under management for continuous or periodic harvesting are cut according to plan. The plan may call for reserving enough of the younger, growing timber in the stand being cut to provide for future harvests, and only a portion of the timber is removed at any one time. In such cases merchantable trees are deliberately left because they will be sufficiently larger and more valuable in the future to justify letting them grow. Or the plan may provide for clear-cutting in strips or blocks, sufficiently small in size so the clear-cut area will promptly reseed from adjacent uncut timber. When the plan calls for clear-cutting larger areas, seed trees are selected and left standing to provide for regeneration of the forest. When merchantable trees are cut in immature stands, such as in thinning, selected trees are removed to provide more space for the crop trees so they will maintain their growth at a normal rate.

The choice of trees to be cut in managed forests is largely based on the plan of management and the silvicultural system in effect. While the selection of trees to be cut is the first step in the harvesting operation, so many technical forestry considerations affect the selection that it cannot logically be considered as a harvesting operation alone, except from the economic aspect. Therefore, considerations affecting the choice of trees to be cut, other than those dealing with the economic aspects and those having a bearing on the harvesting operation itself, are not treated here.

Economic Considerations. Trees are cut for sale or for conversion into salable products, so it is essential that only those trees be cut that will return more than the cost of harvesting, manufacturing, and selling. Since timber grows over such a wide range of topographic and ground conditions that greatly influence the cost of harvesting; since trees vary so much in size and quality as well as in species; and since the selling prices for products and the cost of harvesting vary so greatly from time to time, constant attention must be given to the selection of trees in harvesting to avoid cutting trees too small or too defective to return the cost of manufacturing them.

This is particularly true in the case of saw logs because, on the average, with decreasing tree size the cost of harvesting and milling increases while the value of the lumber obtained decreases. Figure 7 and Table 12 illustrate the relationship between tree diameter, production cost, and value of product. In pulpwood, since the wood is chipped in manufacture and loses its identity in the chip bin, the factor of decreasing quality in the small-sized trees is not so marked. Nevertheless, small trees stack less solid wood per cord, the purchasing unit of measure, so there is an element

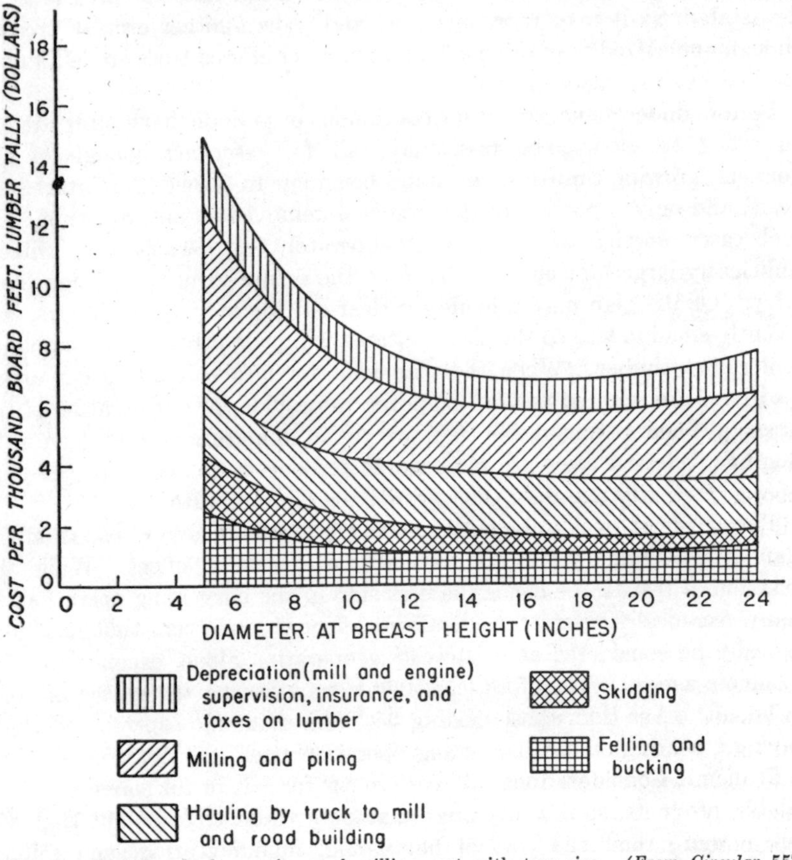

Fig. 7. Variation in harvesting and milling cost with tree size. (*From Circular* 557, *U.S. Department of Agriculture, "Cost of Producing White Pine Lumber in New England," by V. S. Jensen, et al.*)

of lower quality with smaller size. The cost of harvesting small trees for pulpwood, however, does increase rapidly as tree size decreases and there is a minimum size below which, if trees are cut, a loss is incurred.

Special products, such as poles and piles, are cut to specifications, so there is less chance of cutting unprofitable piece products, unless the cutters are careless and cut trees that will not meet the specifications. In some cases trees can be too large for profitable manufacture into piece products. When oversize trees are cut for hewed ties, for example, extra work is necessary to hew the round sections down to tie dimensions and this, of course, adds to the cost. When oversize trees are cut for poles or piles, even though they are acceptable according to the specifications, the extra weight of the excess wood increases the amount of freight to be paid, which reduces their net value.

TABLE 12. VARIABLE COST OF CUTTING PULPWOOD BY TREE SIZE

Tree Diameter, In., at 4½ Ft from the Ground	Cost of Cutting per Cord
4–5.9	$1.40
6–8.9	0.90
9–11.9	0.72
12 and above	0.76
Average of all trees cut	$0.80

* Source: Reynolds, R. R., "Pulpwood production studies in Shortleaf-loblolly Pine Stands," *U.S. Forest Serv., South. Forest Expt. Sta. Occas. Paper* 71.

For every product, under a given set of operating conditions and for a given set of costs and selling prices, there is a class of trees of such size or of such quality that the cost of harvesting just equals the return obtained for them. These are the so-called marginal trees. Any trees smaller or poorer in quality will result in a loss if they are cut and manufactured and all larger or better in quality will yield a profit, increasing with increasing size and quality.

A determination of the smallest and the poorest trees to cut for any product requires a thorough knowledge of the costs of harvesting trees of various sizes, species, and qualities and the returns that can be obtained from them. Adequate cost records combined with time studies or analyses of a particular operation will show the minimum operating limits below which cutting should not be done. This is a subject for more advanced study and it is discussed in this text only to recognize its importance and the necessity for foresters who are to engage in harvesting timber to become competent in making such determinations.

CUTTING INSTRUCTIONS ON MINIMUM SIZE

On operations not concerned with reserving trees for subsequent harvests, instructions to the cutting foremen and crews on the choice of trees to cut are either written or verbal. Such instructions simply state the diameter of the smallest tree of each species and the poorest quality larger trees that are to be cut. Thus, while the instructions may state that no white oak trees smaller than 10 in. in diameter at the stump are to be cut, they should also state how much defect will be acceptable in larger logs. A "rule of thumb" on one operation, for example, stated that no logs of above minimum size that were more than one-third cull were to be cut. In perfectly sound trees, however, the cleaner and straighter trees are cut for saw logs at smaller sizes than the limby or crooked trees, because of the higher sale value of clear lumber.

Since the minimum profitable diameter for an operation changes from month to month with changing costs and changing lumber prices, a more or less continuous state of uncertainty exists as to whether or not certain small trees should or should not be cut. This is particularly true when cutting instructions are verbal only. By the time the cutting foreman gets the crews fully instructed about the minimum tree size and top utilization under one set of conditions, another fluctuation of costs or selling prices brings a change in the minimum profitable utilization size. Effective supervision over cutting instructions can seldom be maintained adequately by verbal or written instructions only, because of the constant adjustments that cutting crews have to bear in mind, which creates a sense of uncertainty and confusion.

Cutting specifications, either written or verbal, but preferably written, should clearly state that the diameters prescribed are either inside or outside the bark, to avoid confusion. While either can be used satisfactorily, it is better, ordinarily, to specify outside-bark diameters, as the crews can soon become accurate at estimating tree diameter at the stump and will need to measure only those close to the minimum size. Calipers, of course, could be carried, but they are cumbersome and probably would be frequently lost or broken. A short measuring stick or marks on the log or pulpwood measuring stick is all that is needed to furnish the crew with a ready means of checking stump diameters for the purpose of conforming to the minimum specifications.

While most cruising and estimating is done on the basis of tree diameter at breastheight, it is more convenient and accurate to specify minimum cutting diameters at stump height. This permits examining an area accurately even after the timber has been cut to determine whether or not the instructions have been followed. The fallers themselves also are better able to conform to the specifications as they can get an accurate measure of the stumps and would have to estimate the breast-high diameter.

MARKING TREES FOR CUTTING

It is extremely difficult if not impossible to draw up a set of verbal or even written instructions for cutting when some of the trees are to be cut and some left, if some of those to be left are of profitable size. Timber varies greatly in density of stand per acre, sizes of trees, age, and soundness, and in forestry treatment required; to cover in writing the ordinary range of possibilities on any one operation alone would be prohibitive and well beyond the grasp of even the best woods workers and foremen.

In lieu of categorical sets of instructions, reliance must be placed on the knowledge and practical common sense of the forester charged with maintaining a forest in permanent or long-time production. Using his knowledge and judgment, either personally or through closely supervised and competent assistants, a forester should choose and mark the trees to be cut or the trees to be left, whichever is the more efficient procedure.

Adoption of the practice of marking trees to be cut is a very forward step initiated within recent years by many lumber companies, pulp mills, and individual forest owners. Marking takes this important decision out of the hands of laborers, who are interested primarily in their daily wage, and places it in more responsible hands.

Marking timber requires that some form of visible evidence be placed on the trees in a conspicuous place to indicate those to be cut or those to be left. This is done either by blazing them with an ax or by putting paint or some other mark on the bark. Blazing the trees that are to be cut is used by the United States Forest Service and by many other forest owners. Since the trees are to be cut, there is no harm done by the blazing method. Blazing trees that are to be left is harmful, however, since the tree would be permanently scarred and insects or rot fungi may gain entrance to the sound wood of the tree through the wound. Even if this does not occur, resinous streaks are formed in coniferous trees, which degrade the lumber that may in the future be cut from that part of the tree. Ax blazes, then, are a satisfactory way of marking trees to be cut but on operations where trees are marked to be left, marking should by all means be done with paint or some other similar substance.

Ax Marking. The U.S. Forest Service in selling timber from the national forests uses the ax-blaze method of marking almost exclusively. In designating trees that are to be cupped for naval stores on some of the Southern national forests, an ax blaze is not used as it would damage the tree for cupping purposes. Instead, a light bark blaze is used or a paint mark is put on trees.

The Forest Service method of marking trees selected for cutting is to make a smooth ax blaze 5 or 6 ft above ground, where it can readily be seen from some distance through the woods. The blaze exposes the wood of the tree so it is unmistakable. Another similar blaze is placed on the root swell of the tree on the same side as the marking blaze. On both blazes the letters USFS are branded by a blow with the hammer side of the ax. These blazes and brands officially designate that the trees so marked are to be cut. The stump blazes serve as a check so when a Forest Service officer inspects the area he can tell from the stumps that the trees cut were actually marked trees. A stump without a branded blaze would indicate that an unmarked tree had been cut. This same method, but often without a brand, is frequently used by private owners selling stumpage to reliable buyers.

Ax blazing of trees to mark them for cutting on the logging operation of one Southern lumber company, for example, is done by the saw foreman, who blazes trees for the fourfold purpose of (1) indicating which are to be cut; (2) laying out cutting drifts for the felling crews; (3) showing how the tree is to be bucked into logs; and (4) keeping a tally of each saw crew's daily production.

Drifts are the areas designated for a particular cutting crew and are indicated by the side on which the trees are marked. The foreman, in blazing trees to be cut, blazes them on the side facing him on an area wide enough and long enough to keep a saw crew busy at least one day. On his return trip through the next drift he offsets the width of a drift and again blazes trees as they face him, keeping a tally of the trees marked on each drift. The trees in the second drift, therefore, are blazed on sides opposite from those in the first drift. For example, if the trees in the first drift are blazed on the north side as the foreman walks south, when he offsets and starts north through the second drift he blazes the trees on the south side. Trees in subsequent drifts are likewise alternately blazed on opposite sides.

Saw-crew earnings on this particular operation are based on the number of logs cut and, as the company cuts long logs, the foreman keeps a tally of the number of trees blazed on each cutting drift. At the end of each day he tallies the day's work of each crew by counting the uncut blazed trees on each drift and deducts them from the total number marked, which is much less tedious than keeping a record of each tree or log cut. This method also makes it unlikely that saw crews will cut unblazed trees, because if they do they get no pay for them, so very few if any trees are cut that are not marked to be cut.

This marking method serves still another purpose in that if a tree is to be cut full length into one long log, a single blaze is put on it, but if it is to be made into two logs to utilize it more advantageously, then the tree is marked for "double stocking," and two blazes are put on it one just above the other. Trees marked for double stocking are tallied as two logs by the foreman, so the piece tally is not affected by the fact that some trees are cut into two logs and some into only one.

The ax is a convenient marking tool because no other material is necessary, such as paint. Marking trees for cutting with ax blazes makes it difficult for the marker to change his mind once a tree has been marked. A blaze cannot be removed from a tree except by making a bigger one or by damaging the tree. Therefore, when there is any possibility that the marking will be changed, as in the case of marking timber belonging to another who may want to make some changes, ax blazing should not be used.

Paint Marking. Marking trees with paint is the method used by a large number of pulp and lumber companies and other forest owners to designate trees to be cut or to be left. Marking with paint has proved particularly desirable not only because it is an effective way of indicating trees to be cut or left, but also because on partial cutting operations paint marks can be removed if it is found that some adjustment is desirable in the choice of trees after marking has been done.

In partial cutting, the volume to remove per acre frequently is specified by the management plan and marking is done with the objective of cutting a certain volume of timber per acre. In practice it commonly is discovered, after the tally of marked trees has been computed, that too much volume has been marked, and the marks on some of the trees have to be removed to bring the volume designated for cutting into conformity with the stipulation of the management plan. Removal of paint marks by lightly chipping off the bark flakes is quite possible, but it is impossible to remove ax blazes without further defacing of the trees already marked. When timber of others is marked with paint for selective cutting, the owner can then inspect the marking and if he does not approve of certain selections, there is but little difficulty in adjusting the marking to suit him. With the rapid spread of selective cutting of timber owned by individuals but marked by the operating company, as well as of timber owned by the operating companies themselves, this change of marks has become an important consideration.

FIG. 8. The shortleaf pine tree being felled in this illustration has been marked with white paint applied with a brush. No stump mark is used on this operation. Note the low stump which gives maximum utilization of the best part of the tree. Alabama.

Another advantage with paint marking is that trees can be marked for leaving without in any way damaging them, as would be the case in ax blazing. In stands where seed trees are to be reserved, many more trees are cut than left, so much time is saved in marking trees to be left instead of marking those to be cut. In younger stands the "crop trees," or those chosen to be kept for the final stand, are sometimes marked with paint or some other permanent noninjurious marking material. In such stands, the chosen trees are marked plainly and the saw crews then may cut any unpainted trees of sufficient size or quality for logs or such other products as may be wanted from time to time.

With paint marking, the first consideration is that a color be used that can be readily seen and that the paint be durable enough to last for several months. This permits marking considerably in advance of cutting, which is helpful to the logging superintendent as he can go over the area and determine the best time and method of logging the marked trees. Colors most in use are white, red, and yellow. Of the three, red is the least conspicuous in the woods and is not widely used. Both yellow and white are distinctive and are in much favor for tree marking.

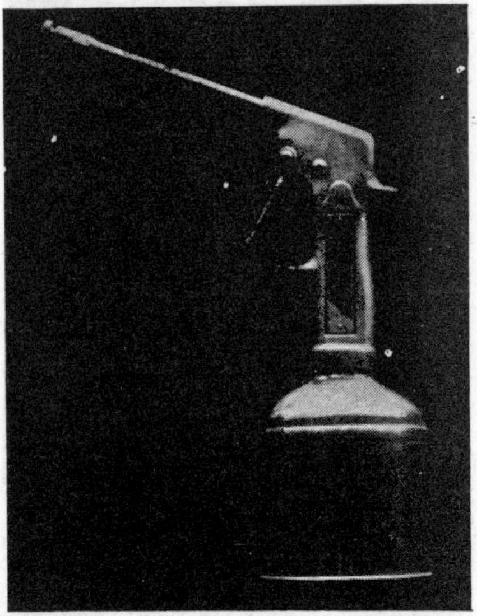

FIG. 9. The type of grease gun used for marking trees with a paint-kerosene mixture is illustrated here. A pull on the trigger ejects a jet of paint through a fine opening in the nozzle that carries accurately for distances up to 20 ft or more. The container can be unscrewed for refilling with paint.

Paint can be applied to the trees either with a brush or by means of a hand-type pressure grease gun, adapted to the use of marking paint. Brushing paint spots on trees, of course, follows well-known painting procedures. Many trees have thick flaky bark and to make a bold paint mark requires that the outer bark be smoothed to some extent. Paint marks brushed on trees over loose outer bark also can easily be lost or deliberately removed and therefore are not as durable as when the rough bark is shaved to a solid base. In the case of tight-barked trees like beech and some maples, no bark preparation for brushing a paint mark is needed. As in the case of ax blazing, paint marks can be and frequently are placed both on the trunk and on the base of the tree or stump to prevent careless or deliberate cutting of unmarked trees.

The paint-marking gun, adapted from a pressure grease gun, is proving convenient and effective for marking trees. The instrument has a pistol grip and, when the trigger is pulled, projects a jet of liquid paint through a small nozzle with accuracy for distances of up to 20 ft. It has a paint reservoir below it which holds approximately a pint of marking paint. A skillful marker using this gun can place a spot on a tree, over rough or smooth bark, at almost any point on the trunk. Another pull of the trigger will put a spot on the base of the tree. No bark smoothing is needed because the stream of liquid paint penetrates the cracks and crevices in the bark, which is difficult to do with a brush.

Ready-mixed paint, as usually obtained, is too thick to use with a paint gun and therefore it is thinned, usually with kerosene, but not sufficiently to dim the color appreciably. One recommended mixture[1] is 1 gal of paint mixed with 1½ gal of kerosene; another is equal parts of paint and kerosene. This gives the necessary dilution without appreciably diminishing the color when put on the trees. Being more fluid than paint normally is, the mixture flows into bark cracks and shows up better than brush marks on loose bark. Tree-marking paint suitable for use in spray guns is now obtainable from several manufacturers in a variety of colors. Figure 9 illustrates the type of paint gun that is widely used for marking trees.

FIG. 10. The tree to the left in the above illustration has been marked with trunk and stump spots to indicate it is to be cut for pulpwood. The tree on the right was marked with a circle on the trunk and a stump spot to show it is to be cut for saw logs with pulpwood to be cut only from the top. Arrows indicate the white tree marks that were applied with a paint gun. South Carolina.

The paint-gun method is used extensively by Southern pulp companies and their contractors because of the large number of trees that must be marked per acre in small pulpwood timber. On saw-log operations, larger and fewer trees are cut per acre, so a quick method of shooting spots on trees is not so essential. One objection to the paint-gun method is that the markers can spot a tree without going up to it to determine its condition. A tree should, especially when marked to be left, be looked over from all sides to check for defect, crook, or other decay that might not be apparent from one side only.

The bark scribe is a tool used for marking trees, but it is not used to any extent on cutting operations as it does not give a prominent mark that can be seen from any distance, and this is essential on harvesting

[1] Bratton, Allen W., and R. M. Fergus, Use of the spray gun in marking timber *Jour. Forestry*, vol. 43, No. 2, pp. 113–117, February, 1945.

operations so the saw crews can judge their felling and so the foreman can lay out the most suitable skidding roads.

Ordinary whitewash can be used with a brush to mark trees if the timber is to be cut within a few weeks; otherwise rain will wash off the marks. Another method of marking trees shortly before cutting is by means of powdered lime in a sock. The sock is partially filled and carried in the marker's hands. To mark a tree he strikes it and the powdered lime sifts through, leaving a distinct mark.

FIG. 11. Method of marking trees for cutting with a paint gun and a paint-kerosene mixture. Virginia.

There are many variations and combinations of methods of marking trees selected for cutting, adapted to the needs of individual owners and operators. Frequently the verbal-instruction method is combined with tree marking, as in the case of one company that sells timber from its forest and specifies that all trees 20 in. dbh and larger are to be cut except those marked. In this case paint is brushed on the trunks and on the stumps of those trees below 20 in. dbh to be cut, so a later inspection of the cutting area will show whether or not any marked trees have been taken.

OTHER CONSIDERATIONS IN MARKING

On more intensively managed forests an inventory-taking operation can be combined with the marking, and the crew not only marks trees to be cut but in addition makes a tally of the trees marked for cutting and trees to be left, sometimes by both their diameter and height. In addition to furnishing the forester with data needed for planning the next cut, this method serves to give the logging department exact information on the amount, location, and size of the timber to be harvested on each designated area and permits the logging superintendent to plan logging operations well in advance and with full knowledge of operating conditions.

When timber is being cut for two or more purposes, different paint marks can be placed on the trees to indicate for what product each is to be cut. Thus, some owners in the Southeast, using the paint gun, mark trees to be cut for pulpwood by placing one spot on them; those to be cut for saw logs by placing a circle of several inches in diameter on them; while trees to be cut for poles, piles, or other specialized products are marked with two horizontal streaks or lines. This effectively indicates to the cutting crews which trees are to be cut for each product.

Good harvesting usually requires that the highest valued products, or those with the most rigid specifications, be cut first. Pole or piles cutters should cut first, then the log cutters, after which the pulpwood cutters remove the remainder to be cut. The final cutting for pulpwood also removes the tops of the saw-log trees, rejected poles left on the ground, and any damaged trees that were not marked for cutting. Hence, designating the trees selected for each product makes for a more orderly series of harvesting operations.

The cost of marking is various, depending on the products the cut trees will be made into, the amount cut per acre, and the size of the individual tree, as well as the wage rates and the method of marking used. The less volume cut per acre, the greater the number of acres that must be covered per day to get the same volume. Likewise, since the cost of marking a small tree is nearly as much as marking a large tree, the cost per unit of product is less for large timber than small timber. Actually, the cost of well-conducted marking work is so small proportionately to other costs that it is not a determining factor. In one instance where saw timber was marked for selective cutting with a paint gun, 1 to 1½ gal of paint mixture marked about 60,000 ft of timber, which was used at the rate of 1½ gal per day covering 10 acres of timber. In this case two spots were placed on each tree, one head high on the trunk and one at the stump. Paint cost was only 13 cents per 100 trees marked, and the total cost of doing the marking was not more than 12 or 15 cents per 1,000 ft of timber marked. It is apparent, therefore, that the cost of good marking with a paint gun can be very reasonable.

Competent marking by any of the methods mentioned, or others, can be done by nontechnical woodsmen if they are adequately trained and closely supervised by technically trained foresters. Such markers should be competent to carry out instructions and capable of exercising reasonably good judgment where set procedures cannot be prescribed in advance. The forester should devote his time and attention, in addition to supervising the marking, to developing and expressing a sound operating policy and to keeping in touch with the constantly changing relationship between cost of cutting and milling and sale price of lumber and the effect of this on the size of the marginal tree. Thus he can revise the marking policy when it is indicated.

So that cutting crews can work through a tract in an orderly manner, all trees marked, regardless of method, should be marked facing the same direction. This permits the cutting crews to see clearly which trees are to be cut and allows them to fell trees in the most efficient way. When timber is marked by strips, as is common on many operations, each drift may be marked facing opposite directions, so when a crew has completed cutting one, the next one can be cut without returning to the opposite end.

Control of cutting is essential for bringing about a sustained yield of timber, and the adoption of efficient marking methods makes it possible to mark timber under practical operating conditions at a very nominal cost. In fact, the cost of marking is largely offset, if not saved altogether, in better timber cutting by the saw crews. The method best adapted to each operation should be the one used. On some the ax blaze will be best, while on others a brush and paint, paint gun, or lime sock will be found more economical and practical. Marking the trees to be cut, or left, as the case may be, should be adopted on all harvesting operations except the strictly liquidating type.

CHAPTER 7

THE FELLING OPERATION

Trees in the forest have only potential value, commercially. They must be removed and manufactured, by hand as in the case of hewed ties or by milling as for lumber, before they become useful products. Obviously before they can be removed they must be felled, or severed, usually at the base close to the ground. While ornamental trees that are dug up and transplanted are also commercial products, they are a rarity in forestry and are not considered here.

Felling the trees selected for cutting, therefore, is the first stage of the actual manufacturing process. It is an extremely important step, as will be shown, because how it is done affects the value of the trees for the products they are intended to produce, the efficiency of the operation, and the condition of the forest itself after the trees have been felled.

The unit crew in felling is the number of men required to operate the saw used to cut trees economically and quickly, varying from one in the case of some pulpwood operations to four or five for saw-log operations on which heavy power saws are used. Saw crews are frequently referred to as "saws," since the saw is the unit of production in timber cutting and log making.

DESIGNATION OF FELLING AREAS

To avoid confusion on the operation and to provide for orderly felling and bucking, as well as to facilitate supervision, each felling crew should be assigned a strip or an otherwise designated area in which the timber is to be cut. This assures that both the hard-to-fell trees and the easy-to-fell trees will be cut; otherwise, especially on piece-work jobs, some crews would cover a large area picking only the large and easy-to-fell trees and leaving scattered trees uncut, so their production and hence their piece-rate wages would be greater. Other advantages of allocating definite working areas to each saw crew are

1. Less danger of felling trees on other workers.
2. Easier scaling when the output of each crew is kept separately.
3. Better placing of tops to avoid brushing up the working area.

Felling, and bucking as well, must be carefully planned and must proceed in regular order if the operation is to be kept economical. Thus, in logging when trees are bucked to log lengths as soon as felled and the logs are promptly skidded out, the felling operation should proceed from the farthest point towards the area closest to the log yard or log-assembling place. Otherwise, the teams, tractors, or other skidding or log-moving machines would constantly be hampered by tops of felled trees. Also, there is much less danger that men, horses, and equipment will be struck by falling trees when the cutting is begun at the back of the strip and progresses towards the yard or loading-out place.

Even when all the trees on an area are felled and bucked before they are moved, care should be observed to fell them so the tops will be out of the way of the skidding operations to follow. When cable skidding is to be used, little attention to the placing of tops is required as the cables pull the logs to the machine with such force that tops are no great hindrance. However, care should be exercised to fell them so subsequent bunching and skidding will not be made unnecessarily difficult.

CHOOSING THE DIRECTION OF FALL

When only a portion of the stand is to be cut, as in selective cutting operations, particular care must be observed to fell the trees in directions that will cause the least damage to other trees not to be cut and to young timber and sapling stands that are needed to grow into future crops. Skillful fallers, who can correctly undercut and backcut, can place trees on the ground with a high degree of accuracy, and only such skilled saw crews should be permitted to work in managed forests, for the protection of the residual stand, if for no other reason. As openings are created in stands as trees are cut, others can be felled into these openings by workers who not only know how to fell trees correctly but who also are willing to be careful.

One of the important considerations, therefore—not only in partial cutting operations but on any harvesting operation—is the direction of fall for each tree cut. This is determined in some cases by a foreman who sizes up the situation and tells the crew on what side to undercut. In other cases the crews themselves determine the direction of fall. The larger and more valuable the trees, the more likely it is that the foreman of the saw crews or the "saw boss," as he is commonly called, will make this important decision.

In deciding upon the direction of fall for a tree or trees to be felled for log making or for any timber product such as poles, ties, veneer bolts, and pulpwood, the following considerations have an important bearing.

To Avoid Lodging in Other Trees. When trees are not thrown clear of adjacent trees and lodge or "hang" in them, much waste time results in getting them to the ground. This is an unnecessary complication and can be avoided by accurate felling work. Furthermore, it is often dangerous to work at dislodging a tree as the tree in which one is hung may have to be cut. Throwing still another tree across a lodged one may bring it down, but it is not unusual to lodge it also, resulting in a tangled straw pile of partially felled trees that will require a tractor to straighten out. This causes added expense as well as lost time, meaning greater costs and lower output.

Prevention of Breakage of the Trees Felled. Trees are heavy, weighing many thousands of pounds when of saw-log size, and the force with which

Fig. 12. By carelessly felling this tree across another one already on the ground the felling crew broke this tree as shown. Although the break was made a bucking point, from 2 to 4 ft or more of merchantable log length was wasted. From an operation in the Piedmont region of North Carolina.

a tree strikes the ground is sufficient to break large trees if they strike irregular surfaces, especially a rock, mound, or abrupt change in topography. Trees felled across wide sloping hollows also may break in dropping into such concave depressions. Logs, stumps, or other obstructions likewise can break the trunks of trees striking them, and thereby cause serious losses of merchantable volume and product values.

The speed of a falling tree is much greater in the upper portion than at the lower end because of the much greater distance the top must travel to get to the ground in the same length of time the butt end does. This greater force in the top, combined with the smaller tree diameter at the top and the weakness frequently caused in tops by limbs and knots, results in more breakage in upper sections of trees than in butt sections. Felling a tree across a stump or log, therefore, except when it is close to the butt, is inviting breakage and waste.

A falling tree, until it reaches an angle of about 45 deg with the ground, exerts great pressure against the stump due to the weight of the tree and also to the downward motion of the tree. The direction of this pressure changes as the tree falls, from a downward vertical pressure, as in a standing tree, to pressure from the side towards which the tree is falling. As the tree passes the halfway mark (45-deg angle with the ground) in falling, the momentum gained by the top, in moving more rapidly than the base, pulls the butt of the tree away from the stump in the direction

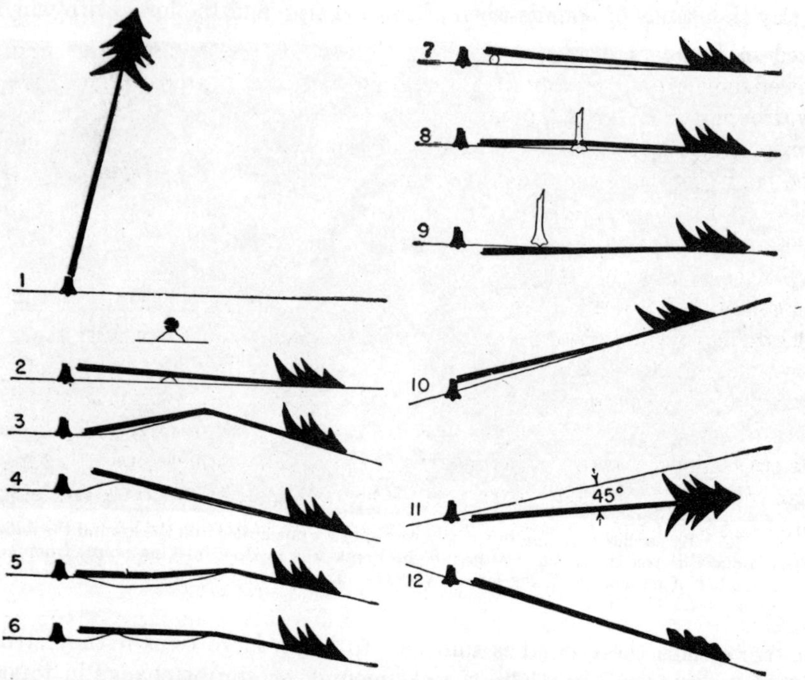

FIG. 13. Principal causes of timber breakage in Douglas fir. 1. The ideal condition is practically level, spongy ground. 2. Dropping trees squarely on rocks or stumps unless relatively close to the stump of the falling tree will burst the trunk at the rock or stump, even when no bending stresses are present. 3. Dropping a tree across a ridge or log about 60 ft or more from the stump so that the butt is not supported will break the tree at the ridge or log. 4. Dropping a tree across a ridge or log 30 to 40 ft from the stump will not ordinarily break the tree at the ridge or log. 5. Dropping the tree across a hollow and over a ridge 100 or more ft from the stump ordinarily will break the tree both at the ridge and over the hollow. 6. Trees may be dropped across two ridges, the first 30 or 40 ft from the stump and the second as much as 100 ft from the stump without breaking at the first ridge or over the hollow between the ridges but will break at second ridge. 7. Hanging the tree on the stump or on a large log near the stump will produce the equivalent of a hollow on otherwise level ground and may be sufficient to break the bole below the center of its length. 8. Trees striking leaning trees or snags near the ground and about the center of the length of the falling tree are likely to be broken at the leaning snag or tree. 9. Trees striking leaning trees or snags materially closer to the stump than one-half the height (60 or 80 ft) will ordinarily have the butt thrown to one side without breakage. 10. Trees thrown uphill are at times hung up on the stump or spur roots and are broken in the bole through bending or the butt split upon a spur root. 11. On moderate uniform slopes throwing trees not directly uphill but an at angle of about 45 deg appears to give best results. 12. Throwing trees directly downhill breaks the tops badly but does not seem to break the boles any worse than throwing the trees directly uphill. Trees thrown downhill never hang on the stump and strike a sliding blow. (*From Timberman, Falling timber with less damage, by J. A. Newlin, June,* 1940.)

of fall, so when the tree strikes level ground the butt is from 1 to 3 ft from the stump, as a rule. These conditions affect the possibility of breakage or injury to workers, as will be shown.

In unusually large timbers such as redwood, giant Douglas fir, cedar, and other Western species, the direction of fall for each tree felled is extremely important. Great waste in broken boles and tops will result unless such trees are felled so they will come to earth on prepared sites, freed of boulders, logs, and even graded to eliminate uneven ground. Power shovels and tractors with bulldozers are used to prepare such beds. By dropping such large and valuable trees in prepared places much reduction of breakage is possible. Redwood, especially, is quite brittle and a tree may break and split open from end to end like a ripe melon if it is felled on rocky or uneven ground or over logs. When a bed has been prepared in which a tree is to be felled, therefore, it is extremely important that it be accurately placed on the ground.

Ease of Bucking the Tree. When trees are dropped on uneven or rolling ground they may be strained so that they will pinch the saw in bucking. Saw pinching due to simple sagging of the tree in bucking can be overcome by driving a wedge into the saw cut, but in severe cases and especially in trees of small diameter, it may prevent the saw getting far enough into the cut to drive a wedge in behind it. To avoid any unnecessary pinching that will require bracing, prying, or "underbucking," saw crews and saw foremen should seek the best lie for each tree, and other factors permitting, drop the trees in places where they can be most readily bucked. When the lay of the land would make bucking difficult, felling crews should have a log skidded across the place a tree is to be felled, usually near the butt, to give it the best stress for bucking.

Prevention of Damage to Trees Not to Be Cut. In managed forests or on any cutting or harvesting operations where merchantable trees are to be left for later cutting or younger trees left to grow to larger size, it is particularly important to fell the trees being harvested so they will not destroy or damage any of the reserve stand. To do this requires that falling crews be skillful and that they be willing to use care in undercutting and felling to direct the trees into open places devoid of trees and young timber. When a partial cutting operation is directed towards salvaging overmature and defective timber, more difficulty is experienced in felling trees without damaging others.

The large size and the large tops of overmature virgin trees require considerable space in falling under even the best conditions. When such large timber is also defective to the extent that the stump or butt of a tree is rotten or hollow, the direction in which the tree will fall is somewhat uncertain. Hollow or rotten trees give way sooner than sound trees when being felled and frequently fall in directions contrary to the intention of the sawyers. Nevertheless careful saw crews can minimize damage to other trees by the exercise of skill and judgment.

Method of Bunching or Skidding. In general, trees should, if possible, be felled parallel with the direction of skidding, especially when team or tractor yarding is carried on simultaneously with felling and bucking. This permits the chokers (fastenings) to be placed on several logs at one time with a minimum amount of backing and stopping by the tractor or team to get a load. This saves valuable time and makes for a more orderly and efficient operation, especially when long logs or poles and

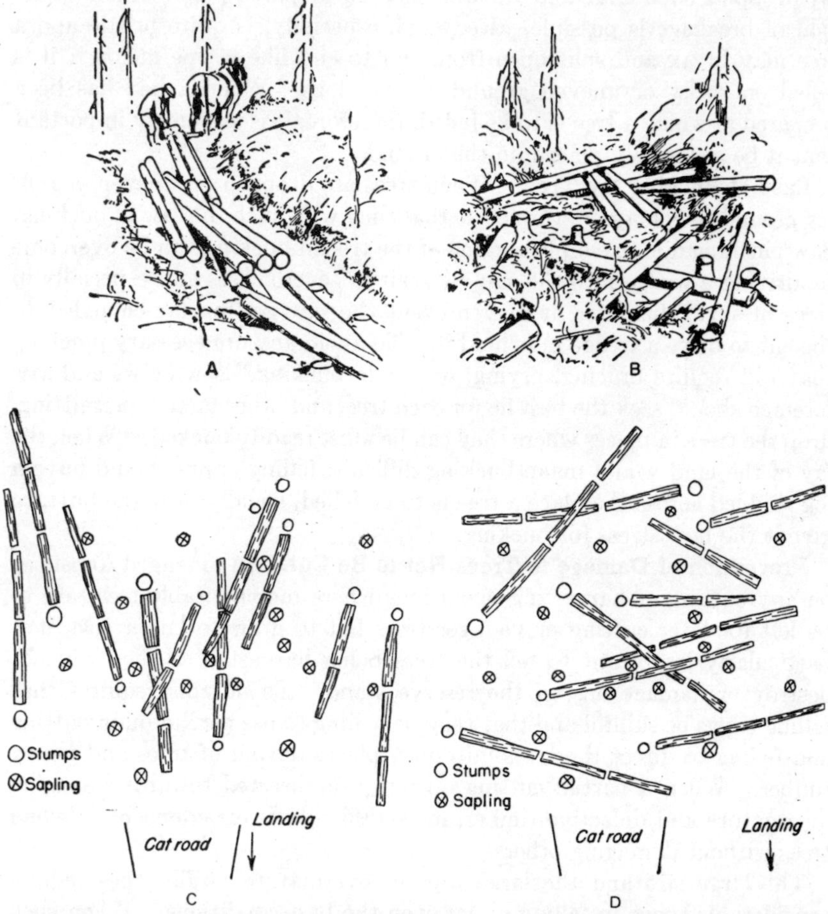

FIG. 14. Trees should be felled to make skidding easier rather than more difficult. *A.* Logs well placed for skidding on a well-cleared road. *B.* Logs badly placed and stumps on the road are too high. (*From "Woodcutters' Handbook," by A. Koroleff, forester, Woodlands Section, Canadian Pulp and Paper Association*, 1942.) *C.* Trees felled according to pattern make it easier to get logs out. *D.* Trees felled in haphazard fashion make it hard to yard logs around saplings. (*From West Coast Lumberman*, 1939.)

piles are being handled. In cable skidding, close attention need not be given to felling the trees parallel with the skidding direction as the powerful drums, blocks, and lines will straighten them out, once chokers are fastened to them.

When logs or pulpwood are to be bunched or assembled prior to skidding or hauling, trees should be felled so this operation can most easily be accomplished. When products such as pulpwood are to be piled in ricks, pens, or piles, much saving in effort can be made by felling trees so that when they are cut up the bolts will be close together. Indiscriminate felling may scatter the bolts so much that assembling them is made unnecessarily difficult. This same principle holds when logs are assembled in small groups for skidding. By throwing the trees close to each other, the logs will be partially bunched and only a minimum of actual bunching with team or tractor will be necessary.

Keeping Skidding Roads Clear of Tops. If skidding trails or roads are blocked by tops of felled trees, they must be opened promptly to keep costs lower on later operations. In cases where cutting specifications call for piling the tops, considerable saving in this cost can be effected by throwing several tops together, so that as little handling as possible will be necessary. On the other hand, in cases where lopping the tops and scattering the branches is required, it is more economical to keep the tops as widely separated as possible.

Slope. In mountainous country or on steep slopes, trees must be felled with particular regard to the topography. Other factors permitting, trees should be felled laterally to the slope of the ground so they will strike approximately horizontally. Felling a tree up steep slopes invites damage to the tree and injury to workers because it strikes the ground while pressure is still against the stump. If the undercut has been incorrectly made or if some interference with the tree during its fall jars it loose from the stump, it may split or jump violently backward, or both, with obvious damage and danger. Kickbacks result from trees striking the ground while they are still exerting pressure against the stump, and the jar of striking the ground before it reaches the reverse stress frequently frees it from the undercut ledge tending to hold it at the stump and causes the butt of the tree to kick backwards.

Felling trees down steep slopes accentuates the pull of the tree away from the stump in the direction of fall. When the tree continues to fall downward past the horizontal, the momentum of the top increases to such an extent that the butt will fly upwards and outwards as the tree strikes the ground, and frequently it will somersault end over end several times or slide 100 ft or more down steep slopes, with obvious damage to the tree itself as well as to adjacent trees and with danger to workers who may be in the vicinity.

FELLING TECHNIQUE

The larger the tree to be felled, the more important it is that the felling operation be properly accomplished. Large trees are more valuable than smaller ones; because of their size they damage or break easier; and the area affected by the fall is greater for large trees than for small trees. Determining the direction of fall, therefore, is the first step in actually felling a tree because this determines where the undercut or felling notch

FIG. 15. Completing an undercut in a Western pine tree in preparation for felling. In this case, the undercut is cut in with an ax only.

will be made. When the several factors having a bearing on where a tree should be felled have been weighed and a decision reached, then the tree is undercut on the side towards which the tree is to be felled. First, however, a working space is cleared around the tree by chopping out all sprouts and other growth that might interfere with the fallers.

A perfectly erect and evenly balanced tree in a windless atmosphere, when properly undercut and backcut, or severed, will fall squarely in the direction of the undercut. Conditions that upset the balance of a tree, such as wind, an uneven crown, or lean, may prevent the direction of fall from actually being in the line of the undercut. Skillful undercutters, however, can judge the combination of factors pulling a tree in other directions and can place the undercut, felling notch, or "bate," as it is sometimes called, at the place where it will influence the other stresses to the degree desired and thus determine the direction in which the tree will fall.

The backcut also can be so made that a tree can be pulled to one side or the other of the undercut, which gives still greater control of felling

(*a*) (*b*)

FIG. 16. (*a*) The stump splinters shown here were pulled from a high-quality butt log and some of the boards sawed from it will have to be cut off 2 ft in the mill. Poor felling is shown by too shallow an undercut and a sloping backcut that went lower than the undercut. This waste could readily have been avoided by more care in felling. North Carolina. (*b*) This badly split tree was incorrectly undercut. Had the sawyers made a standard cut of ¼ to ⅓ the diameter of the stump instead of making only 2 or 3 ax cuts this waste would have been avoided. North Carolina.

direction. By driving one or more wedges in the backcut, a tree can be forced in the direction desired. The pressure exerted by a wedge driven in the cut by blows from a sledge is tremendous and even very large trees can be forced to fall against their natural lean by judicious wedging.

Making the Undercut. After the direction of fall has been determined, the undercut is made to guide the tree in the chosen direction. An undercut is a notch at the base of the tree that removes support for it at that point and thereby increases the tendency of the tree to fall in that direction. As the backcut, or felling cut, is being completed, the narrow bridge of uncut wood that breaks off as the tree falls serves as a hinge that guides the tree in the direction of the undercut and prevents it from rotating on the stump.

The depth of the undercut should vary with the size of the tree and whether or not it is to be felled in the direction of its lean. In straight,

FIG. 17A. The "V" undercut illustrated here has just been chopped in the tree by the two undercutters. The next step is to saw the tree off an inch or two above the top of the undercut, starting from the opposite side of the tree. Alabama.

even-crowned, and sound trees the bottom of the undercut is chopped or sawed in horizontally for a distance of one-fourth to one-third of the diameter of the tree. When the tree is leaning in the direction of fall, the undercut should be deeper to prevent the tree from splitting. Corner notches should also be chopped in on each side of the undercut in leaning trees to prevent pulling side splinters as the tree falls.

When a tree is felled against its lean or natural direction of fall, a shallower undercut is advisable. The undercut in a leaning tree being cut to fall in the direction of its lean or natural pull should have a deeper undercut than a straight tree; otherwise it may break off before it has been severed sufficiently and induce splitting and splinter pulling, causing damage to the butt log, which contains, generally, the wood of highest quality. This is because the portion of the base being sawed is in tension. The more lean in a tree, the deeper should be the undercut, if the tree is to be felled in the direction of its lean.

The base of a tree being felled against the direction of lean is in tension on the side where the undercut is placed and in compression where it is sawed off, so wedging is necessary to prevent the saw from being pinched and to force the tree to fall in the desired direction. A deep undercut would free the side in tension sufficiently to permit the tree to fall prematurely in the opposite direction from that desired.

After the undercut has been sawed in the correct distance, it is chopped out by one or two men with axes or it is sawed out. When chopped out,

FIG. 17*B*. The "V" undercut, when correctly made, gives a clean-cut stump like this and a butt log without splinter pull or splitting. This particular stump could have been slightly lower for close utilization. Alabama.

the angle should be only sufficient for good ax work, or about 45 deg. Too much unnecessary chopping is done when a high angle is used and with a low angle chopping is difficult. It is important that the undercut be chopped or sawed out cleanly and as far into the tree as the saw cut goes; otherwise, the portion left serves as a flat lever that enhances the tendency to pull splinters from the tree when it begins to fall. When two men chop out an undercut, one chops right-handed and the other left-handed. Some fallers prefer to chop out the undercut entirely, without sawing, even in large trees.

Another type of undercut is the "V" undercut, which cannot be sawed, even partially, but must be cut in with axes. Usually two men chop at once, one on either side of the tree, and cut a deep V-shaped notch in the base of the tree, on the side towards which it is to fall. The notch is then cornered, forming two smaller V's at each side of the main V. This kind of undercut is reputed to reduce the danger of splinter pulling, which is especially wasteful in trees being cut for timbers, as the butt end of a sawed timber must be cut off as far up as the pulled splinter holes penetrate, with consequent reduction in its length and value. Splinter holes in lumber logs also reduce their value but not to the extent they do in timbers, as only those boards cut from the log that contain the splinter cavities are cut back in length. Special types of undercuts are made in

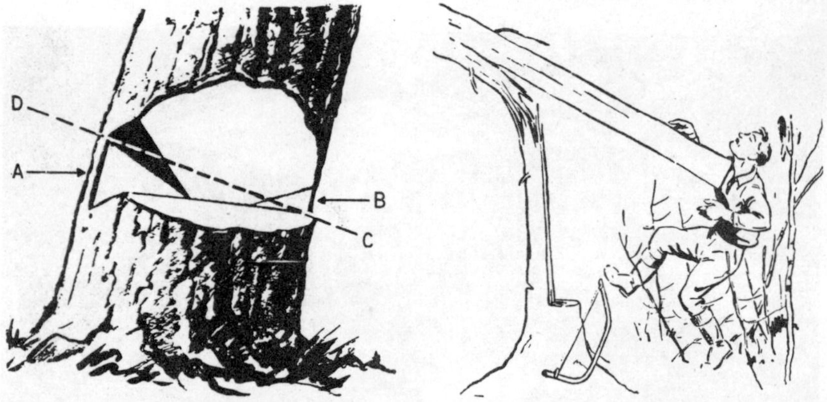

(a) (b)

FIG. 18. (a) Sketch showing special type of undercut used for felling large leaning trees to avoid breakage. Developed by H. J. Ness of Hamilton, Wash. A. Holding wood on upper side. Leave 1 in. of straight face for each foot of lean. B. "Dutchman" on lower side. This straight face acts as a leg when tree starts to fall. C-D. Position of saw. (*From West Coast Lumberman, September*, 1942.) (b) Poor undercutting increases danger of sudden butt splitting in trees with heavy lean and stump rot. (*From "Woodcutters' Handbook," by A. Koroleff, forester, Woodlands Section, Canadian Pulp and Paper Association*, 1942.)

felling trees offering special problems, such as large leaners in steep topography, where it is necessary to fell them at right angles to the slope to avoid excessive breakage.

Large trees can also be felled satisfactorily without undercutting by a method called "matchsawing," which means that the bole is severed from the stump by two saw cuts meeting near one side in the same plane. The tree must be forced to fall by wedging if it does not lean. However, if it leans or if the crown pulls it in one direction, match-sawed trees will nearly always fall in that direction.

Splinter pulling can be almost entirely eliminated by match sawing, as the tree is held erect by the flat sawed surface until the tree is practically severed from the stump. In leaning trees, the first saw cut is made on the under side of the lean until the saw pinches, then the saw is removed and the tree is severed by cutting from the opposite side and in the same plane as the first cut.

The disadvantage of match sawing is that the fallers have but little, if any, control over the direction in which the tree will fall, unless it is so evenly balanced and the air is so still that it must be wedged. This adds materially to the danger in felling by match sawing. Where it is essential or desirable to fell trees in definite places, as in the case of selective or partial cutting, undercutting is always preferred, and, in fact, it is far more

FIG. 19. Making a conventional type of undercut in a Western pine tree with an electric chain saw. Washington.

widely practiced than match sawing. Also, since there is no bearing line, as between an undercut and the saw cut, the tree is much more likely to slip backwards over the stump in falling and, if the crown is uneven, the tree will rotate, with no "hinge" to hold it.

The use of power saws in felling trees has made possible the use of undercuts different from the conventional types. With handsawing, the bottom of the undercut is sawed to the required depth and the upper slanting portion chopped out with axes. With power saws, however, the undercut is commonly made by sawing two parallel horizontal cuts 4 to 6 in. apart. The intervening wood is then chopped out with a special tool with one ax blade and one adz blade. Or, the conventional type of undercut can be used, both the upper and lower cuts being made with the power saw. The reverse of the conventional type can also be sawed in; *i.e.*, the upper cut is horizontal and in the usual position of the lower cut, while the lower cut slopes up, thus placing the notch on the underside and on the stump.

All of the variations are as satisfactory as the usual types. They are possible because of the speed of power-saw cutting and the ability to saw from various positions. Horizontal undercuts or those with the notch below have the advantage of leaving the butt logs square ended, which results in some saving in lumber volume.

FIG. 20*A*. Spring boards on which undercutters and fallers stand are placed in notches chopped in trees with large butt swell or on sloping ground. (*From West Coast Lumberman.*)

Because of the importance of getting large redwood trees to earth on prepared "beds," the undercut is usually "gunned" during its making to be certain it is aiming the tree correctly. A "gun" is an A-shaped wooden frame 6 or 8 ft long. In use, the faller holds both legs of the A, placing one in each corner or end of the undercut. The peak of the A points in the exact direction in which the tree will fall; and the observer, standing within the frame and holding it, can sight through the point of the A and so determine this direction. Such adjustments as are needed in the undercut are then made before starting to sever the tree from the opposite side. Many fallers gun, or sight, their undercuts in smaller timber by placing a double-bit ax in it so the head is against the undercut with the handle indicating the direction in which the undercut is aiming.

FIG. 20*B*. Types of undercuts made with chain saws in small and large trees. (*From West Coast Lumberman.*)

Severing the Tree. When the undercut has been completed, the tree is severed by sawing from the opposite side. When trees with thick fibers or scaly bark are being sawed, the bark should be removed or the loose portion shaved off with an ax to prevent it from being drawn into the cut by the saw. The severing, felling, or backcut, as it is called, is made from 1 to 3 in. above the bottom of the undercut, being the highest in large trees. In an evenly balanced tree the backcut should start at the middle of the

FIG. 21. Making a parallel-type undercut with a gasoline-motor chain saw. The wood between the two saw cuts will be cut out with an adz ax. Oregon. (*Photograph courtesy of the Mall Tool Company.*)

FIG. 22. Gunning an undercut to see if it will cause the tree to fall in the desired direction. This undercut was chopped in by axmen. Note the springboard on which one axman stands, and the long, narrow-bladed double-bitt ax used for chopping undercuts. Washington.

FIG. 23. Southern pine tree just as backcut has been completed by crew using hand crosscut saw. (*Photograph courtesy of the Southern Pine Association.*)

FIG. 24. Making backcut with hand crosscut saw in felling a longleaf pine tree in Flordia. Note low stump and worked out naval stores face.

opposite side of the tree and proceed parallel to the undercut until the final narrow uncut section breaks off and the tree settles and falls. In windy weather and with rotten or hollow-butted trees, sawing is usually done around the stump, leaving until the last the triangular portion directly in the middle of the undercut to prevent a premature fall from

FIG. 25. Felling a tree with a circular-type power saw. Note the low stump that is readily obtained with this type of saw. (*Photograph courtesy of Lowther Saw Co.*)

tearing the sides of the tree or splitting the butt log. Wedges are driven into the backcut, if necessary, to assist in making the tree fall in the direction desired as well as to keep the weight of the tree off the saw.

It is frequently desirable to "pull" a tree being felled to one side or the other from the straight direction of the undercut by sawing the backcut at an angle to the undercut instead of parallel to it. The backcut, by this method, will be completed on one side while the opposite side is still only partially cut. When the tree starts to fall, the unsevered side holds and pulls the tree to that side, while the sawyers rapidly complete the backcut.

Thus, it is seen that saw crews by skillful work can fell trees with great accuracy and even against their natural tendency to fall. In some cases considerable wedging is necessary to force the tree in the desired direction; crews ordinarily prefer to fell in the easiest direction because they can then produce more per day and, when working on a piece rate, thereby earn more. When it is necessary, however, they can be required to go to the extra trouble necessary to fell trees where they should go.

As a tree is about to fall, a sawyer should call "timber" or give some other signal to warn any who might be within the danger area. As the tree

FIG. 26. Making a backcut in felling a tree on an Oregon logging operation. The chain saw shown in this illustration is powered by electricity. Similar saws are powered by gasoline motors and by compressed-air motors.

starts to fall and before it reaches the ground, the saw is removed from the cut to prevent any damage to the blade, and the fallers step back out of danger. From time to time during the sawing operation in both felling and bucking pines or other resinous trees, kerosene is sprinkled over both sides of the saw blade to remove gummy substances that interfere with sawing and increase the friction of the saw in the cut.

Felling Small Trees. When trees are cut into cordwood products such as pulpwood, fuel wood, or distillation wood, some of the precautions observed in felling saw-timber trees need not be observed. Splinter pull from the butt section and splitting are not serious defects, as the wood will be chipped if pulpwood, charred if distillation wood, and burned if fuel wood, and splinter cavities cause no serious losses. Breakage, however, is a cause of loss in that most cordwood products are cut to specified lengths

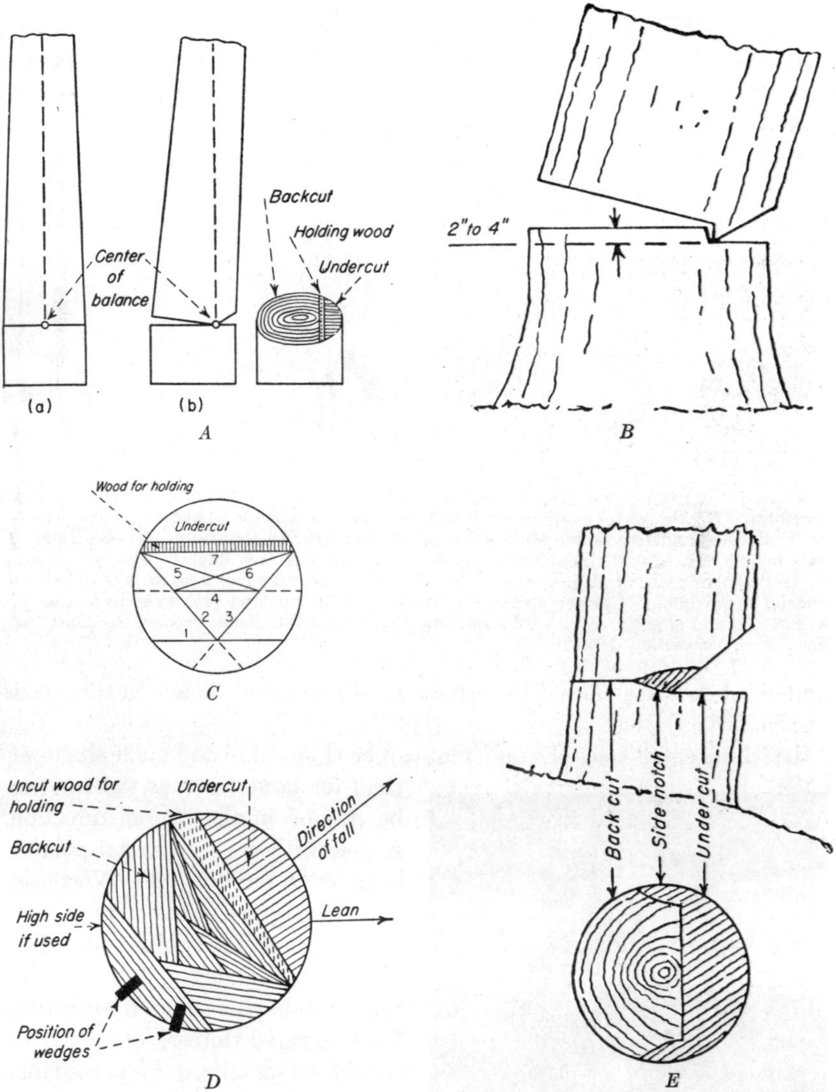

Fig. 27. Felling Procedures. *A.* The undercut removes support on that side of the tree and when the backcut is wedged the center of gravity of the tree moves in the direction of the undercut as shown. *B.* A tree correctly undercut and correctly backcut cannot kick back over the stump, as the above illustration shows. *C.* By backcutting large trees according to the above pattern, sawing is easier as the saw does not at any time cut through the entire diameter of the tree. Also, the direction of fall can be better controlled by varying the amount of holding wood in accordance with the lean of the tree. *D.* When trees are felled in a direction other than their natural lean, they can be "pulled" away from their lean by leaving a wider band of holding wood toward the direction of fall. *E.* Corner notches are chopped in with axes at each side of the undercut in leaning trees to prevent splinter tear when the tree falls. (*From Snag falling, by R. C. Lindberg, Timberman, November,* 1943.)

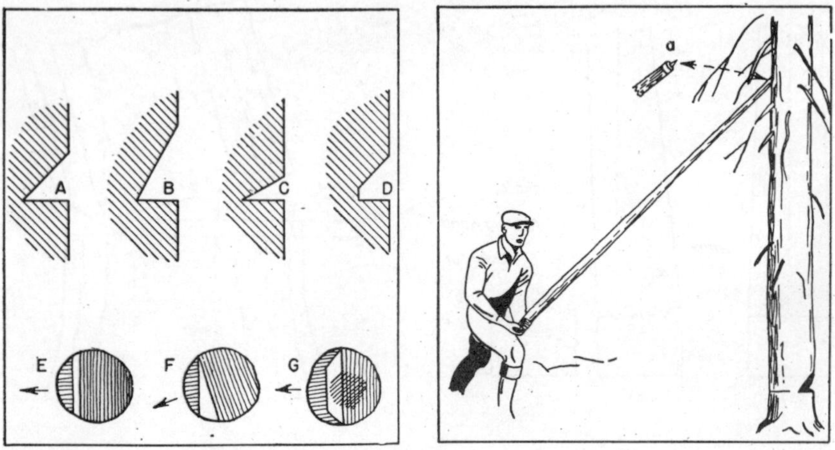

FIG. 28. (*a*) Good and poor undercuts and methods of backcutting. *A*. Good shape of undercut. *B*. Too wide: waste of work and wood, and bad shape of butt. *C*. Too narrow. *D*. Unfinished undercut: some work and wood wasted. *E*. Even "bridge" between undercut and felling cut. *F*. Wedge-shaped bridge to "draw" the tree slightly against its lean. *G*. Undercutting when butt may split. (*b*) Method of pushing small trees in desired direction instead of wedging. Correct slant of pole (45 deg) in pushing tree over from the hip. *a*. Spike at end of pole. (*From "Woodcutters' Handbook," Woodlands Section, Canadian Pulp and Paper Association*, 1942.)

and short sections caused by breaks in felling would be left in the woods or rejected.

In felling small trees the undercut can be chopped in and made shallower than for large trees, as the tree can be pushed in the chosen direction. A pole with a sharp metal point is commonly used in the Northeast and in eastern Canada to push trees being felled for pulpwood. This eliminates wedging and helps guide the tree in the desired direction. Even in small timber, correct felling should be practiced in accordance with the factors already mentioned, although time should not be wasted in meticulous preparations for felling when they are made unnecessary by small tree size or the nature of the products being cut.

FIG. 29. Felling a Southern pine tree for pulpwood with a one-man bow saw. (*Courtesy of the Southern Pulpwood Conservation Association.*)

Trees cut for poles and piles are selected with respect to their size, straightness, and freedom from defects, and particular care is called for in felling to avoid breakage that would disqualify the tree for the product for which it is intended. If this occurs a valuable tree must be converted into less valuable products, such

FIG. 30. Felling trees with power saws; above with a chain saw and below with a circular type. (*Photographs courtesy of the Southern Pulpwood Conservation Association.*)

as saw logs or pulpwood, or be wasted, with obvious loss to the owner and operator.

Trees to be hewed into ties are frequently felled without an undercut with the intention of causing the tree to split in falling. This leaves an unsightly split on the stump extending up 3 or 4 ft or more, but it does flatten one side of the butt cut, which makes it unnecessary to hew as much wood from that side as would otherwise be necessary in shaping the tie.

STUMP HEIGHT

In early logging little attention was given to stump height because the principal objective was to cut the trees as cheaply as possible. Trees, therefore, commonly were cut almost waist-high, the most convenient height for the fallers, which resulted in a great deal of waste from the best part of the tree. Now, with higher priced timber and mobile woods machinery, low stumps should always be required to prevent waste of valuable timber and to facilitate the use of tractors and trucks in the woods. When trees are felled with snow on the ground, the snow should be removed from around each tree before it is felled to permit the cutting of correct stump heights.

The best stump height in sound trees is the natural stump height, or just below the top of the butt swell. This varies from species to species and from tree to tree of the same species, usually varying with the tree diameter in the latter case. The larger the tree diameter, the higher the swell extends up the tree and hence the higher the natural stump.

In cutting saw-log trees very close to the ground and into the butt swell, an actual loss may result, instead of close utilization. The curve of the annual rings follows the external swell of the stump and when lumber is sawed from a log with butt swell, the swell is slabbed off but the grain at the end of lumber from the outer portions of the butt logs is cross-grained. This causes warp and twist in drying and a tendency for such boards to split in use. The supposed saving in the woods by cutting stumps practically at ground level may cause a greater loss in the mill from trimming off warped ends of boards, which means shorter length lumber, often shorter than the extra log length taken from the stump, due to standard lumber lengths being in 2-ft intervals. For example, if a tree were cut 1 ft too low and it was necessary to trim off the lumber cut from the butt log, it would be necessary to trim the lumber to the next standard length, or 2 ft shorter, thus losing instead of saving 1 ft.

Stumps, therefore, can be too low as well as too high in trees cut for saw timber. In cutting trees for pulpwood, such cross grain is no factor, as the bolts will be converted into small chips. In cutting for ties, veneer bolts, poles, and other products, the requirements of each product must be considered in determining stump height. Usually they are cut above the swell to avoid producing bell-shaped bolts, which are difficult to process.

Hardwoods are sometimes defective at the base and when this is the case, the felling cut should be made high enough to eliminate, if possible, excessive amounts of decay or unsound wood in the butt log. When this cannot be done, or when unsuspected butt rot is discovered after the tree has been felled, the felled tree should be "jump-butted," or the unmerchantable piece at the butt end cut off and discarded before any logs are measured or cut from the tree.

When a similiar condition is present in pine trees they also should be cut high enough to eliminate excessive butt decay or damage, such as is caused by naval-stores operations in the case of longleaf and slash pines. As a

FIG. 31. The above illustration shows a well-cut tree with a low, but not too low, stump and a clean undercut and a horizontal backcut that permitted the tree to fall without splitting or pulling splinters. Note that the tree fell exactly in line with the undercut. North Carolina.

rule, however, softwood trees seldom need to be cut with high stumps on account of decay. High stumps often are left in felling large redwood, Douglas fir, and similar Western species because of the wide butt swell. The stumps are often head high and the fallers use springboards inserted into each side of the tree to serve as a platform on which to stand in sawing or in manipulating a power saw.

All the decay should not be eliminated by cutting high stumps or by jump butting, as clear logs with decayed centers or catfaces will yield full-length, merchantable lumber from the sound portions of the logs, provided the sound material is not all in the butt swell, where it will have to be slabbed off, yielding only boards shorter than the full length of the log.

Poles, on the other hand, must be entirely free of decay in the butt as well as in other sections, and stumps are ordinarily cut higher than when similar trees are cut for saw timber. At the same time, they should not be cut higher than necessary; by so doing the tree may be made too short and hence fail to meet the specifications of the cutting order. When doubt exists, the stump should be cut at or just below the point of butt swell and, if necessary, jump-butted after the tree has been felled to eliminate defect. Trees with hollow or unsound centers can seldom be used for veneer bolts as there is no firm purchase for the lathe and hence the bolt could not be turned against the cutting knife.

In large saw timber, averaging 18 in. in diameter and 100 trees per acre, a loss of 12 ft log scale per tree is incurred when stumps are cut 1 ft too

TABLE 13. FEET DOYLE LOG SCALE LOST PER STUMP WHEN STUMPS ARE CUT TOO HIGH

Diameter of stump at cutting point, in.	Excess heights on stumps			
	2 in.	4 in.	6 in.	8 in.
16	1.0	2.0	3.0	4.0
20	2.6	4.0	6.0	8.0
24	3.0	6.6	10.0	13.0
28	5.0	10.0	15.0	20.0
32	7.0	14.0	21.0	28.0
36	9.0	18.6	28.0	37.0
40	12.0	24.0	36.0	48.0
44	15.0	30.0	45.0	60.0
48	18.0	37.0	53.0	73.0
52	22.0	44.0	66.0	88.0

Source: Bryant, R. C., Prolonging the Cut of Southern Pine, Part II, *Yale Univ. School Forest. Bul.*, 2, 1913.

high. This is a loss of 1,200 ft per acre. If the stumpage price is $10 per 1,000 ft, this is a direct loss of $12 per acre to the owner if he cuts his own timber or sells on the basis of volume cut. If the operator purchased for a lump sum, then he loses that amount. In pulpwood-sized timber substantial losses can also occur. The stumps of trees 10 in. in diameter, if cut 1 ft over correct height and averaging 200 trees per acre, contain 1¼ cords, worth about $2.

FELLING TOOLS

Felling is done by the use of a number of tools, which will be considered in more detail in Chapter 10. A saw is, of course, one of the most essential of all felling tools and for a great many years saws were strictly hand tools, being operated by either one or preferably two men. Recently, however, power saws have been perfected to the point where they can be used effectively and economically in logging. They are most widely used on Western operations where the trees are of large size and felling one tree is a job of several hours duration for handsawyers. Regardless of whether handsaws or power saws are used, however, the principles involved in undercutting and felling trees are the same.

CRITERIA OF CORRECT FELLING

It is frequently the duty of a forester to inspect cutting operations to report on the quality of workmanship of the felling crews. As it is not possible for the inspector to be with the felling crews at all times, he must be able to judge the kind of work done by examining cutting areas after harvesting has been completed. The quality of felling work can be accurately judged by noting the stumps and other visible evidence, as explained below. The inspector should, of course, also examine the work being done where the operation is going on.

An efficient and careful job of felling is indicated by the following:

1. Low stumps with undercuts from one-fourth to one-third of the stump diameter; horizontal backcuts 2 to 3 in. above the bottom of the undercuts; and a clean break without splinters or splitting where the holding wood broke as the tree started to fall. The stump of a tree that was a leaner is usually elliptical or the pith is at one side as a result of the tree forming more wood on the side under the lean than on the opposite side, as a result of greater compression. Such stumps should show that a deeper undercut than usual was made. All stumps should have marks on them showing that the trees had been selected for cutting.

2. No excessive breaking of or damage to other trees left standing and no more damage than necessary to reproduction.

3. Tops placed where they were not in the way of skidding and, if piling is required, bunched to facilitate this operation, and not close to such standing trees as may have been reserved for growth and future cutting.

CRITERIA OF POOR FELLING

Indications of poor felling practices or poor workmanship on the part of the saw crews are

1. High stumps, unless they are defective enough to be justified or were deliberately cut high because of the kind of product being cut.

2. Splinter pull, which is evident from splinters projecting up from the holding wood, or unsawed portion, of the stump when the tree fell. These may be relatively short or up to 5 or 6 ft in length. The crews may cut such splinters off after the tree has been felled but the damage has already occurred. When this has been done to conceal poor workmanship, it can always be detected by a skilled observer from the evidence of ax cutting and from the splinters lying on the ground, unless they have been scattered.

3. Split butt logs. Although the split logs may have been removed, the

stumps will show that the trees broke before the backcut reached the undercut and split the butts of the trees.

4. Where the timber has been marked for cutting, unmarked stumps indicate that the crews deliberately violated the marking and cut trees they should have left. In cases where trees were marked for leaving, the stumps should, of course, be unmarked.

5. Trees lodged in others or broken stubs or sections of other trees either standing or lying on the ground are evidence of poor judgment or carelessness in felling. Also, when tops of trees have been chopped away from roads and skidding trails and when tops have destroyed or damaged an excessive number of young trees, it is evidence that no attention was given to saving future timber crops and that the cutting crews caused unnecessary extra work for the swampers cleaning out skidding roads and trails.

DANGERS ON FELLING OPERATIONS

In felling trees in the woods there is danger on every side for the careless or unaware observer and workman. The larger the trees being felled, the greater the danger of serious or fatal injury, but even small trees can seriously injure or kill any who may be struck by them in falling. Good woodsmen are known as much if not more by their manners and behavior around cutting and other harvesting operations than by the words they speak. Hence, all who have business in the woods or have occasion to visit cutting operations or to direct or advise cutting crews will find it very much to their advantage to maintain good manners in the woods. One of the requirements of good woods manners is that of avoiding danger whenever possible by knowing from what quarter it may come and observing due precautions. The person who exposes himself unnecessarily to injury not only annoys the workmen but interferes with the operation and is generally a nuisance.

Some of the dangers that should be avoided where trees are being cut are mentioned briefly in the following paragraphs. Some thought will assist in enlarging upon the observations given.

1. A swinging ax is a dangerous tool and can cause serious injury to a worker or observer if he is struck a full or glancing blow with it. Observers should be alert to keep at a safe distance from swinging axes. Workers ordinarily do not and should not have to take time to look back to see if anyone is standing too close to them. When a workman is swinging an ax in chopping, a close bystander could be seriously hurt. An observer facing a working axman takes a chance that the head might come off or the handle slip from the worker's grasp. When axes get caught in vines, twigs, or limbs, or when a glancing blow is inadvertently struck, the chopper may lose control of his tool and bystanders, in such cases, may get hurt.

2. Frequently trees have large dead limbs or debris from other trees that have fallen in the past lodged in their crowns; portions of their own tops may be rotten or loosely hanging. Undercutting and sawing a tree may jar this material loose and cause it to fall on the crew or on a bystander beneath the tree. Similarly, a tree in falling may brush against another tree, knocking lodged material from it or, if it is dead or fragile, a portion of it may break off and fall. Such material lodged in the top of a tree is commonly known as a "widow-maker." Workers and observers, therefore, should look into the crowns of trees to be felled to determine if such danger exists and take the necessary steps to avoid or minimize it. When it is necessary to fell a tree with a widow-maker in its top, the foreman or a crew member stands off and watches it closely and if it starts to fall, he calls out and warns the crew in time for them to reach safety.

3. If a tree is not properly undercut, does not fall true due to some interference, or is felled uphill, it frequently kicks backwards for several feet because of the downward and backward force it exerts against the stump, and anyone standing directly behind would be struck a violent and heavy blow. Therefore, observers should never stand directly behind a tree while it is being felled and the saw crew should step well back to either side so if the tree should kick back no great harm would be done.

Poor undercutting or misjudging unsound trees may result in the butt splitting instead of breaking off clean at stump height, and in such cases the severed portion usually kicks upward and backward, which likewise is dangerous to anyone behind the tree. If a tree breaks clean at the stump it still may strike other trees, or logs already on the ground in such a way that the butt will move violently to one side or the other. Standing too close at the side, also, is not advisable unless one has observed the situation and determined that there is no likelihood of such danger.

4. A carelessly placed ax or saw may be struck by a falling tree or the tree may strike some other object that will throw the ax or saw, so observers cannot be too careful when standing around sharp cutting tools. Such tools should always be placed flat on the ground when not in use, as this minimizes the possibility of such accidents.

5. Workers and others carrying axes and saws should observe the usual precautions. These require that a crosscut saw be carried on the shoulder with the cutting edge away from the neck, and that an ax be carried in the hand, grasped close to its head, with the edge pointing downward and outward. The sharp edge of the saw or ax would be away from the body, should the carrier stumble. Serious cuts have been suffered by failure to observe these precautions.

6. When trees are being felled, there is always danger of being struck directly by the tree. Particular care should always be taken to keep entirely clear of trees that are falling and away from others that a falling

tree may knock down. One should not depend upon one's hearing alone to detect this danger but should constantly look around to observe any nearby felling crews. With noisy power saws and tractors in use, the "timber" call of the felling crew may not be heard. Frequently trees in falling bend other trees nearly to the ground, and when they are suddenly released they may throw limbs or other heavy material to strike the unwary bystander not looking for danger from this source. With due precautions, however, observers and workmen can avoid the dangers created in felling timber.

7. Workers skilled in their jobs and in the use of their tools automatically observe safety rules. Careless or incompetent workers are more of a liability than an asset and ordinarily are not tolerated on harvesting operations, as they jeopardize not only themselves but their associates as well. Even in handling such sharp cutting tools as axes and saws there is no cause for accidents in the woods except through incompetence or carelessness. All that a workman needs to do to perform his work without risk is to observe good common sense and do a degree of thinking about his job and the use of his tools.

CHAPTER 8

BUCKING FELLED TREES

The merchantable portions of felled trees require some preparation or processing before they are removed from the forest. Since the purpose for which a tree is cut influences the bucking operation, the principles and practices involved in bucking trees are treated separately for saw logs, cordwood products, and piece products.

FACTORS AFFECTING THE BUCKING OF TREES FOR SAW LOGS

When trees are cut into logs for the purpose of later sawing them into lumber at a sawmill, felling and bucking are the first steps in the manufacture of lumber. It is important, therefore, to cut the trees into logs in such a way that they can be sawed at the mill to obtain the maximum volume of lumber from them and also to obtain the best lumber grades possible.

Serious damage and waste can occur in bucking trees. Waste is of two kinds, volume loss and quality loss, both of which result in a reduction of the value yield obtained from the trees, and both should by all means be avoided or held to a minimum. Loss of time by inefficient bucking crews is also a major loss item in that it indirectly affects the quantity of lumber obtained from the trees. Inefficiency results in greater than necessary harvesting cost, which raises the minimum limits of merchantability with respect to size and quality and so reduces the volume of merchantable products that can be obtained from a given stand of timber.

Full Utilization of Trees. For complete utilization of a tree, the top end of the last or top log must be at the limit of merchantability. The lengths of the logs to be cut from each tree can be so chosen that in practically all cases there need be no short merchantable section left in the top. Softwood logs are cut to produce construction and yard lumber in even foot lengths, with 12-, 14-, and 16-ft lengths the most common. Hardwood lumber is used in odd and even foot lengths, so hardwood logs can be cut accordingly, although it is common practice to cut the logs in even foot lengths up to 16 ft. Shorter and longer logs may be cut, but it is preferable to cut standard lengths, except when special orders specify other lengths.

By varying the combination of log lengths to be cut from a tree, it is possible to utilize almost any total merchantable length. As Table 14 shows, three-log trees can be cut in the several combinations of lengths to give total used lengths in intervals of 2 ft. It is evident from the table that not more than 2 ft of merchantable top length need be left in the top of any three-log tree, since if it amounts to 2 ft it can be included in one of the logs, if the combination of lengths is correctly chosen. Similarly, trees with four or more logs can be fully utilized by choosing the correct combination of log lengths.

TABLE 14. TOTAL USED LENGTHS OBTAINED BY VARIOUS LOG-LENGTH COMBINATIONS FOR THREE-LOG TREES

Number of logs			Total length of logs, ft	Total used length, ft, including 1 ft for trimming allowance
12 ft	14 ft	16 ft		
3	0	0	36	37
2	1	0	38	39
1	2	0	40	41
0	3	0	42	43
2	0	1	40	41
1	1	1	42	43
0	2	1	44	45
0	1	2	46	47
0	0	3	48	49
4	0	0	48	49

Source: Wackerman, A. E., Bucking or log making, *South. Lumberman*, Jan. 15, 1942.

To cut a tree into the correct combination of log lengths, it is advisable and usually necessary to measure off the logs first so if adjustments are necessary to include all the merchantable length they can readily be made. If the tree is bucked as each log is measured starting from the butt, it is seldom possible to make the last cut at the limit of merchantability. Because butt logs are the largest and generally turn out the highest grades of lumber, and since lumber in longer lengths is usually more valuable than the same grades in shorter lengths, it is standard practice to cut 16-ft logs from the lower portion of a tree and shorter logs from the upper section, which usually contains more defects, especially knots.

In tall timber with as many as six or more logs per tree, it is not necessary to measure off all the logs before actual bucking begins to utilize each tree completely. When only three or four logs remain in the upper portion of such a tree, the top limit of merchantability should, however, be determined so the remaining log lengths can be adjusted to fit the remainder of the tree. For other reasons, as will be shown later, it often is desirable to mark off the log lengths prior to bucking even in tall timber.

Adequate Trimming Allowance. Because it is impossible, or at least impractical, to make crosscuts exactly at right angles to the tree all the way through and because, furthermore, lumber sawed from logs at the mill should be long enough to permit trimming off the rough ends, logs should be cut several inches longer than the full length of the finished lumber to be sawed from them. Softwood lumber standards require that each piece be exactly an even number of feet in length, so that lumber which is even a fraction of an inch too short, as cut from the log, must be cut back in the mill to the next even foot of length.

Thus, if a log is cut for a length of 16 ft and no allowance is made for trimming, or if the allowance is insufficient, the lumber will be cut off to 14-ft lengths at the trimmer in the mill, resulting in close to a 2-ft loss in length of the entire lumber contents of the log. This obviously is a serious waste of merchantable lumber that is entirely unnecessary. Sawmill managers and superintendents make it a point, therefore, to observe lumber passing over the trimmers in sawmills, because logs coming to the mill short of full length plus trimming allowance can readily be detected at this point.

Usually a trimming allowance of 4 in. is sufficient for logs up to 16 or 18 in. in diameter, if the saw crews are skillful and can make reasonably straight crosscuts; and up to 6 in. or 1 ft for larger logs. If, in bucking, the saw pulls to one side of the cut in excess of the trimming allowance, then a larger allowance is needed. In practice the trimming allowance is measured in addition to the length of each log. Thus, a 16-ft log is measured off as 16 ft 4 in., and the next one similarly should include an adequate trimming allowance.

While sufficient trimming allowance is needed to avoid waste, an excessively large trimming allowance causes waste. It is not necessary in cutting good lumber tc trim more than an inch or two off each end of a board, so when more than this amount of allowance is included, the excess cutoff is waste. The trimming allowance, therefore, should be sufficient to prevent lumber being cut back to shorter lengths in the mill, yet the allowance should not be so large that waste occurs from cutting off more than the minimum allowance needed.

Prevention of Split Logs. Splitting occurs when an unsupported log is being sawed from a felled tree. Trees lying in such a way that in bucking them they tend to split or pinch the saw are said to be in a "bind." When they sag at the point of cutting, the cut closes at the top but it can be held open with wedges and no damage occurs in sawing through the tree. When a saw cut progresses through a tree with one unsupported end, however, it weakens the tree at the point of cutting and the weight of the unsupported end results in a split into the next log before it has been completely severed. A split log is wasteful in that the lumber cut from the split section will contain splits and must be cut off to the next smaller even foot length. The amount of waste from splitting depends upon the length of the split and its location, but it always is considerable and is quite unnecessary.

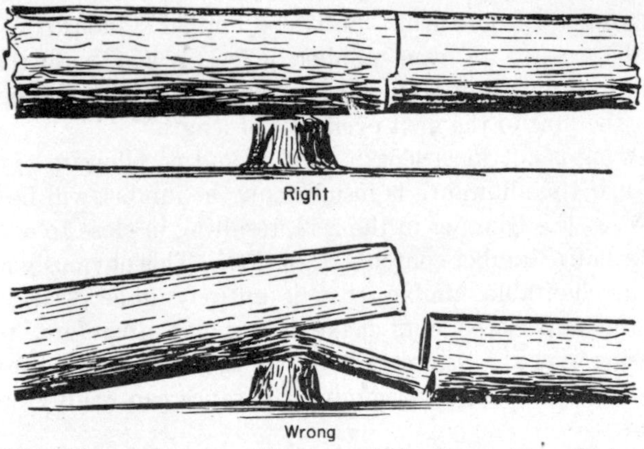

FIG. 32A. Right and wrong methods of underbucking. (*From Timberman.*)

Splitting can be prevented by providing a support beneath the hanging end of an unsupported log or by making the log cuts at other points along the tree until the remainder of the tree lies without stresses. Saw crews should not always buck a tree beginning from the butt end progressing towards the top. Depending upon how the tree lies, they frequently should start from the top or the middle or at some other intermediate point when this will permit the unbucked section to settle and thus eliminate stresses to prevent damaging the logs and also to do the job more easily. If this still leaves an unsupported log to be cut off, it can be braced with a strong upright stick firmly placed under the log or by some other method that will hold the log until it has been severed.

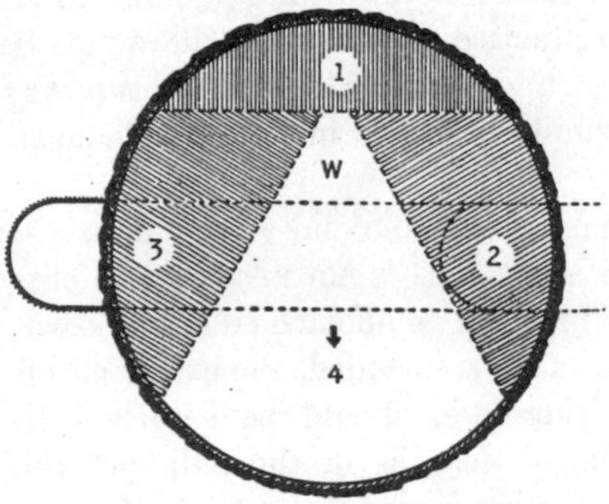

FIG. 32B. Bucking tree with bind. Procedure for bucking a fallen tree lying across stumps or other supports, causing a sag or bind in the mid-section. The one-man power saw is applied flat across the top of the tree, making cut 1. Then cuts 2 and 3 are made in the approximate direction shown, taking care to leave a wedge *W*. The saw is then driven straight into the tree as shown in the dotted lines and allowed to work downward. The wedge *W* serves as a key to keep the tree from sagging and pinching the saw. As cut 4 is completed the increased strain breaks the wedge. (*From Timberman, August,* 1943.)

When large trees lie over logs or on uneven ground in such a way that a hanging log cannot be supported, it is desirable to underbuck it. This is done by sawing up from beneath far enough so that when the log is cut off

FIG. 32*C*. The log shown above split because the end was unsupported in bucking. Virginia.

FIG. 33. Underbucking a large Douglas fir log to prevent breakage. Note that the ax handle supports the saw. A special tool, called an underbucker, is often used for this purpose. Also note the kerosene bottle with hook for hanging it conveniently nearby.

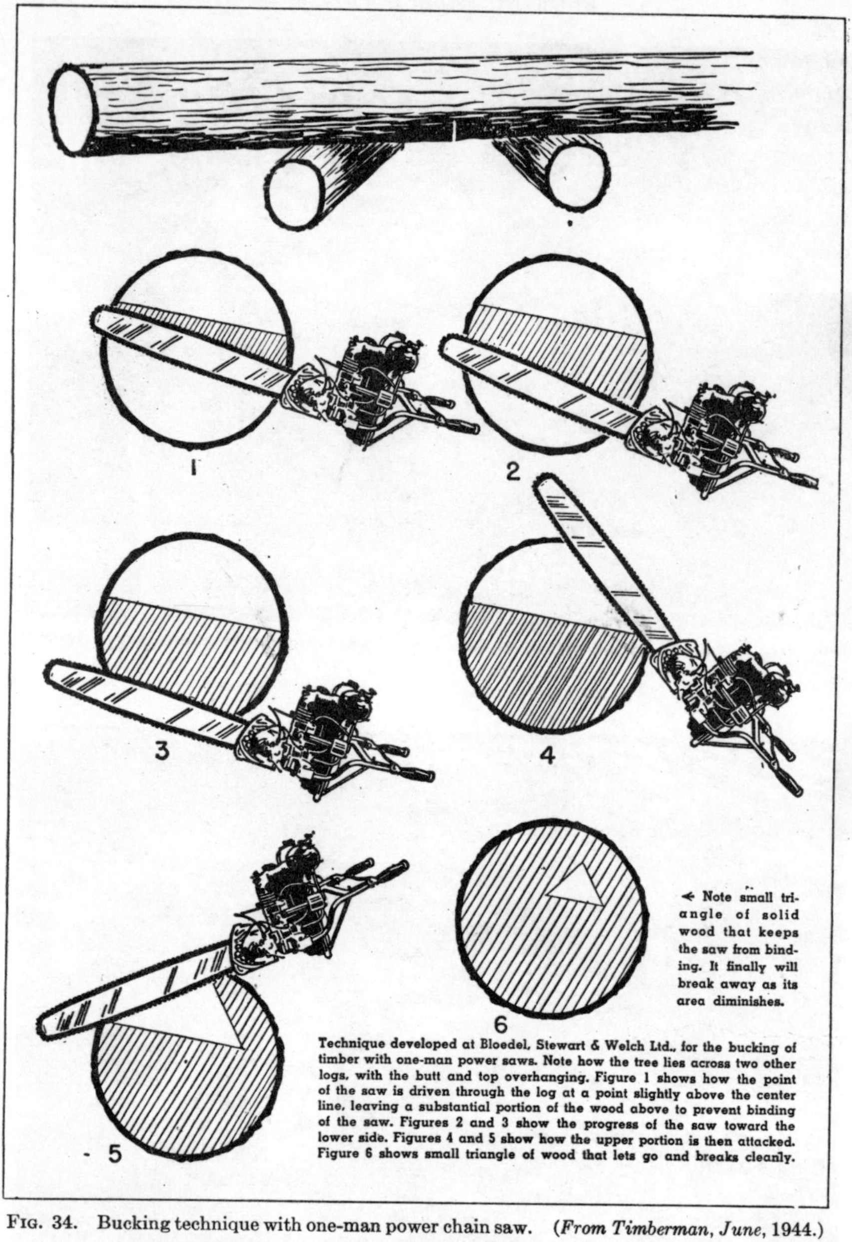

Technique developed at Bloedel, Stewart & Welch Ltd., for the bucking of timber with one-man power saws. Note how the tree lies across two other logs, with the butt and top overhanging. Figure 1 shows how the point of the saw is driven through the log at a point slightly above the center line, leaving a substantial portion of the wood above to prevent binding of the saw. Figures 2 and 3 show the progress of the saw toward the lower side. Figures 4 and 5 show how the upper portion is then attacked. Figure 6 shows small triangle of wood that lets go and breaks cleanly.

FIG. 34. Bucking technique with one-man power chain saw. (*From Timberman, June,* 1944.)

from the top it will not split. In some cases a narrow wedge is cut out so when the log is sawed from above it will break off cleanly in much the same manner as a tree does in falling when correct undercutting and backcutting have been done.

Underbucking is rather awkward in hand bucking, especially with one-man crews, but is not so difficult with the chain-type power saws if there is sufficient space beneath the tree to permit using the saw in this fashion. A recently developed type of one-man chain saw that can be used as a mortiser, cutting at the end as well as along the blade, makes possible a new technique in bucking hanging logs. The saw is driven clear through the log just above the center. The cut to the bottom is then made, after which the top portion is cut from above.

FIG. 35. This large, sound section of a yellow poplar tree in Virginia was so cut by an unskilled saw crew that it had to be left in the woods. At least two-thirds of this section would have sawed out clear lumber had it been cut long enough for a log. The small fork should have been sawed off close to the large section and then cut into logs.

Cutting Forked Trees. The problem of cutting logs correctly from forked trees occasionally arises in pine timber but is more often confronted in hardwood timber. In a forked tree the diameter of each fork is always considerably smaller than the diameter of the trunk below the fork. Therefore, to cut a forked tree so one of the logs includes the forked section is to waste much of the best portion of the log, even if the fork portion is in line with the rest of the log, since a log must be sawed in the mill on the basis of its smallest diameter. Much of the lumber from such logs will be shorter than the log length, which is undesirable, as short lumber even if of high quality is less valuable than full-length lumber.

Whenever possible, the log lengths cut from the portion of the tree below a fork should be arranged so a log cut comes just below the fork. This can be done by adjusting the combination of log lengths so a cut comes at the desired point, as in the case of obtaining full utilization in the top.

Then logs can be cut independently from each forked section. Ordinarily the fork will separate without much difficulty but if not, the forked section should be cut out, or jump cut, to eliminate it entirely from the used portion of the tree.

Jump Butting and Jump Cutting. Serious defects in trees, such as hollow or rotten butts and decayed portions further up the tree, should be cut out to eliminate the defect from the logs if it is of such size that it would cause most of the lumber cut from that section to be culled. Cutting off defective butt ends of trees is called jump butting and is always desirable if the rim of sound wood is so narrow that most or all of it will be slabbed off in the sawmill in squaring the log. Large logs with 6 in. or more of sound clear shell around a defective butt can be taper-sawed in the sawmill to obtain lumber of the full log length from the clear outside layers. In such cases, the lumber produced from the center portion will be shorter; but this is not so serious because it usually is of lower grade anyway.

Jump butting should, of course, be done before the butt log is cut. Simply cutting one end off a log already cut would defeat the purpose of jump butting the tree. The advantage of jump butting is to eliminate a short defective butt section so the first log above the stump will be sound enough to yield full-length lumber at least from the outside portion and avoid producing it in less valuable short lengths.

Cutting a defective section out of the tree at some point above the butt is jump cutting. This removes a rotten, forked, or otherwise defective section to prevent including it in a log. A log with such a defective section of a few to several feet in length will cut out boards with defective middle sections and sound ends, thus yielding only short lengths of usable lumber because the defect must be cut out of the lumber in the mill, reducing its value considerably.

In measuring off logs when a jump cut is to be made, a log cut should come at the lower edge of the defective section. Then a cut is made just above the defect and full-length logs can be cut above the defective jump-cut section. While it is desirable to eliminate all worthless sections from the logs, good judgment is required on the part of the saw crews in jump cutting so usable material is not cut off and left in the woods. Defective centers should not be entirely eliminated by jump cutting as this would waste good wood when rot, for instance, tapers into the center of a log, as the outside if sufficiently thick can yield full-length, sound lumber.

Elimination of Crook. Trees with short crooks need not necessarily produce crooked logs, as much of the crook can be eliminated by arranging log lengths so the tree is cut at the point or points of crook. If the crook is included in a log, then heavy slabbing in the mill is necessary, with resultant waste, and an excessive amount of short cross-grained lumber is produced. Careful attention to bucking crooked trees, therefore, can eliminate much of the crook from the logs. Occasionally, sharp localized crook should be eliminated by jump cutting. Of course, some trees are so crooked or twisted that no ingenuity on the part of the saw crew can prevent crooked logs. Trees with wide sweep from stump to top should ordinarily be bucked into short logs to minimize the effect of the curvature and so reduce the waste in slabbing.

Cutting for Grade. Trees ordinarily vary in quality from the stump to the top because the lower portion of a tree is the oldest, and hence the largest and least knotty. Sapling limbs on young forest-grown trees are soon pruned off naturally as trees grow, and only the upper portions of the matured trunks ordinarily have large limbs or concealed knots near the surface. Hence, the lower portions of most trees will yield the clearest lumber.

Clear lumber, of course, is much more valuable than the knotty common grades, and trees should be bucked for saw logs so they will yield the largest possible amounts of the best grades of lumber. Softwood lumber is graded on the basis of the number of defects in the piece as a whole, and a board that is clear of defect for all but a foot or two of its length must still be graded and sold as a common grade if it is knotty or otherwise defective for the remainder of its length. By cutting off the knotty end of a board it is possible to raise the grade when the remainder is free of defect, but this involves wasting the portion cut off and discarded. Hence, although the board is raised in grade and value, it is reduced in volume. Since hardwood lumber is graded on the basis of the amount of clear surface that can be cut from boards, this consideration is not so important in hardwoods as it is in softwoods.

Trees should always be bucked to keep as much as possible of their clear lengths in clear logs, leaving the common logs with as little as possible of better than common grade, since the better part will be reduced in grade to that of the poorer part. It is better practice to cut a 12-ft clear log if the cut comes at the point dividing clear and knotty sections of a tree than to cut a 14- or 16-ft log and by so doing put some of the knotty section in with the clear. In the latter case, either the knotty ends of the lumber will be trimmed off in the mill and wasted or the lumber will all be lowered in grade. By keeping as much as possible of the clear length of a tree in clear logs and putting the knotty length in common logs, the greatest value yield will be obtained.

IMPORTANCE OF PREVENTING WASTE IN FELLING AND BUCKING

The methods of getting the most out of trees in bucking them for saw-logs cannot always be followed in actual practice because of the great variation in tree size, form, and quality. Nevertheless, those doing the job of cutting logs or supervising such operations must know and understand the

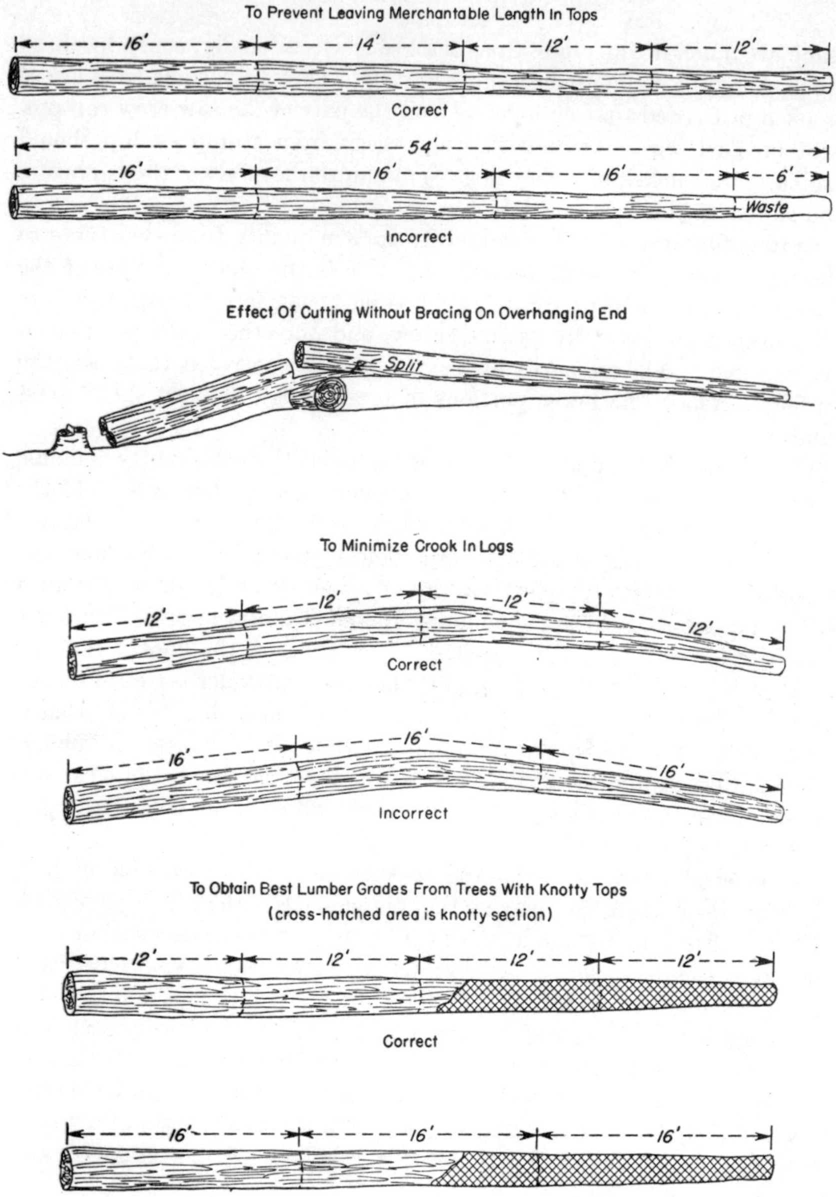

FIG. 36. Methods of cutting logs to obtain maximum utilization.

best methods, if full volume and value are to be obtained from each tree cut. That much waste can and does occur on many logging operations is shown by a study made in the Piedmont region of North Carolina in 1942.[1] Several different logging operations typical of the type supplying small portable mills with logs were investigated and it was found that unnecessary waste in volume alone amounted to 1.95 per cent of the total volume cut. This does not include waste in value from failing to cut correctly for grade. Table 15 shows the amount of loss by kind for the operations investigated.

TABLE 15. AMOUNT OF WASTE IN FELLING AND BUCKING BY KIND AT TYPICAL LOGGING OPERATIONS IN THE PIEDMONT REGION OF NORTH CAROLINA

Operation	Volume of waste, board feet								Ratio of waste to total volume, per cent
	Total volume	Splinter pull	Split	Break	Wasted top	Excess stump	Miscellaneous	Total	
I	6,290	5	11	60	20	63	7	166	2.64
II	5,115	18	6	..	30	17	13	84	1.64
III	12,440	18	10	..	95	19	..	142	1.14
IV	11,220	11	14	30	75	156	..	286	2.55
V	9,885	26	8	..	105	39	..	178	1.80
VI	5,400	6	..	40	54	..	..	100	1.85
VII	26,200	21	30	22	85	241	..	399	1.52
VIII	31,095	119	18	45	100	233	..	515	1.66
IX	5,895	9	7	2	50	95	..	163	2.77
X	14,560	33	8	..	250	170	..	461	3.17
Total..	128,100	266	112	159	850	1087	20	2,494	1.95

Source: Minor, Charles O., "Logging Waste Resulting from Poor Felling and Bucking Practices in the Piedmont Region of North Carolina," Unpublished thesis, Duke Univ. School Forestry, 1942.

In terms of annual production, the waste was about 28,000 ft for the average small mill, or about equivalent to one carload of finished lumber. For the Piedmont region of North Carolina the annual loss was estimated to be about 12 million board feet, or 600 carloads of lumber, or enough to build 850 five-room houses. In terms of stumpage prices, the loss was estimated as $97,000 per year. Since much timber is sold on the basis of lumber sawed, the forest owners actually suffered this loss, as they were not paid for lumber not cut although the volume was in the trees. To make up this loss, about 3,480 additional acres of timber land would have to be cut, which is an unnecessary addition to forest drain.

[1] Minor, Charles O., "Logging Waste Resulting from Poor Felling and Bucking Practices in the Piedmont Region of North Carolina". Unpublished thesis, Duke Univ. School Forestry, 1942.

Too often log cutting is considered a common labor job that requires only a strong back and a weak mind. True, it is hard work, but it should be done by skilled workers to get full value from the timber being cut. The saw crews themselves will profit by learning the correct methods of sawing logs, since they will find it possible to cut more timber in the same period of time by doing it correctly rather than incorrectly. Both workers and employers profit when felling and bucking are done skillfully, carefully, and correctly. Operators profit by getting more volume and greater value from the same amount of timber and the workers profit by being able to cut more logs with the same amount of effort and time. Furthermore, with skilled workers using good judgment in bucking trees correctly, employers will be justified in paying higher wage rates.

Timber owners, mills, individuals, and farmers have a direct interest in good timber utilization. Now that timber is a crop being consciously grown, it is poor business to waste it in the cutting. Good harvesting of any crop, whether wheat, corn, cotton, potatoes, or timber, means getting to market all of the usable product grown. Skill and good judgment in bucking logs are attained only by careful training and supervision of the bucking crews. Waste of volume or value, however, is too costly to tolerate and the time and expense necessary to obtain good bucking is more than justified by the lumber saved and higher lumber values obtained.

BUCKING TREES FOR CORDWOOD AND PIECE PRODUCTS

Pulpwood. Trees cut for pulpwood seldom exceed 15 to 18 in. in diameter on the stump and may be as small as 7 or 8 in. in diameter. Hence, bucking trees for pulpwood is a much simpler operation than bucking trees for saw logs. Also, pulpwood is cut to a uniform length and there is no opportunity or need to vary bolt lengths to obtain the best utilization. Nevertheless, the bucking operation is one of the important steps in producing pulpwood because the considerable number of cuts that must be made to sever the trunk into the usual short bolt lengths require a large amount of time and effort as compared to other steps.

When pulp logs are produced in connection with a saw-log operation or on a strictly pulp logging operation, the trees are cut into logs in the same general way as for saw logs except that the log lengths usually vary from 12 to 20 ft, in multiples of 4 ft. When large trees are cut for pulp mills, as in hemlock forests of the Lake states and on the Pacific coast, the trees are ordinarily cut into logs and the usual machine logging methods followed in delivering them, because even if cut in short lengths, the bolts would still be too heavy to handle by manual labor.

The length of pulpwood bolts varies from mill to mill and from one section of the country to another. In the Lake states and the Northeast, lengths are either 4 or 8 ft, while in the South the most common length is 5 ft 3 in.

In the northern areas wood that is to be driven down streams is more conveniently handled in 4-ft lengths, as there is less tendency for it to jam in narrow water. Wood to be trucked may be either 4 ft or 8 ft long, depending upon the local custom. Wood that is to be rafted across large lakes is cut in 8-ft lengths, as it is less likely to be lost in transit. Actually the bolts are usually cut 50 or 100 in. in length, because in driving, the ends frequently "broom" from striking rocks and other obstacles in the water. The extra length allows for this waste.

In the South practically all pulpwood moves to the mill by truck, rail, or barge, and none is floated by stream driving or rafting. Pulpwood bolts, therefore, must be cut short enough for loading and unloading by hand, which is the reason bolts longer than 5 ft or 5¼ ft are not cut. In some instances where power equipment is available for loading, pulp logs are cut and handled in the same way as saw logs and are cut to shorter lengths mechanically at the mill before being conveyed into the chippers.

Although these variations exist in the length of pulpwood bolts, the length is uniform on any one woods operation, and usually for all of the wood brought to any particular mill. Therefore, the problems met in bucking saw-timber trees with respect to correctly placing the bucking cuts are not involved in pulpwood bucking. Furthermore, since the wood is to be chipped, there is no loss due to crook or sweep, unless it is so severe as to prevent a bolt from entering the chipper. Likewise, no value loss is involved in mixing clear and knotty portions in a bolt or a pulp log, as the wood loses its identity once it has been chipped.

Bolts. The principles involved in bucking saw logs apply, in general, to bucking trees for bolts of various kinds. Bolts are, in effect, short logs, usually not over 4½ ft long, and are cut for such products as veneer, handles, staves, and hardwood dimension stock.

Most bolts are cut from hardwoods and generally only from specified species. Thus, handle bolts are cut only from ash and hickory trees of suitable quality, and most stave bolts are cut only from oak, principally white oak in the case of tight cooperage. Bolts cut for slack cooperage may be obtained from a wider range of species. Veneer bolts usually are cut from a variety of species, for example, yellow poplar, gum, sycamore, maple, oak, basswood, and similar woods, as well as pine in some cases, such as for vegetable hampers.

Rigid specifications as to diameter and freedom from defect must be observed in cutting most bolts, since they are produced for a particular product. Premium prices are paid, generally, for clear sound bolts of larger diameters, but second-grade bolts are also accepted. Otherwise it would be difficult to interest producers in cutting bolts, as not enough of each tree could be utilized.

Most manufacturers using bolts purchase them in various lengths. This permits producers to buck trees so as to localize or minimize defect and to choose the correct combination of lengths to obtain the best volume and quality utilization in a tree, in the same manner as described for saw logs. Lengths vary from 24 to 50 and 60 in., depending upon the product. For any one product the range usually is narrower, and in 2- or 4-in. intervals. Some length specifications include a trimming allowance while others do not and one must be added. The producer, therefore, should be certain on this point to avoid cutting short bolts.

It is very desirable in harvesting timber for bolts to cut other products at the same time. This permits each tree to be fully utilized, as those portions that do not meet bolt specifications can be sawed for saw logs or, in some localities, for pulpwood.

Some veneer and handle manufacturers buy logs and cut them into bolt lengths at their plants. In such cases the entire log must meet the specifications for bolts. This is done primarily to interest those producing saw logs in cutting bolt stock. Since saw-log operators already have their log-handling equipment in the woods, they can handle logs more readily than bolts. Because such species as are often required for veneer, shingles, handles, and other specialty products occur only in minor proportion in a stand, it is not feasible to organize a logging operation or, in some cases, even bolt operations for such a small and scattered volume. By removing it along with other products, however, no excessive cost is incurred.

Care should always be observed in cutting bolt products to prevent damage and waste. It is important, also, that the full length required be cut, although too much trimming allowance would be wasteful. Splitting either results in a deduction from the scale or culling the entire bolt, so it is very wasteful. Occasionally jump butting and jump cutting must be done to obtain the maximum volume and quality utilization of a tree. Since bolts are much smaller than logs and in most cases can be handled at least to some extent by hand, it is not difficult to brace hanging cuts and trees in a bind and so prevent splits and other damage in bucking.

The principal factors that must be considered in bucking trees into bolts, therefore, are similar in most respects to those involved in bucking saw logs, in that damage due to splitting and cutting them short of specified lengths and incomplete utilization means unnecessary actual volume waste, and lack of skill in cutting for grade means quality loss.

Piece Products. When trees are bucked for piece products such as poles, piles, ties, mine props, and fence posts, care must be observed in cutting correct lengths and in preventing damage in bucking at the required intervals. Piece products must be cut to meet rigid specifications with respect to dimensions at butt, top, and length, as well as with respect to defect.

Any breakage or other flaws caused by careless bucking or measuring means the loss of the entire piece or requires that it be cut back to a smaller and less valuable class. Hence, great care must be taken in bucking trees selected and felled for particular products because of their special form or quality, to prevent ruining them for those products.

The precautions and measures that should be taken in bucking saw logs apply equally to bucking piece products. A split tie or a short tie is a cull tie. Likewise, a tree felled for a pole or pile that is split in the top or cut too short means that it is either disqualified for that specification and must be cut off in length, perhaps at both butt and top, to fit it into a less valuable pole class with obvious loss, or culled altogether.

THE BUCKING OPERATION

Bucking is almost entirely a sawing operation except for the removal of limbs, which is done with the ax. The sawing is done by crews of one or more men, depending upon the size of the timber, the tools used, and local practices. One-man bucking crews are required for bucking the very large timber of the West with handsaws, because it is impossible for two men to work at opposite ends of the same saw at the same time. In northern pulpwood operations bucking small trees is commonly done by one-man crews using bow saws or one-man crosscut saws.

Most bucking, however, is done by two-man crews, one man at each end of the saw. A third man may work with the crew and do the limbing and undercutting. The same crew frequently, but not always, does the combined operation of felling and bucking. On many Western operations felling and bucking are done independently by separate crews. When power saws are used, the crew size varies greatly, depending upon the kind of power saw and local custom.

Trees felled for conversion into saw logs are sometimes left full length and later bucked into shorter lengths after being skidded to a loading point, or at the mill, or, as is commonly the case, they may be severed into single- or multiple-log lengths in the forest before being bunched or skidded. Similarly, trees cut for pulpwood and other cordwood products and trees cut for veneering may either be removed full length or sawed to piece lengths in the woods before they are removed.

When poles and piles are produced, the entire tree is usually required for each piece, so the trees are left full length and the only processing necessary is to remove the tops and limbs to the upper limit of utilization. The pieces may or may not be peeled of bark, depending upon their intended use. When ties are cut, the trees are ordinarily bucked into tie lengths for removal either in the round for sawing or are hewed in the forest to be removed as finished products, except for preservative treatment in the case of nondurable woods.

When markets for various forest products exist, two or more products are frequently cut from the same trees. The lower portions of trees, for instance, may be cut for veneer bolts, or saw logs, or both, and the upper portions for pulpwood, fuel, or other cordwood products. Or, more rarely, a saw log may be cut from the butt and the remainder of the tree used for piles. Whenever possible, such integrated use is highly desirable since it means much closer utilization than when only one product is cut, and a greater monetary return for the timber. Close utilization of the tops of trees for cordwood products also reduces the subsequent fire hazard, as most of the limbs are left close to the ground and so decay rapidly.

Removal of Limbs. The limbs must be removed from that portion of a tree to be used, regardless of the product. This is a comparatively simple operation, usually done with the ax except when exceptionally large limbs are encountered, as in large hardwood trees and large Western conifers. Limbing, or swamping, as the operation often is called, may be done before trees are bucked into logs or it may be done as bucking progresses from the butt of a tree towards the top, either by the bucking crew or by a separate crew. It is generally preferable to complete the limbing operation before the bucking has proceeded so far that log lengths cannot be adjusted to fit the top limit of merchantability.

In some cases it is preferable to leave uncut some of the limbs on the under side of a tree until bucking has been completed to hold the bole off the ground for easier bucking and to reduce saw pinching and prevent splitting the logs. One or two "flags," or limbs extending upward, are often left on logs cut in the winter in regions where deep snow is common to aid the skidding crews in finding them after a heavy snowfall.

All limbs and dead branches cut off should be trimmed close to the trunk so the logs will handle readily on later operations. If this is not done at the time of swamping it must be done later, which is an unnecessary added expense. When an ax is used in limbing, the cut should be made from the lower side of the limb to avoid tearing the wood and to remove the limb with the minimum effort and danger. In larger limbs, cuts may be made from both sides, but in any case the limbs should be wholly or partly cut from the under side.

Bucking with Handsaws. The felled trees are bucked into log or bolt lengths by the comparatively simple though strenuous process of pulling a saw back and forth across them at the indicated intervals. Bucking may be accomplished by any of a variety of crew combinations. The two-man crew that felled the tree often does the bucking also, or bucking and felling may be by entirely separate crews.

Frequently on pulpwood operations in the North, bucking is done by one-man crews but two will combine to do the felling. On many Western operations bucking does not start until all of the trees have been felled, and one-man bucking crews are very widely used. Because of the large diameter of many of the trees and because they frequently are several feet off the ground across other trees, it would be quite inconvenient if not impossible for two men to handle a crosscut saw. One man, however, can stand beside a tree that may be higher than his head and buck it with a one-man saw of sufficient length. Felling often is done by power saws with bucking by one-man crews using handsaws.

When trees of small diameter are bucked, it is seldom necessary and often impossible to wedge them, as not enough space is available behind

FIG. 37. Bucking a Southern pine tree with the popular and widely used two-man crosscut saw. (*Photograph courtesy of the E. C. Atkins Co.*)

the saw to wedge the cut. Cuts that tend to bind can be opened up by prying up on a stick placed beneath the tree near the bucking point. It is usually held by one of the sawyers. In large timber this is not possible because of the great weight of the logs, but a bucking cut can be kept open with wedges to prevent its closing and pinching the saw. At least one member of a saw crew and sometimes both wear holsters in which wooden or iron wedges are carried. It is poor practice to drive wedges or do other pounding with the flat side of a double-bit ax, as it loosens it and bends the eye, throwing the ax out of balance. Even with a poleax only light pounding should be done. When heavy wedging is to be done, a sledge should always be carried. One of the advantages of the bow saw for bucking small trees is that it is seldom pinched, since the blade is so narrow and thin that it is not affected by the cut closing up at the top.

FIG. 38. Method of bucking large Western trees with a one-man crosscut saw. Note the steel wedges used to keep the cut open, which prevents pinching the saw.

FIG. 39. Trees may be skidded full length to a yard and there bucked into log lengths as shown in the above illustration of a harvesting operation in South Carolina.

The precautions noted previously should all be observed in bucking, to prevent damaging logs or wasting volume or value. When trees are bunched at a yard or landing, sometimes called a "brow," after they have been bunched or skidded, fewer problems arise, since the landing can be located on flat ground or the logs can be yarded in such a way that they do not lie in a bind requiring special efforts in bucking. When

FIG. 40. Above, bucking trees for pulpwood with a bow saw with a two-man crew and below, a one-man crew. (*Courtesy of the Southern Pulpwood Conservation Association.*)

logs are bucked in yards, separate saw crews do the limbing and cut out the tops.

Bucking with Power Saws. The method of bucking with power saws varies with the type of saw used. Mechanization of bucking is not as widely practiced as mechanized felling in Western forests. In the South and East, where the timber is smaller, this difference does not exist and, in fact, on many operations trees are felled by hand and bucked with power saws.

FIG. 41. Bucking trees with three types of power saws; top, chain saw; middle, circular saw; and below, a rim-driven circular saw. (*Courtesy of the Southern Pulpwood Conservation Association.*)

The reason for lack of progress in bucking large Western timber is due, at least in part, to the difficulty of getting power saws spotted at the places where trees are to be bucked. The difficulty arises from the practice of felling all trees on an area before bucking is done. This makes power saws unwieldy as they must be carried over logs and used on trees that may be several feet off the ground.

FIG. 42. Bucking with a gasoline-motor-driven chain-type power saw in Oregon. (*Courtesy of the Mall Tool Company.*)

In large timber, chain-type saws cannot be used satisfactorily for the same reason that two-man crosscut saws are impractical. When power saws are used for bucking large Western timber, they are of the drag-saw type that oscillates back and forth in the cut in the same fashion as hand-saws. In smaller Western timber and throughout the East, North, and South where two men can operate a chain saw, bucking can be readily done with this type of saw. The chain-type saw can be most effectively used where tree lengths have been skidded to yards or decks, as it is not necessary to move the saw over uneven ground, logs, felled trees, tops, and other obstacles.

Circular saws mounted on light wheels are used extensively in bucking small timber for pulpwood in the South where the ground is generally level. Felled trees may be bucked either at the stump where they were felled or they may be skidded full length, after being bunched and topped, to a brow or yard and bucked at that point. The latter practice is necessary where the ground cover is dense or the topography too uneven to permit moving the saw readily from tree to tree.

One-man chain saws have been recently developed but as yet have not come into general use. Some types are limited to cutting trees no larger in diameter than the cutting bar, while one type can be used to buck trees with diameters up to twice the length of the cutting bar. The latter type can be used to cut halfway through one side of a log and then the opposite side can be cut. Likewise, such a saw can be used to cut trees in a bind

FIG. 43. Bucking a large ponderosa pine tree in California with an electric-motor-driven chain saw. The tractor power unit generates the electricity which is usually sufficient for two or more saws.

that would split if cut from above in the usual way. Since the saw cuts at the end opposite the operator as well as on the top and bottom edges of the cutter bar, it can be forced directly into the side of a log to cut the bottom half first, finishing from the top, as previously explained and illustrated.

Power saws can be most effectively used only on operations large enough to justify maintaining a competent filer and a shop in which they can be maintained in good running condition. As in the case of most mechanical equipment, power saws need to be adjusted and lubricated frequently. When five or six or more saws are in use on an operation, a spare saw can be provided so one can always be repaired without slowing production. When only one or two are in use, the extra investment for a spare saw and the cost of necessary maintenance facilities would be excessive for a small output.

CHAPTER 9

HEWING, RIVING, AND PEELING

Most forest products are manufactured in mills and factories but the primary steps in their manufacture are performed in the forest. Cutting trees into logs or pulpwood, for example, actually is the first step in the manufacture of paper or lumber, although one of the primary reasons for taking these steps in the forest is to obtain small enough tree segments for ready handling.

Some forest products are manufactured by hand either wholly or in part in the forest where the trees were felled. Those whose manufacture is practically completed in the forest are not transported to manufacturing plants except for such processing as treating with wood-preserving compounds, seasoning, or further cutting to obtain exact sizes.

Products finished in the forest are handled so differently from those manufactured in mills that they merit separate consideration. Furthermore, the crews or individuals who produce these products also fell the trees, so they are engaged in actually harvesting certain forest products as well as in manufacturing them. Whereas the major forest products usually are harvested by highly organized crews using special and powerful equipment, most of the products which are hewed, rived, or peeled are generally produced by individuals either working alone or in small crews. In the aggregate, however, their output is substantial in volume and value and supplies a number of important manufacturing and processing plants throughout the United States.

HEWING

Hand-hewed ties are produced throughout the forest areas of the Eastern part of the United States, and in some portions of Western forest areas where the trees are not too large for hewing by hand. Several species of trees are suitable for ties, including those with nondurable woods as well as those with durable woods. The species listed by the Railway Tie Association as generally among those specified for crossties by purchasers are:

Ashes	Cherries	Firs (true)	Larches	Pines	Sycamores
Beech	Chestnut	Gums	Locusts	Poplars	Walnut
Birches	Cypresses	Hackberries	Maples	Redwoods	
Catalpas	Douglas fir	Hickories	Mulberries	Sassafras	
Cedars	Elms	Hemlocks	Oaks	Spruces	

All ties of nondurable wood are shipped or trucked to treating plants where they are impregnated with creosote oils or other preservative compounds. Some ties of durable woods also are treated to add to their natural life, but in many instances they are placed in use without further processing.

In hewing ties, the trees are felled usually by a two-man crew using a crosscut saw. Because ties must be composed of sound wood, the stumps of tie trees often are cut higher than they would be for saw logs, to get above butt rot common to many species of trees. Any decay in the butt cuts of saw-log trees can be sawed around in the mill and finally discarded by trimming the defect from the boards, leaving the remaining portions of the boards merchantable. Rot cannot be accepted in ties as it has a tendency to enlarge, especially in wood in contact with the ground, hence all rot should be eliminated from a tie. Also, since the bases of most trees are enlarged or swelled to various degrees, tie cutters—or tie hackers, as they are sometimes called—prefer to make felling cuts just above the butt swell to save the extra hewing that would be involved in cutting the butt tie bolt down to tie size. This is not uneconomical, unless it shortens the bolt sufficiently to prevent complete utilization in the top.

Good felling practice for most products requires that trees be felled without splitting. In felling trees to be hewed into ties, however, splitting is sometimes deliberately induced by not undercutting. When this practice is followed, sawing starts on the side opposite from the lean of the tree and is continued until the tree splits or until it has been completely severed. When the tree falls before it is completely severed, a split portion is left attached to the stump and the tree is then cut from the split with an ax. There is some saving of effort in hewing due to one side of the butt tie cut having been split off, but it is of doubtful advantage as extra effort is required to cut the tree free from the stump portion and the butt cut must be hewed in conformity with the split side, which may or may not be the best procedure. Also, much less control is possible over the direction of fall when undercutting is not done and, in addition, the split stumps add to the difficulty of maneuvering trucks in the woods, so it is not a recommended practice.

A felled tree may be hewed on two opposite sides before it has been bucked into tie lengths, or tie cuts, as they are often called. The tree is steadier for hewing when the top is attached and before it has been bucked, which is the reason this practice is sometimes followed. Tie hewers use a broadax that has a blade from 6 to 7 in. long with an offset handle, which clears the face of the bolt being hewed. The faces to be flattened are first scored by cutting into them to the desired depth along the entire length with a conventional ax or with the broadax. Then the scored face is hewed smooth until a flat face of the desired width is obtained. The same procedure is followed on the opposite side until the desired thickness has been obtained. If the bole was not bucked into tie lengths before hewing started it is bucked at this point. After two faces have been completed each tie cut is turned face down and the opposite rounded sides are hewed to form the maximum-sized standard tie it is possible to obtain from the tie bolt. Most tie specifications require that the top and bottom faces of a tie be parallel although the sides are not required to be parallel.

Tie cutters often stand on top of the tree or tie being hewed and hew first one side and then the other. From this position they can best

FIG. 44. A tie cutter hewing crossties from a Southern pine tree in Georgia. Note the high stump which, in this instance, is not a waste because the extra hewing would more than offset the saving in length by cutting a lower stump.

apply their skill in hewing smootn, straight, and parallel tie faces with a minimum of time and effort. Some tie hewers split off the scored sides to minimize the amount of hewing when they are producing ties from species that split readily and are straight-grained. A good tie cutter can hew from 20 to 30 ties per day.

Most ties are simply bolts flattened on four sides, although some railroads prefer ties hewed only on the top and bottom with the sides left rounded but with the bark removed. Ties of this type are known as boxed-heart ties, which means that the pith is included in the tie. When only the top and bottom faces of ties are hewed, they are known as pole ties, while those hewed on all four sides are called square ties. Some ties are split from bolts of large diameter but tie cutters prefer to hew boxed-heart ties rather than to split larger trees, as they are able to produce more per day. Trees of larger diameters that can be split into ties ordinarily are more valuable for saw logs, veneer bolts, or other products, so smaller trees are much more commonly used. Ties sawed from large logs in sawmills, which is the method of manufacturing ties throughout most of the Western forest regions and in parts of the East and South, are not boxed-heart ties, unless one is cut from the center of the log with the pith in it. When sawmills produce ties they saw the outer clear portions of the logs into lumber and use only the center portions for ties.

Trees larger than 14 in. dbh are not ordinarily used for hewed ties, since too much hewing would be necessary to produce even the largest sized tie, requiring undue effort and causing unnecessary waste of the most valuable portion of the tree. Trees felled for ties should be tall enough to produce at least three and preferably more ties to justify cutting them. This is important as ties are classed as to size and a substantial proportion of the larger sizes are usually specified in any purchase order issued by tie dealers or railroads. If a tree is only large enough to yield one or two ties, they very probably would be of the smaller sizes. It is necessary that tie buyers purchase a range of sizes, so a tree that will produce one or two large ties and two or three or more smaller ones can be completely utilized.

Ties must be either 8, 8½, or 9 ft long, depending upon the specifications of the user. Most ties are 8½ ft but a considerable number of users specify the 8- or 9-ft ties. Switch ties are longer and range up to 12 or 14 ft, but such ties are always sawed, as they are used at switches and ties of uniform size are required for the satisfactory operation of switches.

After the ties have been hewed, they are loaded on wagons or trucks in the woods and hauled to shipping points or direct to a preserving plant. Bunching and skidding are seldom necessary or advisable, since the volume to be moved per unit of area is generally too small to justify such operations. Ties are loaded by hand on wagons, sleds, or trucks and are unloaded at yards or sidings and there piled so that all faces and both ends of each tie can be seen by the tie inspector who grades each one and so marks it. Tie grades range from number 6, which is the largest and best grade to number 0, which is the lowest grade. Most ties produced fall into grades 5, 4, and 3. Ties smaller than a number 3 are used for very light-weight track such as in mines, quarries, and certain other types of light or temporary railroad construction. The Railway Tie Association has adopted the following specifications for tie sizes and manufacture as standard:

TIES: SAWED OR HEWED

Size	Top, Bottom and Sides
0	5 in. thick by 5 in. wide on top*
1	6 in. thick by 6 in. wide, 6 in. face
2	6 in. thick by 7 in. wide on top
3	6 in. thick by 8 in. wide on top
4	7 in. thick by 8 in. wide on top
5	7 in. thick by 9 in. wide on top
6	7 in. thick by 10 in. wide on top

* None accepted for standard-gauge railway ties.

Size	Top and Bottom
0	5 in. thick by 5 in. wide on top
1	6 in. thick by 6 in. wide on top
2	6 in. thick by 7 in. wide on top
3	6 in. thick by 8 in. wide on top
	7 in. thick by 7 in. wide on top†
4	7 in. thick by 8 in. wide on top
5	7 in. thick by 9 in. wide on top
6	7 in. thick by 10 in. wide on top

Manufacture

Except as hereinafter provided, all ties shall be straight, well hewed or sawed, cut square at the ends, have bottom and top parallel, and have bark entirely removed.

† Buyers who specify both 6- by 8-in. and 7- by 7-in. ties manufactured on top and bottom only and who desire to separate the 6-in. from the 7-in. ties will designate the 7- by 7-in. ties as Size 3A.

Only a small volume of timber products other than ties is hewed in the forest. Some mine timbers are hewed square or flattened on two sides and some beams or timbers for special uses are hewed from felled trees in the forest. Crossties, however, are the principal hewed products harvested from forests.

RIVING OR SPLITTING

A number of forest products are split or rived from felled trees. They include handmade shingles or shakes, split posts, fence rails, stave bolts, distillation wood, acid wood, and fuel wood. Some of these products, such as posts, rails, and shingles, require no further manufacturing, whereas others, such as stave bolts and distillation wood, are hauled to manufacturing plants as a raw material.

Shingles. Only large, straight-grained, and clear sections of trees that split readily can be used for shingle riving by hand. Most commercial shingles are manufactured in shingle mills and are sawed rather than rived. For local use, however, some hand-rived shingles are still produced, chiefly in sections of the South and West. Shingles must be of durable wood to be of much use, so the principal species used for shingles are the durable group. Softwoods are most commonly used because of the greater ease of splitting, although some of the durable hardwoods such as white oak, cherry, and walnut can be and occasionally are used for shingles. It is quite uneconomical to do so, however, because of the high value such woods have for other products, when of a quality suitable for riving shingles.

Trees felled for hand-riving into shingles are either hauled in log lengths to a shop where the shingle making is done or are rived in the forest. Shingles made by hand riving are usually from 18 to 32 in. long and from $\frac{3}{8}$ to $1\frac{1}{2}$ in. thick. The usable portion of a tree, therefore, is cut into blocks of the same lengths as the shingles or shakes that are to be produced.

The bolts are then split and the sapwood and center portions discarded. The usable portions are then placed vertically on a bench or frame and a special knife tool, called a froe, is driven into the block with a wooden mallet and shingles or shakes of the desired thickness are successively split off. Shingles may be rived from one side of the round block to the other instead of quartering it first, producing mostly flat-grained shingles. Edge-grain shingles rived on the quarter grain are of better quality because there is less tendency for them to cup and split after they are placed in use.

After they are rived, the shingles are smoothed on the surfaces and edges with a drawknife and also often tapered. The fresh-rived shingles are flat-piled in such a manner that air can circulate freely around each one to hasten drying, either close to the riving place in the forest or after they are hauled to a more suitable location for piling.

Posts and Fence Rails. Posts and fence rails are cut from both durable and nondurable species in all sections of the United States. Posts of nondurable wood are treated with wood preservatives. Posts may or may not be peeled, depending upon the type of post produced. Small pine trees are cut in some sections of the South for posts, and as they must be treated with preservatives before being used, they always are peeled of all bark. Cedar and cypress posts, being very durable, on the other hand, are often used with the bark left on.

Durable-wood posts are either left round or are split from larger trees. Red cedar posts are produced through the range of this species in the Eastern and Southern states. The tree usually is of comparatively small size so most red cedar posts are left round. The bark generally is not removed unless the post is to be painted or placed in a location where smooth clean posts are desirable. White cedar posts produced in the Lake states and the Northeast and cypress posts obtained in Southern swamps likewise are produced from small trees and so are left round and are ordinarily not peeled until placed in use. White cedar and cypress trees reach large size but the larger trees are more valuable for logs or poles than for split posts.

White oak, catalpa, locust, and other trees of the durable-wood group are preferred for posts. Only the heartwood is durable, so the best posts are obtained from large trees which are felled in the usual manner and bucked to post lengths, 4 to 8 ft, or longer for special purposes. They are then split into segments by means of wedges, sledges, and axes. Wedges are driven in until the log splits and the cross fibers in the open split are then cut with an ax.

The halves are split further along and across the radius until each segment has been broken down into posts of suitable size. Small trees with sufficient heartwood may be split only in halves or quarters. Trees of larger diameters yield up to 30 posts per cut, depending upon the size. They may or may not be piled for seasoning before being hauled, depending upon whether they are to be used promptly or not. If they are to be used locally there is no particular advantage in seasoning them, but if they are to be shipped their weight is greatly reduced by drying, which lowers freight cost.

Fence rails are now seldom used for general fencing, even locally, and only to a limited extent for ornamental fencing on estates, because of the high labor cost of making them and the comparative cheapness of wire fencing. When rails are produced they are rived from felled trees bucked to the desired lengths, much in the same manner as split posts.

Chemical Wood. Certain of the heavy hardwoods, particularly beech, birch, maple, and oak, are cut for destructive distillation to yield acetic acid, methyl, or wood alcohol, charcoal, and tar residues. The wood is used in the form of cordwood usually cut to 42 in. in length to fit the standard type of metal buggies used to place the wood in the retorts for distillation.

Only trees not suited for saw logs or those that will not grow into suitable saw timber should be cut for chemical wood, as there is an abundant supply of poorer quality trees of suitable species and it is not necessary or desirable to cut trees of good form, unless a young stand is in need of thinning. Suitable wood is also obtained from the upper portions of trees cut for saw logs, veneer bolts, or other higher uses. Limb wood does not pile closely enough to give full yields from the retorts and knotty sections are too difficult to split, so such portions of trees are not used. There generally are, however, two or three bolt lengths above top saw-log utilization that can be used for chemical wood, so it is good woods practice to follow logging operations in hardwoods with chemical-wood production, whenever possible. However, much if not most of the chemical wood used is produced independently of other products by small, widely scattered crews.

In producing chemical wood the trees are felled and bucked into bolt lengths. Then each bolt is split so that no piece is more than 5 in. through at its thickest part. This is necessary because the wood is to be carbonized by the application of high temperatures in the absence of air, and wood thicker than 5 in. does not carbonize completely. Splitting is accomplished by the usual procedure using wedges, malls, and axes. Two men can work together efficiently, as they can saw more effectively than one; and when they are splitting bolts, one works at each end with the larger sizes, while small bolts can be split individually. The wood is hauled as soon as it is split or it is ricked in the forest where cut and hauled later.

Chemical wood is produced in areas within shipping distances of distillation plants in the Lake states, Allegheny and Appalachian regions, and, to a limited extent, in the central Mississippi Valley section.

Acid Wood. In the Appalachian Mountain section considerable acid wood is produced. This is cut from dead chestnut trees which, being durable, are still standing and sound after having been blight-killed 20 years or longer ago. Chestnut cordwood is used by extract plants throughout the Appalachian region in the manufacture of tannic acid used in tanning leather.

The dead chestnut trees are too large, as a rule, to use in the round so most of the bolts are split to facilitate handling. It is not necessary to reduce the bolts to pieces as small as 5 in. in thickness as in the case of chemical wood because the bolts are chipped before entering the extractive process. The wood splits readily and the operation requires only the use of axes, malls, and wedges to split the round bolts, using the usual technique.

Stave Bolts. Stave bolts are important split products harvested from forests in some sections of the Eastern portion of the United States. Most split stave bolts are used for tight-cooperage products that will contain liquids, and so only certain woods can be used. White oak is the best tight-cooperage stock and commands a premium price. Some red oak is also used, but other species do not meet the specifications for tight-cooperage stave bolts. Staves are sawed from the split blocks and the heading is usually sawed from round blocks or bolts.

Trees felled for riving or splitting into stave bolts must be 16 in. in diameter or larger and the portions used must be clear of defects. Special care is necessary in felling to avoid splitting and splinter pull. Cavities in a butt cut left by splinters attached to the stump prevent the use of such damaged portions, and because of the high value of clear butt sections suitable for staves, special care is justified to prevent splinter pull.

The central portion of a tree is often unsuited for staves because of small knots, shakes, or other defects. The outer portion, therefore, yields the best stave material and fallers often saw completely around the tree without undercutting before cutting it off, so that if there is any splinter pull it will occur at the center instead of in the best part of the butt bolt. Since undercutting would shorten one side of the butt bolt it is avoided whenever possible, which makes sawing around necessary.

Staves are made in various lengths, so the felled tree is bucked into segments of the lengths being produced. When several lengths are produced by the same crew, more complete utilization of the timber is possible, as the bolt lengths can be adjusted to utilize the clear portions of the trunk to best advantage. The round bolts then are split open and the usable portions split into segments along the radius with the largest width from 6 to 8 in. Narrow edges of the segments towards the pith of the tree are split off and left, as they contain defect and are too narrow for staves. The split stave bolts are then crib-piled, either where cut or at an assembly yard, and left to dry, after which they are hauled or shipped to stave-manufacturing plants, where staves are sawed from them.

Stave cutters work in crews of two or three men and fell the trees as well as rive out the stave bolts. The usual splitting tools, axes, malls, and wedges, are used and the actual splitting is done in the usual way, except that care must be exercised to avoid splitting the bolts into segments that are too small or that include unsuitable wood.

Another type of stave produced from high-quality oak timber usually is partially shaped before being exported and then later manufactured into the finished product. The trees are felled and bucked into bolts, as before, but the bolts are split into individual stave blanks instead of being left as split bolts. Each segment is broken down into stave-size blanks by splitting along the radius and the blanks are smoothed on all four sides with drawknives. The blanks are cut thick enough, and are left large enough so that shrinkage in drying will still leave enough wood for final finishing to exact size. A tree 20 to 24 in. in diameter often yields up to 100 or more stave blanks of this type. The blanks are then piled in open cribs and dried before being shipped. Only skilled workers should split first-quality timber into stave bolts or blanks, as poor workmanship can result in considerable loss by inefficient use of the timber and by producing culls unnecessarily.

Fuel Wood. No particular care is necessary in splitting fuel-wood bolts, as the only purpose of this operation is to reduce the size of the wood to readily usable dimensions and to hasten drying. When small trees are used, sections up to 6 or 8 ft long may be split and dried and then later sawed to length with a power saw. Larger sizes are bucked first and then split either with malls, wedges, and axes or by axes alone.

PEELING

The bark is stripped from forest products such as poles, piles, mine timbers, and some pulpwood before they are removed from the forest. In some cases, bark is a commercial commodity and is peeled from the felled tree so it can be assembled and sold. Products that are to be treated with preservative compounds, such as pine poles, are peeled to hasten seasoning and because the wood itself must be exposed to obtain adequate penetration of the preservative. Also, of course, poles of both durable and non-durable species are always peeled before being placed in use, as the bark would loosen and gradually fall off, making an unsightly pole.

Pulpwood. Pulpwood is peeled when the mill to which it is to be shipped is not equipped with mechanical bark removers, or for the purpose of hastening seasoning to reduce its weight when wood is to be shipped by railroad, trucked, or floated down streams. Some attempts have been made to devise mechanical bark removers that can be operated in the forest but no satisfactory type has yet been developed, or, at least, put into general use, except at pulp mills. Hence, most bark peeling, except at mills, is done by hand labor.

In all species of trees, the bark is much looser and more easily removed in the spring and early summer months than at any other time. As soon as growth begins, the cambium layer just beneath the bark becomes soft and spongy, as that is the location of new wood formation. As a result, the bark can be readily removed so long as springwood formation is going

FIG. 45*A*. Peeling pulpwood in Wisconsin. Note the rack for holding the bolts and the drawknife used to remove the bark.

on. By midsummer the bark "tightens" and removing it is more difficult. This tight condition of the bark prevails until the following growing season, when it again becomes loose.

Most harvesting operations that involve the removal of bark, therefore, are geared to this seasonal condition. Pulpwood that is peeled, such as hardwood pulpwood in the southern Appalachian region and spruce and balsam in the Lake states and the Northeast, is peeled to the fullest extent possible during the 6- to 8-week period during which the bark can most readily be slipped off. Wood that is produced during this spring peeling season is called "sap-peeled" wood.

The bark can be removed from pulpwood either before or after the tree is bucked. During the sap-peeling season the bark is generally removed from the tree as soon as it is felled and before it is bucked. The bark is severed along the upper side of the tree with an ax, after which it is shelled off by inserting a hand spud into the cut and forcing it beneath the bark. As the spud is pushed under and raised, the bark comes free, generally remaining intact in large segments.

FIG. 45*B*. Peeled aspen pulpwood in Wisconsin after cutting and peeling, but before bunching.

It is common practice in the Northeast and Lake states region to fell and peel the trees only during the peeling season and to buck them into bolts during the latter part of the summer after the bark tightens. In this way the full amount of wood needed is produced but the peeling is all done during the period when the bark can most readily be removed. In the Appalachian region, however, where the wood is produced by small scattered crews, it is usually felled, peeled, and bucked at the same time, but peeled-wood operations are conducted only during the peeling season.

In northern states wood can be cut in the winter and hauled to sidings, stream banks, or other assembly points and there peeled during the early spring, when the wood thaws out. The bark loosens as soon as it thaws and during a short period of not over 2 or 3 weeks, the bark can be removed almost as readily as from trees freshly felled during the sap-peeling season. Because of the short period during which winter wood can be peeled easily and because this is the time major sap-peeling operations start, this practice is followed chiefly by farmers and small producers who take care not to cut more in the winter than they can peel in the spring.

FIG. 45*C*. Peeling an aspen tree in northern Minnesota on a pulpwood operation.

Some peeled pulpwood is also produced in the Appalachian region out of peeling season, but the bark must be removed with a drawknife. It is not necessary to remove all the inner bark when peeling pulpwood, so only the outer bark is removed with the drawknife. It is used because it is easier and faster than a hoe spud would be on small bolts and better peeling can be done.

Peeled pulpwood piled off the ground where air can circulate freely through the stacks dries so quickly on the surface that decay and insect attack are negligible for the first two to three months, even in summer. Peeled wood can be stored during the spring for use later in the year. During the winter months cool weather prevents deterioration, so peeled wood can be kept from one spring until the next, thus making unnecessary the production of peeled pulpwood when the bark is tight, except in emergencies.

The hand spuds commonly used to remove bark during the sap-peeling season often are halves of automobile spring leaves flattened at one end. A peeler often uses two at a time, one in each hand, or he may use one spud and his ax. Manufactured spuds usually have handles to make their use easier but the working part of the tool is not much different from half a spring leaf.

Poles and Piles. Products such as poles and piles cannot be harvested from the forest during the peeling season only, as they are cut on order and inventories ordinarily cannot be accumulated very far in advance of need for the poles. Poles from durable woods are an exception in that they do not deteriorate if correctly stacked. However, Southern pine poles, which constitute a large proportion of the poles and piles produced, cannot be kept untreated for more than a few weeks at most without incipient decay of the sapwood occurring.

Pine poles, therefore, and much of the pine piles produced must be peeled while the bark is tight. This requires a somewhat different peeling procedure than for sap-peeling. Tight bark seldom shells off the tree but must be removed in strips. This is done with a long-handled spud with a broad blade sharp enough to cut through the bark and slip over the wood as it is pushed lengthwise along a tree. A common garden hoe with the gooseneck straightened makes a satisfactory spud. The peeler can stand over the pole on the ground and strip off the bark, rolling the pole over to reach all surfaces. It is not necessary to clean-peel poles to be creosoted, as rossing machines at the plants complete the peeling operation and also shave off excessive surface irregularities.

Pine posts and other posts of nondurable wood that are to be creosoted but not rossed at a plant must be clean-peeled and have all inner bark removed. When they cannot be sap-peeled this requires the use of a drawknife or some other similar sharp tool that will shave the surface. The posts are placed on a rack that holds them steady while the peeler removes the bark with a spud and then completes the operation with a drawknife, preferably one with a curved blade rather than a straight one.

Redwood Saw Logs. The great thickness of bark on redwood trees makes it necessary to remove the bark from redwood logs before they are sawed in the mill. Redwood bark is often a foot or more thick and it makes milling difficult when left on the logs. It is removed from the logs either where the trees were felled, or at an intermediate rehandling point, or at the mill. The more common method is to peel the bark by hand labor with spade-like spuds and leave it in the woods. When the bark is removed at the mill it is used for special products, being shredded and treated, in some cases, and sold as an insulating material. Formerly the wasteful practice of burning the bark off after felling was common. This resulted in so much loss of merchantable wood volume, however, that it is no longer done.

Mine Timbers. Much wood is used underground in mines for timbering galleries and shafts to prevent cave-ins. Sawed, round, and split products are used. Round and split products are usually peeled but some unpeeled timbers are used in galleries that are to be worked for only a short time. When longer life is required the timbers are always peeled and in some cases they are creosoted. Bark is removed from the larger sized timbers in the same manner as when poles are hand-peeled, or by the use of long-handled spuds. Small-sized pit props and lagging are peeled with a drawknife, unless the timbers are sap-peeled in the spring.

Peeling for Bark Products. Formerly the bark of Eastern hemlock, chestnut, tanbark oak, and certain other species was widely used as a source of tannin, but more recently synthetic tannic acid and exotic species have supplanted domestic sources. Nevertheless, hemlock and, to a considerable extent, tanbark-oak barks are still produced in appreciable amounts.

In cutting tanbark, as from hemlock trees, the trunk is girdled at 4-ft intervals and the bark is then split lengthwise and pried off in from 8- to 18-in. widths with an ax or bark spud. The segments of bark are then piled against the log or other support to dry the inside. Care needs to be taken not to overdry the bark at this stage, as to do so would cause it to break too readily into fragmentary pieces.

The bark-peeling season in the Northeastern states is usually from May to August, but logs cut in the winter months can be readily peeled in the spring.

One cord of bark is, as a rule, obtained from 1,000 to 15,000 ft of logs. Two experienced men can fell trees and peel from 2 to 3 cords of bark a day.

Mechanical Peeling Devices. Several types of machines have been devised for removing the bark from bolts, logs, or poles in the forest or at concentration points near the harvesting operations. Such machines must be portable so they can be readily moved from place to place as harvesting operations are relocated. As yet mechanical peelers, except where they are permanently located at mills or heavy concentration points, are not widely accepted. In periods of labor scarcity, as during the Second World War, they were rather widely experimented with but as yet are not entirely practical. In time it is quite possible, if not likely, that mechanical bark-removing machines will be in common use.

CHAPTER 10

TOOLS USED IN FELLING, BUCKING, AND PROCESSING TREES

The tools most widely used in felling and bucking trees into required lengths and further processing them in the forest are the ax and the saw. Peeling spuds, cant hooks, and other tools are also used, but axes and saws as a group are undoubtedly the most universally used. Both are of long standing, having been in use for well over a hundred years. Axes are, in reality, the modern expression of the cumbersome stone hatchet of primitive man that has now become a finely balanced tool with a steel head and a wooden handle.

Early American axes used in this country for logging were made locally by blacksmiths who modified the European type brought over by immigrants and made the pole portion larger, giving better balance. Handles also were modified by curving the ends for a better handhold. Cutting edges of steel were welded to the iron heads. The diversity of present ax patterns is probably due to the variety of individual designs created locally by blacksmiths in the days when axes were handmade, and accepted as patterns that had merit. Basically, though, the ax has maintained the same general form that it originally had, and it is still an indispensable woods tool, especially for felling and bucking trees, and for hewing, when that operation is performed.

Saws are of much more recent origin than axes. Until 1790 there were still no crosscut saws made in this country and those that were imported from Europe were both expensive and difficult to obtain. Before the crosscut saw was first used in America, probably during the 1800's, trees were not only felled by chopping them down with axes, but the boles were cut into log lengths also by chopping. The introduction of crosscut saws greatly speeded up felling and bucking and eliminated the large amount of waste in chips. While handsaws are still very widely used, power saws of several types have been developed and are used extensively for felling and bucking trees.

THE AX

Regardless of the great variety of ax shapes, all are characterized by having heads of steel and handles of wood, the head having an eye in which the handle is fitted, or "hung." The handle is held firmly to the head by means of wedges driven into the end of the handle that is fitted into the eye, to spread it and hold it tightly in place. There are two basically different types of axes used on harvesting operations and in general woods work. They are the single-bit ax and the double-bit ax, so named to designate axes with single or double cutting edges. Thus, a single-bit

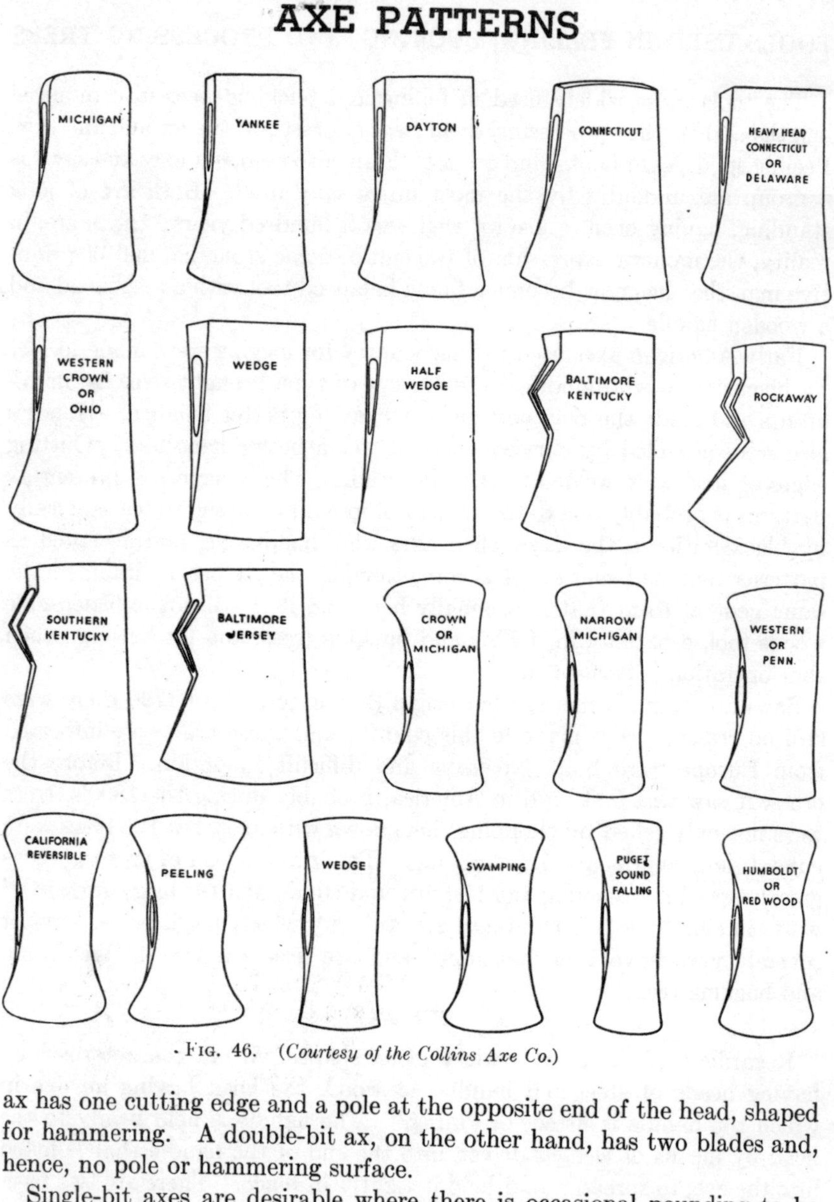

FIG. 46. *(Courtesy of the Collins Axe Co.)*

ax has one cutting edge and a pole at the opposite end of the head, shaped for hammering. A double-bit ax, on the other hand, has two blades and, hence, no pole or hammering surface.

Single-bit axes are desirable where there is occasional pounding to be done, although they should not be used as sledge hammers, as to do so will

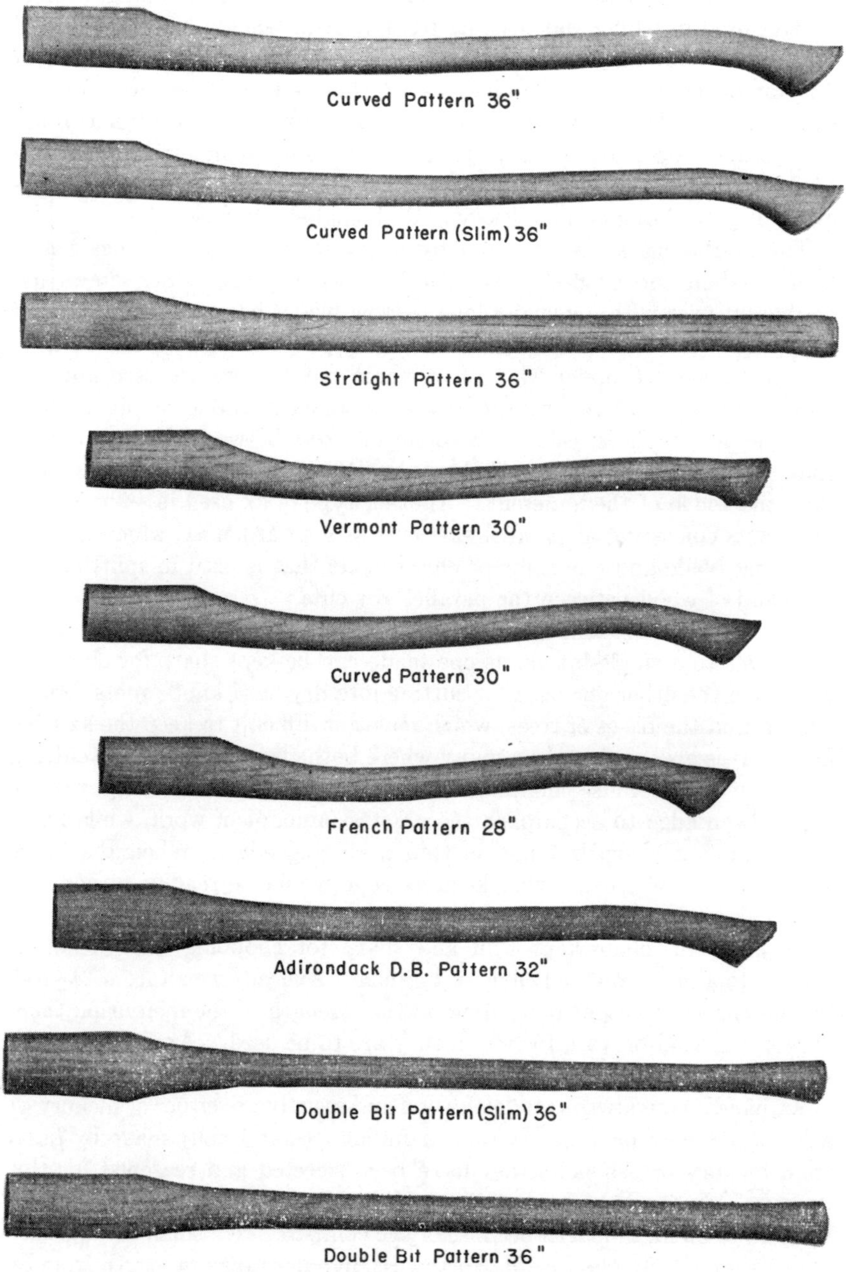

FIG. 47. Types of ax handles. (*Courtesy of the Collins Axe Co.*)

misshape the heads and loosen the handles. For heavy pounding, even of wedges used in felling and bucking trees, a sledge should be used to avoid unnecessary damage to the ax. Single-bit axes are hung on curved or unsymmetrically shaped handles, as all chopping must be done by the same cutting edge. Double-bit axes, because they may be turned to cut with either edge, are usually hung on straight, symmetrically shaped handles; that is, handles that are not curved. In some few cases, however, they are hung on handles typical of single-bit axes.

The double-bit ax is preferred by many fallers because it has better balance than the single-bit ax. In Western regions, felling crews use double-bit axes with unusually long narrow blades because a great deal of chopping is necessary in undercutting and axes that can be driven deep into the wood are preferred. When chain power saws are used and the undercut consists of two parallel saw cuts, an ax with one chopping blade and one adz blade is used. The band of wood between the saw cuts is split out or cut out with the adz blade, which, of course, must be narrower than the width of the undercut. Another type of ax used in splitting out undercuts consisting of parallel saw cuts is a picaroon ax, which has one chopping blade and one spike or chisel blade that is used in splitting out the band of wood between the parallel saw cuts.

For cutting out skid roads or other similar work, a double-bit ax is preferable to a single-bit ax, as one blade can be kept sharp for chopping wood and the other one used for cutting into dry hard knots, roots, brush, and around the bases of trees, which makes it difficult to keep the ax edge keen. It is also preferred by many where both chopping and limb cutting are to be done because chopping operations, such as undercutting, require a thin keen edge to accomplish the greatest amount of work, while limbing requires a sharp but not so thin a cutting edge. When the limbs to be cut off are hard, a thin keen ax edge would be readily turned and dulled or nicked. Hence, in such cases the double-bit ax is frequently used with one blade kept thin and sharp for chopping and the other ground to a more rounded edge for limbing. Axheads weigh from 2½ to 5 lb, the choice of weight depending on the strength of the men using them and on the kind of work for which they are to be used. Axes weighing at least 3½ or 4 lb are used on most logging operations.

Ax handles are always made of wood and usually are either of hickory or ash. They may be either factory manufactured or locally made by hand from hickory or ash bolts that have been selected and seasoned for this purpose. In the Eastern part of the United States, handles are ordinarily from 28 to 30 in. in length because of the comparatively small average size of the timber. In the Pacific Northwest, handles range in length from 38 to 40 and even 42 in., because of the large size of the timber. Throughout the rest of the United States the average length on saw-log operations is 36 in.

On the western slopes of the Cascade Mountains from northern California to Alaska, loggers prefer the swamping type of double-bit ax. In the Inland Empire region most woods workers prefer the straight-back double-bit patterns. There are various other types of the same general pattern more or less generally used, but the Pacific coast is the only region where the swamping, peeling, and falling types are commonly used.

Another type of ax, the broadax, is used in woods operations for hewing ties and, occasionally, hand-hewed timbers. The broadax has a large keen blade 6 to 7 in. long, beveled on one side only and with an offset handle or eye that permits a tie cutter to stand on the felled tree and hew parallel vertical surfaces on both sides of it. The pole portion is small and serves only to balance the blade. Broadax heads weigh from 6 to 7 lb. They are never used in felling or bucking.

The axes used on pulpwood operations in the Northeast and in eastern Canada are noticeably smaller than those used on logging operations in the West and South. This is because of the much smaller average size of the trees being cut. Single-bit axes weighing from 2½ to 2¾ lb with handles from 26 to 28 in. in length are the most common. On Southern pulpwood operations axes and handles of saw-log sizes are in general use probably because the trees are larger and because Southern pine wood is of greater density than that of spruce and fir. Also, the limbs of spruce and fir trees cut for pulpwood are smaller than pine limbs and much of the limbing can be done by using only one hand when the ax is not too heavy.

CARE OF AXES

An ax, in common with all cutting tools, must be kept sharp and correctly ground and hung to the handle for most effective use. A stone

Gage for correct grinding of 3 lb. single or double bit for soft wood

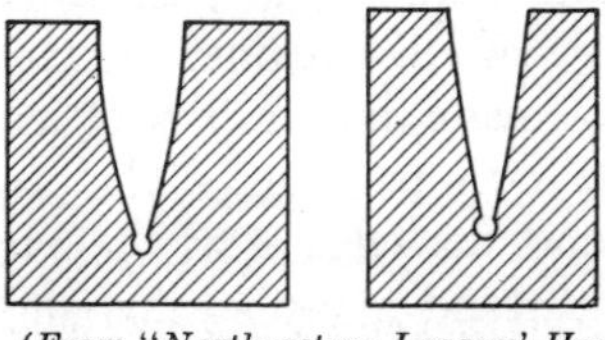

Gage for correct grinding of 2½ lb. single bit for soft wood

FIG. 48. Shape of ax blade. (*From "Northeastern Loggers' Handbook," by Fred C. Simmons, Northeastern Forest Experiment Station of U.S. Forest Service,* 1946.)

grinder, cooled with water, is best for grinding and sharpening an ax, as there is but little danger of overheating the edge and causing it to lose its temper. A dry grinding wheel should never be used, as the thin cutting edge would heat quickly and lose its temper to such an extent that the edge would dull or turn easily. Some axmen use a file to keep a keen cutting edge on their axes, but this is too slow and laborious a process except when an ax blade needs touching up during the day due to cutting into a knot or other dulling object. Filing is only a temporary expedient in keeping an ax in condition. Care should be exercised to keep the profile of the blade from becoming too thick and rounded or too thin or concave. A thick blade tends to bounce out of the tree when chopping instead of making a deep incision. A thin or concave blade wedges into the tree and is difficult to remove for the next stroke. Also thin blades break readily in frozen or dense wood or when knots are struck.

Correctly and skillfully used, an ax is an efficient tool and should be treated as such. In winter an ax should be warmed slightly before using to reduce its brittleness. The handle should be so hung that when the ax blade is placed edge down on a level surface with the end of the handle touching, the blade touches the surface at its center. When hung in this manner an ax will have balance, and when it is correctly ground and sharp, much more work can be accomplished with less effort and danger than with a poorly hung and dull ax.

THE SAW

A crosscut saw of some type is almost universally used as the tool for felling and bucking trees large enough for commercial products. Only when saws are not available are axes used for felling trees larger than saplings. Until comparatively recent years handsaws were the only type in use, but more recently several types of power-driven crosscut saws have come into wide use. Crosscut saws, as their name implies, are used to sever trees, first at the stump and then into shorter segments after they have fallen to the ground. Handsaws and power-driven dragsaws cut by being pulled back and forth across the trunk and, as this action takes place, the teeth cut into the wood, severing the fibers and leaving a clean cut across the grain. Chain-type power saws cut by forcing the teeth contained in the moving chain against the tree, while circular power saws cut by forcing the revolving saw edge against the tree.

Handsaws. Saws are made in a variety of lengths, widths, and types, and with many different tooth patterns. The principal part of a handsaw is the blade, or metal band, one edge of which is toothed for cutting. The blade, since it must withstand many strains and yet hold its sharpness, must be made of the best quality of steel.

The teeth in crosscut saws are of two kinds, cutting teeth and rakers, or cleaners. The cutting teeth are, in effect, a double series of sharp and pointed double-edged knives that cut into the wood, severing the fibers. They are so filed that one tooth cuts on one side and the next on the other. The thickness of the saw cut, which is the distance between the cutting points of the two rows of cutting teeth from side to side, is called the "saw kerf." The raker teeth, interspersed at regular intervals with the cutting teeth, act as chisels and, being slightly shorter than the cutting teeth, cut out the shallow layer of wood in the bottom of the cut that has been severed on both sides by the cutting teeth. This wood is carried out of the cut in the gullets of the saw between the teeth in the form of sawdust.

Hand crosscut saws may be equipped with two handles, depending on whether one or two men are to use the saw. Detachable handles are used with standard types of two-man saws. One-man crosscut saws have a permanently attached handle similar to, but larger than, the handle on a

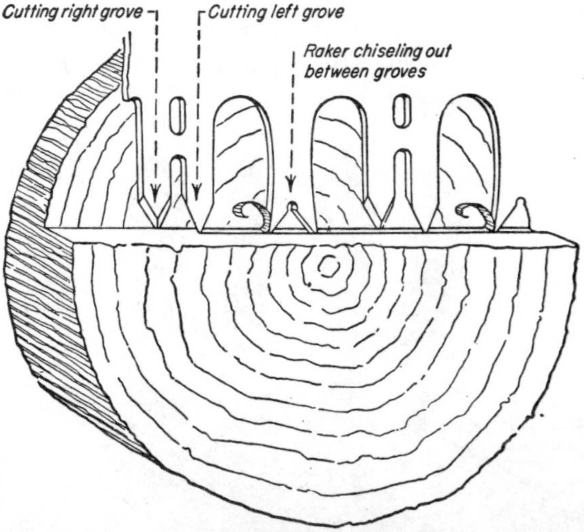

FIG. 49. There are two kinds of teeth on a crosscut saw, cutting teeth, and cleaning or raker teeth. Cutting teeth are like small knives. The blades of these knives are so filed that one tooth makes a cut from the right and the next from the left. The raker teeth, cutting like a plane, rake the chips and sawdust into the hollows or gullets between the cutting and raker teeth and carry them out of the cut. (*Courtesy of the Simonds Saw & Steel Co.*)

carpenter's saw, and also a detachable handle for the opposite end so it may be used as a two-man saw when desired.

Saw blades range in length from 6 to 6½ ft when trees up to 3 ft in diameter are to be cut. In the Pacific Northwest, longer saws, up to 8 and 10 ft in length, are commonly used because of the large size of many of the

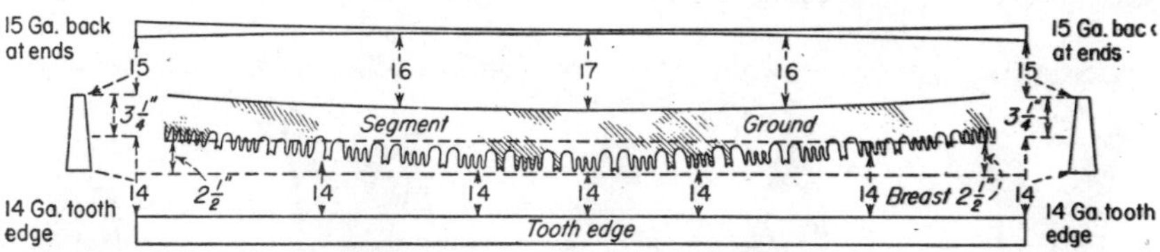

FIG. 50. Details of a crosscut saw exaggerated for the sake of clearness. (*Courtesy of E. C. Atkins and Company.*)

Western trees. The unusually large size of redwood trees in California makes it necessary in some cases to use hand crosscut saws as long as 18 ft.

The width of saw blades varies from 4 in. for saws designed for small trees and light work up to 8½ in. for saws used in cutting saw-log timber. The toothed edge is curved, with the greatest width at the middle of the saw. Thus, when a saw is pulled back and forth in the cut, teeth that are cutting get all the pressure, which makes for faster cutting and also reduces resistance. Also, a minimum of pressure against the cut is required, as the curved blade automatically keeps the teeth cutting as the saw is pulled first to one side and then to the other.

Saw blades must be thick enough to give them rigidity and be as thin as possible to avoid excess weight. Blades taper in thickness from the thicker toothed edge to the thinner back to keep as much of the blades as possible

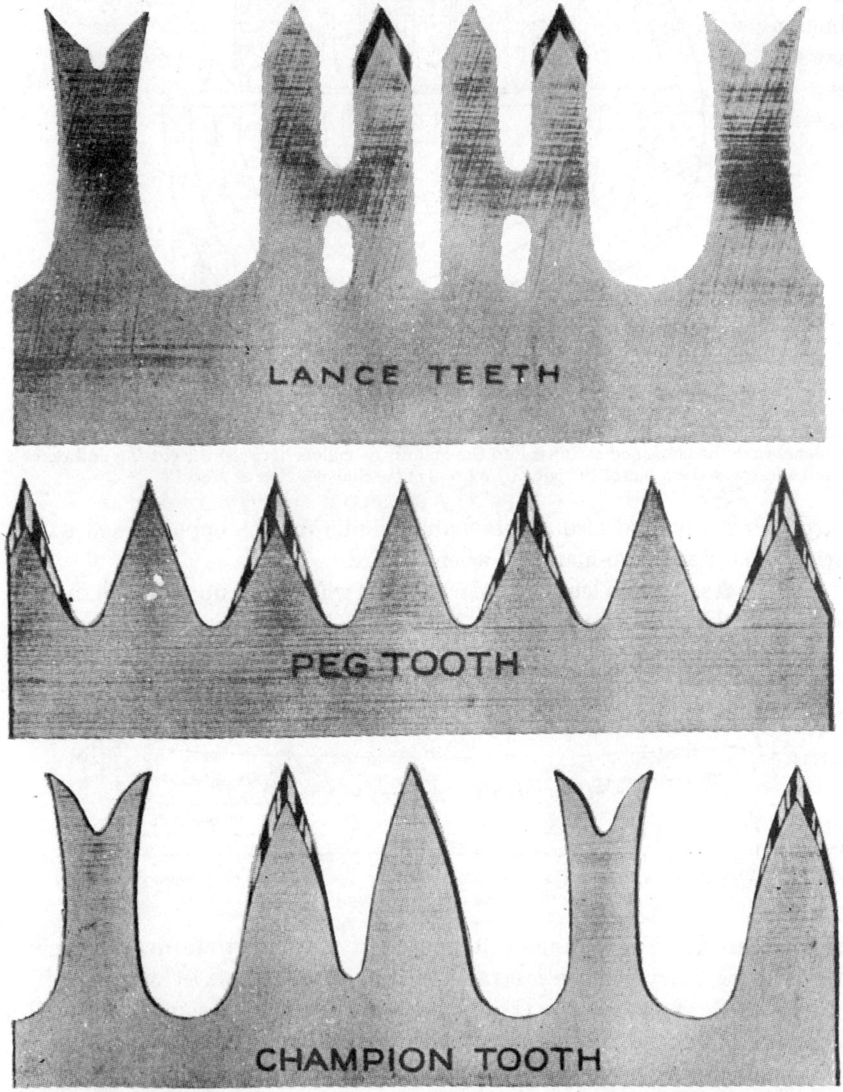

FIG. 51. Types of crosscut saw teeth. (*Courtesy of the Simonds Saw and Steel Company.*)

from rubbing the cut. Some types are also thicker at the ends than in the middle for still greater rigidity.

In many parts of the country the same saws are used for both felling and bucking, but in Western forest regions in large timber where felling and bucking are separate operations, a somewhat different type of saw is used for bucking than for felling. Felling can be done by a two-man crew, one man on each end of the saw, but the large diameter of such trees prevents two men from using the same saw at the same time for bucking. Therefore, bucking is, of necessity, a one-man job in large Western timber when handsaws are used. Greater rigidity is needed in saws being worked from one end only and one-man bucking saws are, therefore, thicker than felling saws.

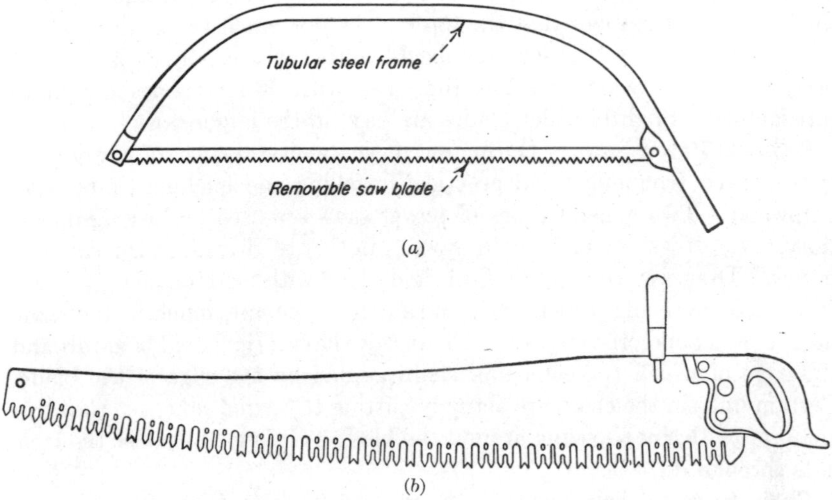

FIG. 52. Types of one-man handsaws. (a) Bow saw for small trees. The blade is tensioned in the tubular frame. (b) One-man crosscut saw. (*From U.S. Department of Agriculture, Farmers Bulletin*, 1907.)

There are various kinds of crosscut saw patterns, which vary with the manufacturer and users' preferences. One-man bucking saws generally have a straight back, and two-man saws for either felling or bucking have a slightly concave back to eliminate as much weight as is consistent with the required rigidity. Narrow saws with a concave back corresponding to the curve of the breast of the saw are preferred in small timber so a wedge can be inserted in the cut behind the saw while the cut is still shallow. With wider types of saws there would not be enough clearance for a wedge until the tree was practically severed.

In small timber, such as is cut for pulpwood in the Northeast and in eastern Canada, bow saws are popular and commonly used for felling and bucking by one-man or two-man crews. There are several types of bow saws but all use the principle of holding a thin narrow saw blade taut by means of tension applied to either end. In metal bow saws the blade is held in tension by the natural spring of the hollow handle, whereas with wooden types a tightening cord, as in the common woodcutting bucksaw, is used to hold the thin blade taut. With a type of labor that likes or is willing to work alone in the woods, and in timber that does not exceed 12 or 14 in. in diameter, bow saws are excellent tools. Their almost exclusive use is on pulpwood operations in small timber. They have long been used in the Northeast and in eastern Canada and are coming into wider use on Southern pulpwood operations, although in the South they are generally used by two-man crews. One important advantage of the bow saw is that the thin narrow blade is seldom pinched even when a tree with a pronounced sag is bucked. The blade is so thin and narrow that even after the bucking cut has closed at the top it does not pinch the blade. When only one man uses a bow saw, the fixed-length type, usually 36 in. long, is preferred. For two-man crews the saws with 48-in. frames are more satisfactory. Slightly wider blades are used in the larger saws.

Crosscut Power Saws. During recent years several types of power saws have proved economical and practical in felling and bucking timber and pulpwood. Two general types of power saws are used on logging operations, *i.e.*, dragsaws and chain saws. Both are driven by mechanical power. Dragsaws consist of a firm rigid blade with teeth on one edge, and they sever by being drawn back and forth in the cut, much in the same manner as a hand crosscut saw. Chain saws have a rigid oval blade around the edge of which travels a link chain guided by the edge of the blade. Certain links in the chain are actually cutting teeth and rakers so that this chain of teeth, by traveling around the blade as it is held against the tree, cuts through it.

Chain Saws. Chain power saws are used for both felling and bucking and are small enough for two or three men to carry conveniently. In felling, the saw is held against the tree at the stump for the undercut and the backcut. In bucking it is placed against the tree where it is to be severed.

Power is furnished by a self-contained gasoline motor, by a compressed air motor, or by an electric motor built into the saw. The saws with self-contained power units are much more portable than those dependent upon other equipment for motive power. Types operating with compressed air or electric motors require that a power unit with a compressor attachment or an electric generator be in operation at the same time for the sole purpose of providing the saw with actuating force.

Saws powered by small gasoline motors, however, are not dependent upon auxiliary equipment and hence can be used as independent units.

Power units for air-driven and electric-driven saws are mounted on tractors, skids, or other portable devices that can be moved to various parts of the operation where felling and bucking are to be done. Diesel motors are the units usually provided and they in turn furnish power for the operation of electric generators or air compressors. Electric cables or air hoses lead from the power units to the saws. Several saws can be powered from each unit and they can operate up to 100 or 200 ft from the power unit.

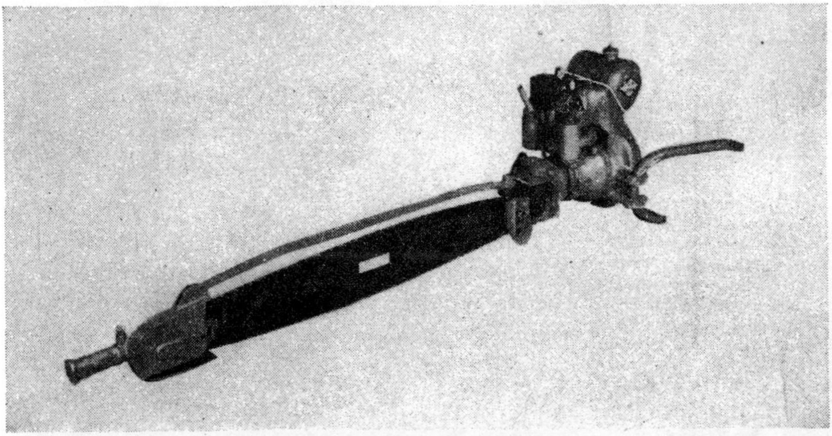

FIG. 53. A gasoline-motor-powered type of chain saw, showing the rigid blade around which the cutting chain travels and a gasoline motor for driving it. (*Courtesy of the Mall Tool Company.*)

Chain power saws are available in various sizes and lengths, with cutting bars ranging from 24 to 48 in. in length and from 2½ to 5 hp. Several types of one-man power saws for use in small timber have recently been developed. They are operated from the motor end only and the cutting chain is exposed on top, end, and bottom. Such saws can be used to cut straight into a log by forcing the end against the log. One-man types are much lighter than two-man types but are adapted only to cutting trees of smaller diameters. Because the one-man power saws can be mortised into a log from each side, it is possible, but of doubtful practicability, to buck logs nearly twice the diameter of the length of the cutting bar.

The chains on all saws of this type travel away from the power end on the bottom of the cutter bar, and as sawing usually is done from the top in bucking, the cutting is done with the bottom of the blade so the sawdust is thrown out of the cut on the side opposite the operator. This direction of cutting-chain movement also serves to hold the power end of the saw against the log, relieving the operator of strain in holding the saw in place. When it is necessary to undercut in bucking, the saw cuts with the upper edge of the bar so the operator must hold the saw in the cut to keep it from working out, as no mechanical restraint is available.

Power saws are better adapted for felling and bucking when a considerable volume of timber is to be harvested per acre. In light scattered stands too much time and effort is wasted in carrying the saw from tree to tree. This is particularly true with saws powered by electricity or compressed air because the power-generating unit must be kept close to the operation.

FIG. 54. An electric-motor-powered chain saw showing the tractor unit with generator and a saw mounted on the tractor. (*Courtesy of the Atkins-Hassler Company.*)

Dragsaws. Dragsaws operated by small gasoline motors are used only in the West in large timber. They consist of a frame containing a motor and the gears and belts necessary to operate the straight saw blade in a reciprocating manner. In bucking, the frame is spiked to the tree with one end on the ground. The saw blade is pulled back and forth in the cut in much the same manner as a hand-operated crosscut saw. The weight of the blade may hold it in the cut firmly enough to operate it or, if not, tension is applied at the motor end. When necessary, this type of saw can be used for underbucking by applying tension to the blade with firm rubber belts that hold the cutting edge upwards in the cut until it is deep enough to permit severing from the top without splitting the log. In felling, the frame is placed in a vertical position against the side of the tree and spiked to the tree and is also supported by a framed platform. The blade is held into the cut by rubber belts or by springs dogged to the tree and moved as the cut progresses.

Dragsaws are adapted only to felling and bucking large timber because the time required to place the saw for cutting would not be justified for only a short setting. The dragsaw can be used to fell and buck trees too large for chain saws, as blades up to 18 ft in length can be used, and they can be operated from one side of the tree, whereas chain saws must be operated from both sides. These saws are widely used on redwood harvesting operations and in other timber too large for practical power felling and bucking with chain saws.

Circular Saws. Portable circular saws driven by small gasoline motors mounted on two bicycle-type wheels are widely used on pulpwood operations throughout the South. The saw is mounted on the end of a beam that projects outward from the axle between the wheels. The motor is mounted over the axle and a handle extends outward on the side opposite the saw. The beam and handle are rigid on the axle, so when the handle is raised the saw is lowered, and vice versa.

This sawing device is moved from place to place by pushing on the handle with the beam approximately horizontal to keep the saw off the ground. In some types the motor is geared to the wheels to assist in moving it. When the saw has been spotted at the point where a bucking cut is to be made, the handle is raised, bringing the revolving saw blade downward and in contact with the felled tree. Considerable skill is required to prevent pinching the saw, which kills the motor. When there is a tendency towards pinching, the operator often makes two or three cuts before severing the tree completely. One man can operate the saw but a second man with an ax usually assists in limbing, marking bolt lengths, clearing a way for the saw, and in holding small trees steady while they are being cut up.

The saw and its mounting are adjustable so that the blade can be operated vertically for bucking or horizontally for felling. It is better adapted to bucking, however, and is more widely used for this purpose. Trees are bucked either where they lie after felling or they are skidded to a brow or yard and bucked at concentration points, where the saw does not have to be moved as much as when it is taken to each tree at the felling point. When circular saws are used for bucking only, the trees usually are felled with hand crosscut saws.

Pulpwood Mills. As a result of efforts to mechanize pulpwood production, various types of portable pulpwood cutting and loading machines have been developed and a number have been placed in use, particularly in the South. Since pulpwood is cut from small trees, usually not over 16 in. dbh, and because the trees are bucked into short bolts from 4 to 5 ft long, except when double lengths or multiple lengths are cut, a great deal of hand labor is necessary in bucking, bunching, and loading pulpwood when mechanical methods are not used.

The object of pulpwood-cutting machines is to eliminate hand bucking and loading, so they are designed to cut and load the bolts on trucks in the forest. In general they consist of a feed roll, or conveyer, a powered circular cutoff saw of the swing type, a bolt conveyer, a power unit, and the necessary auxiliary equipment.

Trees are felled by hand or mechanical saws, limbed, and skidded full length to the machine, where they are placed on the conveyer as needed.

FIG. 55. A pulpwood mill in a Southern forest showing the dolly on which the tree-length log is placed while it is fed to the saw, the swing saw in an upright position, the loading conveyer, and a partially loaded truck in the background. The tractor-mounted log loader at the right is not a part of the pulpwood mill.

One man then feeds the tree along the conveyer and the saw operator pulls or pushes the powered circular saw against it, cutting off bolts of the desired length. The bolts then fall or move onto the loader conveyer and are carried upward broadside onto a waiting truck.

Power for the saw and conveyer is furnished by a tractor, usually of the wheeled agricultural type with special mountings for the saw and conveyer. The power is obtained from the power take-off usually incorporated in such tractors. The entire machine must be portable so it can be readily moved from place to place as the timber in each location is cut. Either tractors or teams are used to skid the tree lengths to the machine. Such machines hold much promise of successful operation when felling, skidding, cutting, and trucking can be closely coordinated to prevent loss of time and when the volume per acre is sufficient to permit operating the machine at one location for approximately a week. Some machines of this type are capable of cutting and loading as much as 60 cords of pulpwood per day when in uninterrupted operation.

CARE OF SAWS

Saw teeth are the working parts of all saws. They function by actually severing the wood when the saw is moved back and forth across the trunk and with sufficient pressure against the cut. Keeping the saw teeth in good working order, therefore, is one of the most important jobs on a timber-harvesting operation. When as many as six or seven saws are used on an operation, it is advisable to provide a saw filer, whose sole job it is to keep the saws sharp and cutting correctly.

Handsaws. A good sawyer earning his living at felling or bucking trees insists that his saw be kept in good working condition, as otherwise a considerable portion of his effort would be wasted. For best results saws should be sharpened after each day of regular use, and on well-conducted operations the company usually provides two saws for each felling crew so one can be sharpened every day. When saws are sharpened after each day of use, they can be kept in good condition with much less effort than if they are permitted to get dull and worn before they are filed and set.

Special saw-fitting tools are used to keep saws in good cutting condition. They are generally called saw-filing, or saw-fitting, tools. The steps in filing, or fitting, a saw are

1. All cutting teeth should be maintained at the same length so each will do its share of cutting. This is accomplished by "jointing" the saw, or placing a guide over the teeth and filing off the ends of any that are too long, until all have the same length. The jointing tool holds the file slightly bent to conform to the curvature of the saw.
2. The rakers are then "swaged," or hammered, to spread them slightly at the points so they will cut as chisels in the bottom of the kerf, or saw cut. The opposite cutting edges of each raker tooth are for cutting on each directional pull of the saw. When swaged and filed, the cutting edges of the raker teeth are from $\frac{1}{64}$ to $\frac{1}{100}$ in. shorter than the cutters, which keeps them from cutting below the depth to which the fibers have been severed by the cutting teeth.
3. The cutting teeth are then filed and sharpened by beveling the points so that each is, in effect, a two-edged knife with the correct shape to the point and bevel to the edge, for the work the saw is to perform. The heavier and harder the wood to be sawed, the less the bevels and the more rounded the points should be to prevent weakening the teeth so much that they may break or bend.
4. The final step is to set the saw by bending the cutting teeth slightly outward, usually $\frac{1}{64}$ in. so the width of the cut is somewhat larger than the thickness of the saw. This is done alternately, one tooth being bent to one side and the next to the opposite side, so the beveled side of the tooth is on the inside of the cut. A saw that is not correctly set is difficult to pull in the cut because of side friction.

The condition of a saw can be accurately judged by the kind of sawdust it makes. A saw in good condition pulls long stringy curls of sawdust from the cut, showing that the cutting teeth are severing the wood fibers on both sides and that the rakers are chiseling out the wood so severed. Fine, granular sawdust is a sign of dull teeth, incorrectly sharpened and set.

The teeth of a saw gradually are shortened in filing so that it is necessary to file or grind the gullets deeper if the saw is still in good enough condition to justify this. Saws should never be leaned upright against trees or walls, as they will sag and become bent, especially if left for any length of time. They should preferably be hung in slotted brackets, toothed edge upwards, as there is then but little possibility of a saw becoming knicked or bent. In carrying saws to and from the woods in cars or trucks, they should always be placed in a box or tool chest in prepared brackets that will keep the teeth from coming into contact with other tools or other saws and becoming dulled or damaged. It is also a safety measure, as loosely placed tools would be thrown around in the event of a wreck, with possible serious injury to the men.

Bow saws are sharpened in essentially the same way as the conventional crosscut saw, but a special set of fitting tools must be used, as the teeth are much smaller and ordinary files are not suitable. Set is put in the teeth with a pair of special pliers that bends the teeth individually. This is possible in thin bow saws although it is not practicable for thicker handsaws.

Power Saws. Dragsaws, since their blades are similar to those of hand crosscut saws, are filed and fitted in much the same way. The cutting teeth of chain saws, however, require a different procedure because the teeth are quite different in design and cut differently from hand crosscut saw teeth. Since the chain is flexible, each segment of it must be held tightly in a vise of special design when the teeth are being sharpened. Each of the several kinds of teeth should be of the same height. All side cutters, for example, should be the same height, and all center rakers should be the same height, although the height of each kind is slightly different. The offset rakers, for example, should be about 0.015 in. lower than the side cutters and the center rakers about 0.025 in. lower than the side cutters.

The teeth in a new chain are formed so that the front of the tooth is farther out than the rear. This gives a minimum of friction in the cut as well as a better cutting point. When a chain develops "timber bind" in use, additional set is required, as a spot behind the cutting point is rubbing in the cut, preventing the point of the cutting tooth from cutting. Only the side cutters require setting and not necessarily every time a chain is filed. Set is put in a tooth by striking it a sharp blow with a hammer, but care must be taken to give all teeth the same amount of set.

Jointing is required when the points of the teeth become uneven in height due to careless filing or the saw coming in contact with a metal wedge or from some similar cause. A jointing tool similar to that used for hand crosscut saws, but with openings to joint three tooth heights instead

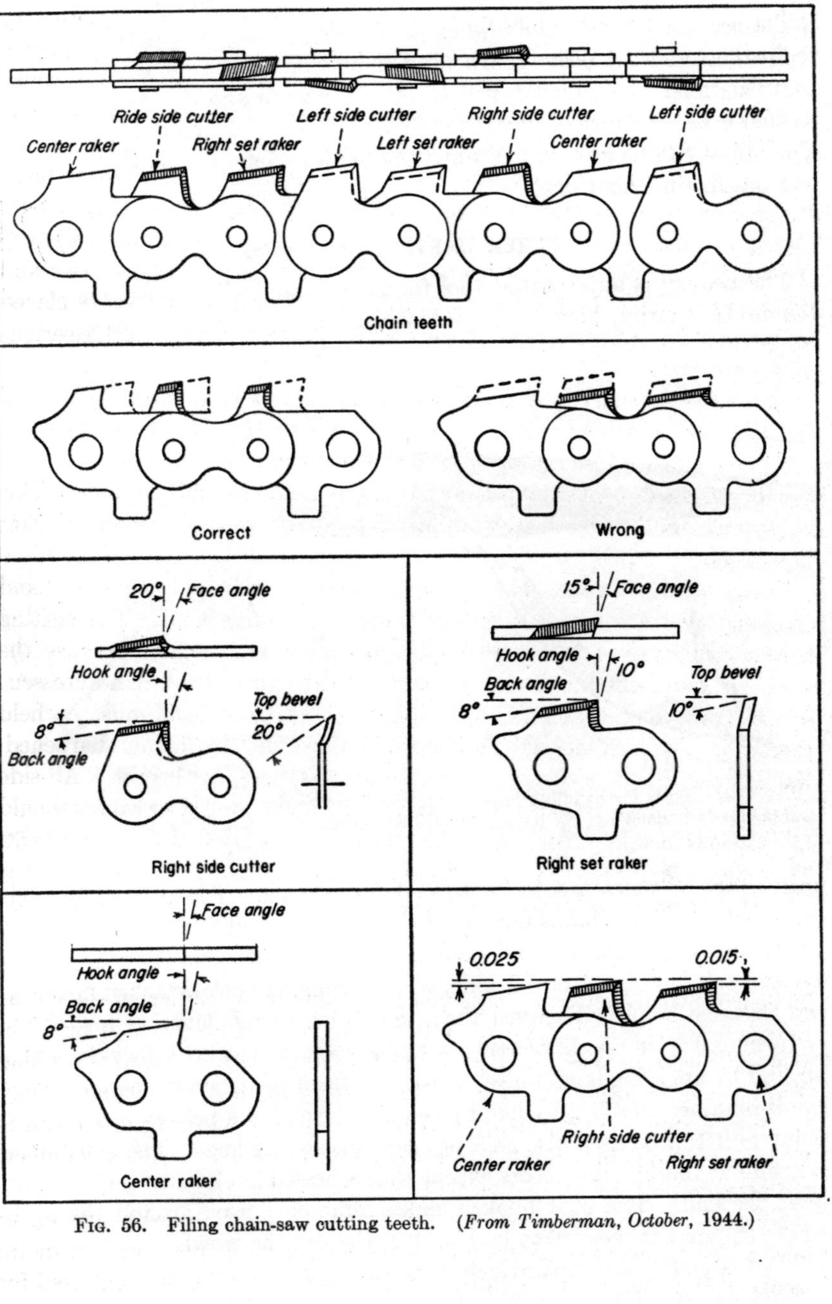

FIG. 56. Filing chain-saw cutting teeth. (*From Timberman, October, 1944.*)

of two, is used to joint chain saws. The tips are filed to the correct height and then sharpened. Only when teeth are jointed should they be filed from the top. All other sharpening should be on the point of the teeth only, as filing from the top will shorten the teeth and decrease the space available for sawdust to be carried out of the cut. In hand crosscut saws the gullets can be ground out when the teeth become too short, but this is not possible in chain saws.

OTHER HARVESTING TOOLS

The peavey is an essential tool found on practically all logging operations. Ordinarily, however, it is not used by felling and bucking crews

FIG. 57A. Tools for handling logs. Note the free-swinging pointed hook and the steel shod point of the handle of the peavey illustrated above. (*Courtesy of the American Logging Tool Company.*)

unless logs or poles are to be turned for peeling. The peavey is very commonly used by bunching and loading crews and for placing logs at yards and on trucks during loading. The peavey consists of a heavy wooden handle, some 5 or 6 ft long, with a swinging hook attached near the end that can be placed against the side of a log to give leverage for rolling or turning it. In operation, the swinging hook fastens into the side of the log and by applying hand power to the handle of the peavey, enough leverage is obtained to roll it by hand power. The end of a peavey has a metal point so it can be used also as a pry pole in moving logs. The cant hook is a similar tool in all respects except that a heeled hook instead of a point is at the end. The cant hook is more commonly used around the mill, where square timbers are to be handled, than in the woods.

The log jack is a comparatively new tool that is coming into general use in bucking saw timber, particularly for pulpwood or other similar products, to short lengths. As the name implies, this tool is designed to jack up the tree being bucked to permit bucking without the saw's pinching. It is particularly well adapted to circular-power-saw bucking, where saw pinching can only be overcome by raising the tree off the ground if it lies in a bind. The log jack is similar to a peavey in many respects, but it has a flattened steel loop on the under side so when the hook fastens into a tree on the ground and the handle is lowered, the tree is jacked up and will remain elevated on the jack. Bucking cuts can then be made without saw

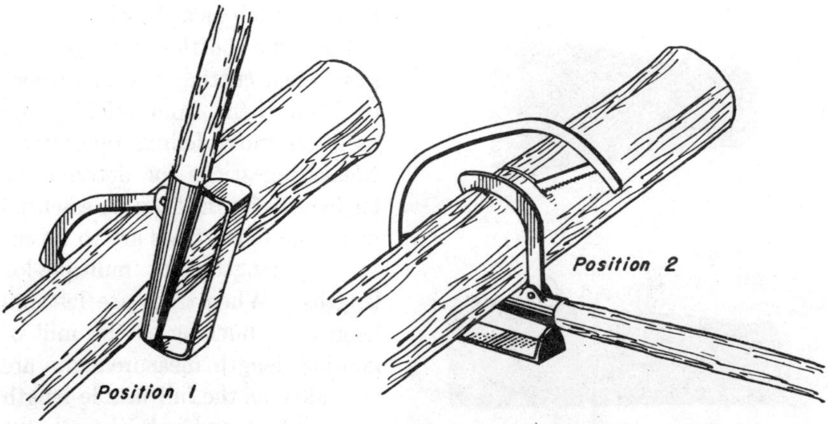

FIG. 57*B*. Tools for handling logs. The tool shown above, known as a "log jack," is used to raise the end of a small log and hold it steady for bucking. (*From "Northeastern Loggers' Handbook," by Fred C. Simmons.*)

pinching and, when bow saws are used, the sawing position is much more convenient. Another advantage is that trees can be bucked without the saw teeth striking the soil and being dulled by rocks and grit.

Wedges are used both in felling and in bucking and are an integral part of a saw crew's equipment. In felling, the wedges are inserted in the cut to prevent the weight of the tree closing the cut and pinching the saw and also to force the tree to fall in the desired direction. In bucking, wedges are used to keep the cut open and thereby prevent the saw from being pinched when the logs settle. Wedges are either of metal or some tough and strong hardwood, such as persimmon or dogwood, that will withstand driving blows with an ax or mall. Sawyers commonly carry several wedges, often in a holster attached to the belt. Wedges are also used for splitting. Malls or sledge hammers should always be used for driving wedges, even wooden ones. Driving wedges with an ax damages the eye and misshapes it and destroys its balance. The malls or sledge hammers are of the usual type, with iron heads and wooden handles.

Several types of spuds, or bark-peeling tools are used when products are to be peeled. One popular tool for this purpose is a common hoe with the eye straightened and the blade sharpened. This gives a long-handled instrument convenient for stripping the bark from poles and other tree-size products. For sap-peeling pulpwood a hand spud similar to a tire iron used in changing casings is commonly used. In stripping bark from redwood logs in the woods, large spade-shaped spuds are used. These instruments and others similar to them only remove the rough bark. When the tight inner bark is to be removed after the peeling season has passed, some sharp knife peeler or shaver must be used, such as a drawknife.

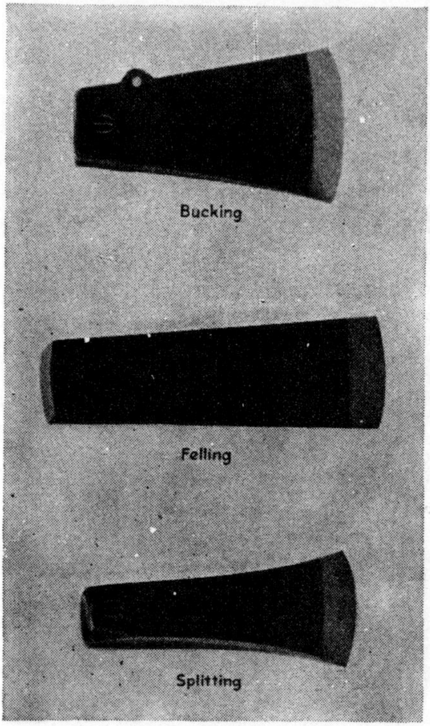

FIG. 58. Types of wedges used in felling, bucking, and splitting. (*Courtesy of the Young Iron Works.*)

Kerosene bottles for spraying saws when cutting resinous woods to keep them from sticking are used on most felling operations. Measuring sticks for determining log lengths for bucking are essential when the felled trees are to be cut to log lengths or multiple-log lengths. When trees are left full length for bucking at the mill or landing, length measurements are not taken as the full usable length is left intact and only the top cut out. For measuring the circumference and length of felled trees for poles or piles, a tape measure is ordinarily used. Measuring sticks usually are simply split or squared straight sticks 8 ft long, marked at 2-ft intervals so lengths from 12 to 16 ft or longer can readily be measured. It is essential that the sticks be full length to avoid cutting short logs, which would mean wasting the length back to the next even foot length. Measuring sticks should be shod with metal tips to reduce the chances that they will be accidentally shortened by cutting them off as the logs are marked. Too much care cannot be exercised to insure that logs are cut full length plus a few extra inches for trimming allowance in the mill.

In riving shingles or shakes, a tool called a froe is used to split off the individual shingles. It consists of a wedge-shaped blade that is held against the end of the block from which shingles are to be rived. The handle is at right angles and extends upwards, so after the blade has been driven into the block a pull on it will turn the blade in the cut and cause the shingle to split from the block. A wooden mallet is commonly used to drive the froe into the block. It may be of the manufactured type or of simple homemade construction.

In the Northeast and eastern Canada, poles are often used to push small trees over as they are being felled. Such pushing poles are from 6 to 8 ft

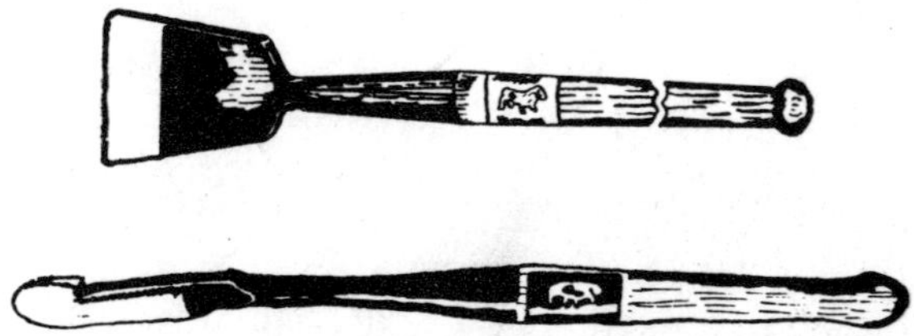

FIG. 59A. Two types of peeling tools, or spuds, are shown above. Another common type is half of an automobile spring leaf flattened at one end. (*Courtesy of the Warren Axe & Tool Co.*)

long and are steel-pointed so the end of the pole will not slip off the tree as it is being pushed. The steel point is short enough so it does not hang in the tree when it is to be removed.

Pike poles are commonly used for handling logs in water, to assemble them in rafts, or to move stranded logs during stream drives. The poles

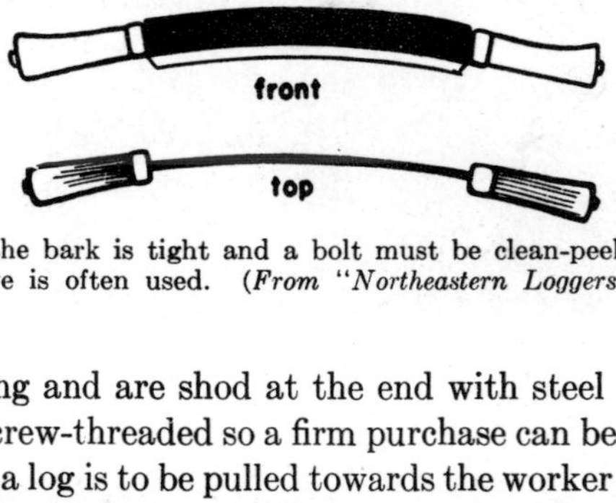

FIG. 59B. When the bark is tight and a bolt must be clean-peeled, a timber shave of the type shown above is often used. (*From "Northeastern Loggers' Handbook," by Fred C. Simmons.*)

are 12 to 16 ft long and are shod at the end with steel points and hooks. Some points are screw-threaded so a firm purchase can be obtained by turning the pole when a log is to be pulled towards the worker instead of pushed. Pike poles are also almost universally used for manipulating logs in millponds.

PART THREE

MOVEMENT OF PRODUCTS FROM THE FOREST

Those operations in timber harvesting that involve the movement of products from the forest are the most costly because green forest products are bulky and heavy and the application of either animal or mechanical power is necessary. Therefore, much attention is given on harvesting operations to locating correctly roads, logging railroads, or other transportation routes that need to be constructed and to the choice of equipment for skidding, loading, and hauling forest products. Transportation of products does not influence their quality or volume yield as do felling and bucking, but it is an important series of operations because of the large proportion of the total harvesting cost that is involved, which, if excessive, reduces the net income received for the timber and makes the forest crop less valuable.

CHAPTER 11

BUNCHING PRODUCTS IN THE FOREST

Bunching is included as one of the steps in the harvesting operation when it is desirable to assemble the products into small groups for more efficient loading or skidding. While the bunching operation resembles skidding in many respects, it is a separate step in harvesting, distinct from skidding, in that products are moved only very short distances. Bunching is done by hand labor in the case of small products and by the use of teams or tractors when logs or other large products are handled. The logs or other products are moved only far enough to assemble them into convenient groups for the next stage of the transportation operation.

Many harvesting operations do not include bunching the cut products, as they are skidded directly from the point of felling. Bunching is seldom if ever done, for instance, with cable skidding, as the cable progressively reaches all portions of the cutting area and it is not necessary to assemble the products for skidding. Also, with cable skidding and with tractor skidding when the power available is sufficient to handle large loads, it is a common practice to cut the trees into long logs that later will be cut into standard log lengths for sawing in the mill. When the trees are not too tall, they sometimes are left full length with only the tops cut off and are skidded as trees. In effect, bunching is accomplished, in such cases, by not cutting the trees into individual logs.

When the products are large enough so that each one constitutes a skidding load, there is, of course, no need to bunch them. Forest products vary so much in size, however, even on any one operation, that while the largest butt log may constitute a capacity skidding load, the other logs in the tree and smaller logs from other trees do not, and two or more must be skidded at one time to obtain full utilization of skidding power, since the skidder must be powerful enough to handle the largest single log or other product. Some tractor skidding is such that even though several logs are skidded at one turn, it is not necessary to bunch. This is the case when the skidding tractor is driven to each of the logs to be attached to it and the chokers, which are already fastened to the logs, are attached to the tractor as it reaches them.

Bunching, therefore, is done only when the products are not large enough so that each constitutes a load for the skidding equipment, and also when the products are to be loaded on trucks or, in some cases, on wagons, throughout the cutting area. In the latter case the skidding operation is eliminated, but bunching is still necessary so the hauling vehicles can be loaded with a minimum number of stops.

BUNCHING LOGS

When logs are to be skidded with carts or sleds and the logs are not large enough so that each constitutes a load, they are bunched with teams or

FIG. 60. Southern pine logs in North Carolina bunched with a single mule and driver.

tractors into groups, usually comprising at least one load at each bunch. Bunching, in such cases, involves dragging the logs together and building small piles over which the skidding wheels can be spotted to pick up a load. When a sled is used instead of wheels, the front ends of the logs are rolled onto the bunk by means of peaveys or with teams, the sled being spotted at one side of the bunch.

Tongs are usually used for fastening onto logs to bunch them, as they are only moved short distances and tongs are more easily hooked and unhooked by the drivers than chokers. The distance logs must be moved depends on the number of trees cut per acre and their size. In scattered small timber it is obvious that more area must be covered to obtain a bunch than in a dense stand of large trees.

When log-hauling trucks and trailers are driven off roads and into the harvesting area to be loaded, the logs are pulled together parallel but are not necessarily piled up as for some types of skidding wheels. It is always desirable to bunch enough logs at each place so that a full load can be placed on a truck or picked up by a skidding device, without moving to other bunches to complete loading. Logs may be bunched before hauling begins when skidding is not done or, as is more common, bunching may be

FIG. 61*A*. A single horse with driver and helper is often used for bunching logs for tractor skidding as in this illustration of a South Carolina operation.

carried on while hauling is in progress. In the latter case, when animals are used, one of the teams doing the bunching also does the loading.

When logs are skidded on the ground with teams, it is only seldom that more than one log is skidded at a time, as the distance that logs can be ground-skidded with teams is so short that it would not be economical to take time to attach to more than one log, unless it so happened that two small ones were lying adjacent to each other and could be tonged or choked together. Bunching would not be economical in such cases, as the team doing the bunching could just as well be engaged in skidding. Furthermore, any log of the usual lengths (12, 14, or 16 ft) and of moderate diameter, say 14 to 18 in., is a load in and of itself for a team doing ground skidding, and the effort required to separate the small logs from the large ones for bunching would require an undue amount of time and effort.

Because a chain must be passed beneath logs bunched for skidding with some types of wheels, they often are so placed that one end of the bunch is over a small log or unmerchantable bolt. This makes it easier to pass the chain beneath the load after the wheels have been spotted over the bunch. In the Middle Atlantic region the chain used to hold a bunch of products to skidding wheels is temporarily fastened to an iron rod two to three feet long with an open hook at one end and a handle on the other end. This rod, locally known as a "loading rod," is passed under the load to lead the chain beneath it. Once under the load, the chain is attached to the axle of the wheeled skidding device and the load is raised beneath the axle of the wheels as the skidding power unit moves ahead.

FIG. 61*B*. Each bunch shown above includes only logs from one tree. South Carolina.

When skidding power and equipment are such that large loads can be handled and logs ordinarily would be bunched, it is not unusual to find that saw-timber trees are left in tree lengths or at least multiple-log lengths. In the West, with both cable skidding and heavy tractor skidding, log lengths commonly range from 32 to 40 ft and even longer. Tractors and cable skidders can be loaded to capacity by attaching to two or three such logs without the necessity of bunching. However, the logs in reality have already been bunched, as they are left in multiple-log lengths to avoid the necessity of bunching them with teams or tractors. Thus, by not cutting trees into the standard log lengths until they reach the mill or some intermediate assembly point, the bunching operation has been eliminated although the logs actually have been bunched by being left attached end to end, unsevered.

Tree length or at least multiple-length logs are sometimes bunched for skidding when the skid haul is by tractors or teams using certain types of wheels and is comparatively long, ½ mile or longer. Bunching in such instances may be by tractors or teams, making the long logs into skidder loads to avoid excessive travel by the skidding tractor. Whenever possible, however, tractors should be loaded without bunching by attaching the already choked logs to the tractor as it passes them in making a turn.

With tree-length logs, even when there are enough cut per acre so that bunching ordinarily would not be necessary, it is sometimes advisable to move the logs so that they can be readily attached to the tractor as it makes a turn. When long logs lie on the ground in all directions it is difficult to maneuver a tractor in picking up a load, as the long logs tangle with each other and with stumps and uncut trees. Therefore, much time is saved in skidding when the long logs are moved into position where they can readily be attached to the tractor as it reaches them. Either a team or a small tractor can be used for such partial bunching since the logs are moved only short distances or only enough to straighten them out with respect to the path of the tractor. Ordinarily when skidding pans or arches are used, tree-length logs are choked at the butt end to reduce ground resistance, so in bunching skidder loads the butt ends are always forward.

The amount of bunching can be greatly reduced on most operations or sometimes eliminated altogether by felling trees so the logs cut from them lie close together and as nearly parallel as possible. Felling trees, therefore, should always be done with due regard to the subsequent operations of bunching or skidding so these operations will not be made unnecessarily complicated. In most stands the trees can be felled in chosen directions except for occasional leaning trees and trees with hollow or badly decayed butts. Thus, two trees within a few feet of each other can be felled in the same direction instead of in opposite directions and the logs cut from them will be adjacent and parallel instead of scattered. Trees can also be felled so the tops lie in piles or rows, with the logs well bunched around the piles, simply by arranging the felling with thought to the next steps in the harvesting operation.

BUNCHING PULPWOOD AND OTHER SMALL PRODUCTS

Small products cannot be economically skidded or transported individually, so the first step in their removal from the forest is usually bunching. Pulpwood and other cordwood products are generally bunched by hand, but occasionally animal or mechanical power is used. When pulpwood is produced in the form of logs, the operation is essentially logging and the methods of bunching and subsequent handling are similar or identical with those for logging operations. Pulpwood is bunched in tree lengths on many operations and then bucked with power saws. Teams and small tractors are used for the bunching, which in the Northeast is locally known as "twitching." The trees are bunched parallel on the ground at a convenient spot and are bucked either as they are bunched or after bunching has been completed.

It is only when pulpwood is cut into individual bolts that bunching is done before skidding or hauling. Pulpwood is bunched in a number of different ways. In the Lake states and the Northeast much wood is cut during the summer or fall and not skidded and hauled until winter, when snow or iced roads are available. As the wood is cut, either rough or peeled, it is bunched into small piles of from 8 to 12 bolts, perhaps more if the bolts are small and the volume cut is heavy per acre. The piles may be in ricks between trees or stakes, consisting of ¼ to ½ cord or more, or the bunched bolts may be laid together in the form of half cribs. In other

FIG. 62. Pulpwood obtained by thinning a stand of loblolly pine in South Carolina bunched in "pens" to obtain a measure of work accomplished and to facilitate loading.

cases the bolts are piled together parallel in a heap of convenient size. In any case, one stick usually is placed on end leaning against the bunch so that it can be readily located even in deep snow. If the wood is to be skidded and hauled soon after cutting during the months without snow, there is, of course, no need to flag the bunches.

In some parts of the South, pulpwood is bunched by penning, as it is called. A pen is simply a crib built by laying two bolts on the ground and building it up with layers of two bolts to form an open crib or pen. This is done for the purpose of assembling the bolts for easy loading on trucks as well as to obtain a measure of work done by the cutters.

Pens are usually either 5 or 6 ft in height, depending upon local custom. A certain number of pens of a given height is calculated to constitute a cord, the number depending upon the length of the bolts. A widely used factor for 4-ft bolts is 5 pens of 6 ft in height to the cord, with proportionate adjustment for bolts longer than 4 ft, as the length of the bolts does not affect the height of the pens.

Pens are often built to random heights when it would be necessary to carry the bolts too far to build all pens of uniform height. A measure of wood penned is obtained by recording the heights of the pens in inches and dividing by the height of a standard pen (72 in.) to obtain the number of full-sized pens.

Penning is not so widely used now as formerly since it is not always necessary to obtain a measure in the woods of the wood cut per crew. Payment now is often made on the basis of the truckload or carload or on truck or car measurements rather than on pen count. The work required to pen wood is often greater than the benefits resulting and whenever possible should be dispensed with in favor of other methods, unless there is some special advantage to be gained by penning. Penning was practiced

Fig. 63*A*. Peeled pulpwood bunched along a cutting strip in Minnesota.

and became an established custom throughout most of the South when pulpwood production was almost altogether a hand-labor operation. Penning has the advantage of making truck loading easier as trucks can be driven from pen to pen and the bolts readily placed on them.

More recently, mechanical methods of producing pulpwood have been introduced and widely adopted. As much hand labor as possible is eliminated, because of both labor shortages and increasing wage rates. Power saws, tractors, and mechanical sawing and loading machines are used to a considerable extent and when such equipment is used, penning is abandoned and it is quite likely that it will, in time, become an obsolete practice.

Pulpwood is also bunched in the South by throwing bolts together into small straw-piles to facilitate the loading of trucks as they are driven through the cutting areas. If this were not done, more time would be required in loading. It is essential for an economical operation to load trucks as quickly as possible so they can make a maximum number of trips per day. The fewer the number of cords produced per acre, the more important it is to bunch the bolts and so prevent truck delays.

Frequently no bunching of pulpwood is necessary. When the volume of wood cut per acre is heavy and the trees are felled and bucked judiciously, the wood lies so thickly on the ground that no advantage would be gained. As in the case of saw logs, pulpwood trees can be felled so that the wood will be well bunched after bucking instead of scattered. This can be done by felling the trees so their tops lie in piles or rows, with most of the bolts readily accessible around the piles of tops or between the rows of tops. Felling according to either plan has the added advantage of reducing the loss of bolts buried under tops and other brush. When trees are not carefully felled according to plan, this loss frequently occurs both because the

Fig. 63*B*. Pulpwood in North Carolina bunched in windrows by felling to accomplish this. Only a few bolts needed to be moved by hand to open the truck lane.

bolts actually are concealed by tops and also because loading crews deliberately overlook the bolts not convenient to load.

In the Lake states and the Northeast, pulpwood is frequently skidded in bunches to a loading point. The skidding is done by teams or tractors using sleds or arches and the bunches may be placed directly on the sled or over a binder chain on the ground. In the South and the Appalachian sections pulpwood is seldom skidded after it is bunched. Instead, it is loaded directly onto trucks, as the cutting areas are usually accessible enough so that this is possible. When pulpwood trees are skidded before they are cut to bolt lengths, no bunching is necessary as the bolts are already concentrated at a yard or brow and can be readily loaded.

Stave bolts, chemical wood, acid wood, mine props, and stulls are bunched in much the same way and for the same reasons as pulpwood. Stave bolts often are bunched in crib or pen stacks not only to assemble them for loading but to hasten drying. When they are open-piled, of course, the freer circulation of air dries the wood more quickly than would be the case in close-piled ricks.

Bunching is an important step in the harvesting operation because, when it is done, it constitutes the first step in the transportation of the products from the forest to the consuming mill or to market. It is often omitted, since on some operations it is unnecessary. When it is omitted, the products are usually skidded directly from the point of cutting, and skidding becomes the first step in their transportation. Skidding, however, is sometimes eliminated on harvesting operations when the products can be loaded in the forest after they have been bunched. When the cut per acre is heavy enough and the products are bunched by felling the trees together, both bunching and skidding may be eliminated when trucks can be driven throughout the cutting area.

CHAPTER 12

SKIDDING FOREST PRODUCTS

Because loading can be much more efficiently performed when quantities of forest products have been assembled along transportation routes, skidding, or yarding, as it often is called, is a usual step in harvesting operations. Skidding, or yarding, means the comparatively short-distance movement in the forest of single products or small bunches of products from the points where the trees were felled to the points of loading. In some instances skidding may be in two stages, so loading does not necessarily follow every movement of products by skidding.

Skidding is not always included in a harvesting operation as, under certain conditions, products often are loaded onto trucks or other vehicles directly from the point of cutting or after they have been bunched into small groups for loading in the forest and off roads. Some harvesting operations are so organized that skidding occurs as an intermediate step between felling and bucking; the trees being felled, limbed, and topped and then skidded full length to the loading point and there cut into logs, pulpwood, or other products.

Skidding obviously requires the application of power to the product or bunch of products to be moved to the transportation system. Power is used in skidding either by the principle of attaching the products to a mobile power unit and then both power unit and product moving to the destination, or by the principle of a stationary power unit pulling products to it by means of a cable attached to the products. The modern tractor with a winch combines both these principles in that it can be attached to a product and move with it or it can remain stationary and draw products to itself by means of the cable and winch.

SKIDDING BY GRAVITY

The force of gravity is used to a limited extent for skidding some forest products. This force can be used only when the products are to be skidded down a slope steep enough so they will move freely without the application of other power. In the southern Appalachian region small logs, pulpwood, and acid wood are skidded down steep slopes by rolling them or placing them by hand in an earthen trough down which they slide unassisted and accumulate at the bottom, where they are loaded on trucks. This method of skidding without the use of power is locally known as "ball-hooting."

Obviously, it can only be used in limited areas where the trees are cut on, or close to, steep slopes and when transportation routes are available in the valley.

Prepared chutes are also used in certain localities in the West for skidding logs down steep slopes into a river. Such chutes consist of logs placed end to end and firmly fabricated in the form of a trough, with the side logs high enough to prevent the logs moving down the chute from leaving it. The method is not extensively used because of the limited areas suitable for such skidding. Because of the speed with which logs reach the bottom, it is not advisable to use chutes unless they discharge into water. If they discharged on land, the logs would be badly damaged. In all cases log slides are only one stage of skidding, as the logs must be skidded to the upper end by tractors or teams. When logs are hauled by truck to the slide from a harvesting operation some distance away, the sliding operation is one stage in the transportation system and is not skidding.

Wooden chutes and portable steel chutes are used to some extent in the Northeast and in eastern Canada on pulpwood operations to skid bolts down slopes into rivers or lakes for rafting or driving. Stretching strong wires down a slope and several feet above the ground is another method of gravity skidding of pulpwood. A nail is then driven into each bolt near one end and bent to form a hook. When the bolt is hung on the wire by this hook it slides downhill, where it is taken from the wire by hand and the nail removed. This method also does not have very wide application, being practical only when the slope is too rugged for ground skidding methods and the bolts light enough for such handling.

SKIDDING WITH MOBILE POWER

The earliest as well as the most recent developments in applying power to products for skidding are by means of mobile power units. Oxen probably were the first power units used in skidding, other than hand labor. They were rather widely used, along with horses and mules, until woods-worthy tractors were designed. Now oxen have practically disappeared from harvesting operations because of their extreme slowness. Other animals also have now largely been replaced by tractors, which are the most modern application of mobile power units to skidding.

Animal Power for Skidding. Draft animals have been bred and trained as mobile power units for use by man from ancient times. The most common in woods work are horses, mules, and, to a very minor extent today, oxen. Certain other animals are used in other parts of the world, such as elephants in Burma, but the most important in America now are horses and mules. The use of animals in skidding has declined markedly with the comparatively recent introduction of mechanized methods. Nevertheless, on small or intermittent operations where, for example, the operator both farms and produces forest products, animals are still very useful.

The comparatively small investment required for a team and equipment as compared to a tractor or skidder is one distinct advantage animals have over mechanized units. However, the lower daily output per team and the constant care required by working draft animals is a factor in favor of a tractor when an output is required equivalent to that which several teams could put out. The use of teams requires more labor per unit of

FIG. 64. Type of barn used on logging operations in the South where severe weather does not require tightly constructed shelters to house logging animals. The barn shown above was used in North Carolina.

output than does the use of tractors, trucks, and other motor-powered equipment. Hence, when labor is scarce or wages high, the use of animals in the woods becomes more costly than tractors or trucks, and operations are converted to mechanized methods.

Where animals are used, barns must be provided and they must be fed and cared for whether the operation is going or idle. Barns range from small temporary lean-to shelters for mild weather to log or frame winter barns as warm and windproof as possible. As it is advisable for animals to start fresh on the job each morning, it is necessary to locate barns close to the working area, preferably within 1 mile. A minimum of travel distance to work saves energy and time of both men and teams and means a larger daily output.

The effective work life of a horse or mule is from 4 to 5 years in skidding, so there is a constant turnover in animals to keep them all at maximum efficiency. Considerable skill is required by team foremen in matching animals for teams. Correct pairing of horses and mules is essential for a good working team. Draft animals are matched with regard to strength, temperament, size, and gait. Correctly broken and handled, animals often become so accustomed to skidding work that they know what needs to be done almost as well as the driver and can be handled with a minimum of effort.

Horses can stand cold weather and ice and snow conditions much better than mules and, hence, are the only animals used in winter logging. Mules, however, can stand heat better than horses and so are much more widely used in the South for draft purposes than are horses. Mules also are less excitable than horses and seem to respond readily to negro drivers, who in turn seem to understand mules better than many white men do.

Mules are at a disadvantage on soft ground as compared to horses, as they have smaller feet and bog down more readily. Both horses and mules, however, may damage themselves severely by struggling in deep mud, and they must be handled carefully on soft ground. One very pronounced advantage that favored oxen in logging is that they have large cloven feet and are not excitable. Their great strength and susceptibility to being hitched into teams of from six to ten head make it possible for them to pull almost unbelievable loads through hip-deep quagmires.

An added advantage of ox teams was that once their useful life had been completed they could be fattened, after a fashion, and butchered for the camp crew, making for complete utilization of the investment. Also, they required but a minimum of attention and were watered and fed only at night. Harness for oxen was very inexpensive, consisting of a wooden yoke and chains, whereas mule and horse teams require rather costly leather collars and harness. However, being so slow as compared to other and more modern equipment, and since tractors now can do the same type of work, the use of ox teams has practically disappeared from commercial operations.

Horses must be fed and watered three times a day when working and are often permanently disabled if they become overheated. Hence, careful drivers are needed who will avoid overexerting the animals, while still obtaining maximum output from them. Mules are less susceptible to overheating and do not need as careful attention as horses. Disease occasionally will wipe out an entire barn of horses or mules or render them valueless as work animals, requiring disposition of them, fumigation of the barn and premises, and complete replacement of the teams. They also are subject to harness sores and attacks by flies, gnats, and screw worms, so almost constant veterinary care and attention are needed when teams are worked steadily in skidding logs and other forest products, as the work is hard and is performed often under difficult conditions for footing.

Skidding with animals requires that they exert strenuous effort and hence they cannot skid effectively for more than comparatively short distances. Dragging logs or trees over unimproved ground or pulling them partially raised on various types of wheels or skids requires greater effort than pulling a wagon or sled over a road. Because they return to the woods empty for the next load they obtain a degree of rest and relief from the strain of skidding effort. This makes possible short periods of strenuous exertion during the actual skidding. This is quite different from the sustained but not so exerting effort required in hauling loads on wagons or sleds. Maximum effective skidding distances for teams range up to 1,000 ft, depending upon the equipment used and the slope and ground conditions.

When skidding is only for short distances, teams often are more efficient than tractors and other mechanized methods. However, short-distance skidding is possible only when roads or railroad spurs are close together and the added cost of constructing transportation routes at intervals close enough to favor animal skidding usually is greater than the saving due to cheaper skidding with animals. Tractors of various sizes and kinds, therefore, are used much more extensively than teams on all larger harvesting operations except, perhaps, in the Northeast and in eastern Canada on pulpwood-producing operations, where teams still are rather extensively used.

Tractor Power for Skidding. For heavy skidding, the crawler type of tractor is widely used and is, in fact, the only type used for skidding in the big timber forests of the Pacific coast and throughout most of the West. They also are extensively used throughout the entire country in skidding saw logs, poles and piles, and even pulpwood in operations where the entire tree is skidded. Where ground conditions are not too rough or brushy, small timber products, such as pulpwood, can also be skidded effectively with wheeled tractors of the type used for agricultural purposes.

Crawler tractors used in harvesting timber are of the same type used on heavy earth-moving jobs or similar operations. The principal feature is the self-laying track consisting of metal plates fastened together with pins and grooved to fit the wheels which revolve along the inner surface. This principle of wheeled transportation gives the utmost mobility to the power unit, as the track prevents the wheels from coming into contact with the ground, and effects much firmer traction. Nevertheless, on extremely soft ground, tractors will bog down and literally dig themselves into the ground, so care must be used in operating them even though they are much more versatile than the usual types of wheeled vehicles.

Crawler tractors used in skidding range in size from 18 to 130 drawbar horse power, so practically every size of log cut in the forest can be moved by a tractor. The largest logs, of course, are found in the Pacific coast region, and in the forests there tractors are used to skid logs up to 6 or 8 ft or more in diameter and weighing up to 15 tons.

For such large loads and where the timber stands heavy per acre, temporary roads are constructed to facilitate the skidding. Some tractors on an operation may be worked full time at road construction. Also, it is not uncommon for a skidding tractor to be equipped with a bulldozer mounted on the front so it can also be used to prepare roads to the logs to be skidded.

Where the topography is comparatively level, few if any roads are constructed for skidding, as the tractors beat down saplings and brush and themselves make a road. Even with small to medium-sized timber, however, it is necessary on steep ground to prepare skidding roads or trails when the products must be skidded horizontally to the slope to keep the tractor and its load from upsetting downhill.

Tractors of the crawler type are powered with either gasoline or Diesel motors, the larger and heavier tractors usually being Diesel-powered and the smaller ones either gasoline or Diesel-powered. The great advantage of Diesel motors is that the fuel cost is much less than for gasoline motors of the same horsepower. The disadvantage formerly was that they were much heavier, although this fault has now been practically overcome. Wheeled agricultural-type tractors are powered by light-weight gasoline motors.

The usefulness of a tractor in skidding is greatly enhanced when it is equipped with a winch connected to the power take-off of the tractor. The winch consists of a sturdy drum, varying in size with the tractor and the purpose for which it is to be used, attached to the rear of the tractor. Wire rope or cable is wound on the winch, which can accommodate from 100 to 400 or 500 ft, depending upon the size of the cable and the size of the winch. By means of a clutch and gear lever, the driver can engage the winch, and as it winds the cable, a powerful pulling force can be exerted.

Most skidding tractors are equipped with winches; in fact, practically all heavy-duty tractors are winch equipped, as well as many small and medium-sized machines. A tractor with a winch can be operated to exert either a line pull or a drawbar pull, so logs can be drawn to the rear of the tractor for skidding without the necessity of driving by each one separately to attach it to the tractor. All skidding wheels and arches with "fair-lead" blocks require winch-equipped tractors, as the winch cable extends through the elevated block and is used in assembling loads and in elevating the forward ends of the logs during skidding. Winch-equipped tractors are sometimes anchored in place and used for high-lead skidding or ground skidding, in which case a double winch is mounted on the rear, one for the skidding line and one for the rehaul line.

When skidding roads must be built, as is common on many Western operations in mountains, the skidding tractors are also equipped with bulldozers for road construction. A bulldozer is a sturdy blade attached to the front of a tractor that can be raised or lowered and, hence, can be adjusted to the correct height in grading temporary skidding roads. The bulldozer can also be used for pushing over stumps, moving rocks, and moving logs, and adds much to the usefulness of a tractor when road work or other heavy pushing is required. On large-scale operations where a large amount of grading for truck roads and railroads is done, it is advisable to provide one or more tractors for such work only and not use skidding tractors for grading.

Maintenance of tractors is an important part of a tractor-skidding operation, and when several or more are used, adequate shop and repair facilities must be provided. They should be greased and oiled frequently; because of the heavy work they are steadily used for, treads must be rebushed and other moving parts replaced or repaired at frequent intervals. With adequate care and maintenance, tractors have a useful life of 4 to 5 years.

A recent improvement in motor-powered units, including tractors, is the torque drive. This is a means of transmitting the power from the motor to the driving mechanism through a set of nonmeshing gears in the form of disks enclosed in a housing filled with heavy oil. When the driving gears revolve from the power of the motor, the force is transmitted by the movement of the oil to the opposite gears connected with the driving or operating mechanism, causing them to revolve also. This results in smooth power in that there are no metal contacts between the driving mechanism and the source of power. When sudden jolts or jerks occur, as is common to skidding, slippage occurs in the oil and not where metal is in contact with metal. The torque drive, therefore, gives internal-combustion motors the aspect of steam power, as sudden jerks and overloads due to logs hanging on stumps or other obstructions are quickly absorbed by the torque drive, and breakage of skidding lines and chokers and gear stripping is greatly reduced.

ATTACHMENTS FOR APPLYING POWER TO PRODUCTS TO BE SKIDDED

Skidding power, whether mobile or stationary and regardless of the kind of skidding unit used, must be applied through attachments to the logs or products to be skidded. This is done by fastening cables or chains to the products and to the skidding unit or to other cables leading out from it. Cables are used for heavy products, and chains more commonly are used when small to medium-sized logs and similar products are to be skidded.

Where a chain is used, it often is attached to skidding tongs that are placed over one end of the log and bite into it when the chain is pulled. The other end of the chain is attached to the power unit by hooking it into a ring attachment. When tongs are not used, the chain with a hook on the end is used as a choker. With this method the chain is wrapped around one end of the log and the hook placed over one of the links so that as the pull comes on the chain, it binds and drags the log.

When the products are to be raised or partially raised off the ground with some of the skidding methods to be described, the chain choker, or, in some cases, tongs, are attached near the middle of the product or group of products to be raised and skidded. In the case of small products as many as 10 or 15 may be handled as a group, with the chain passing under and around or partially around them. Choker chains vary in length depending upon the size of the skidding load. They should be long enough to reach around the load and then reach the skidding unit, leaving 4 or 5 ft between the load and the unit. Ordinarily, from 10 to 15 ft of chain is sufficient for a choker; less if tongs are used.

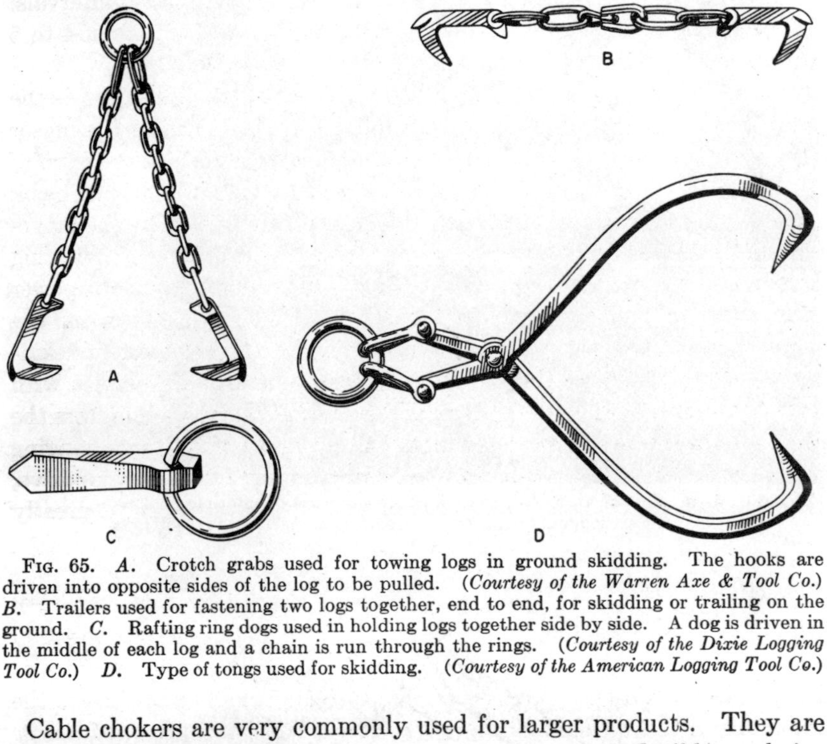

FIG. 65. *A*. Crotch grabs used for towing logs in ground skidding. The hooks are driven into opposite sides of the log to be pulled. (*Courtesy of the Warren Axe & Tool Co.*) *B*. Trailers used for fastening two logs together, end to end, for skidding or trailing on the ground. *C*. Rafting ring dogs used in holding logs together side by side. A dog is driven in the middle of each log and a chain is run through the rings. (*Courtesy of the Dixie Logging Tool Co.*) *D*. Type of tongs used for skidding. (*Courtesy of the American Logging Tool Co.*)

Cable chokers are very commonly used for larger products. They are not so adaptable to smaller products as they are not as flexible as chains and so are seldom used with team skidding. Tongs are seldom if ever used with cables because with large heavy logs the tongs would split out of the log and because tongs cannot be effectively used to fasten to logs very much larger in diameter than the distance between the points of the open tongs.

Cable chokers are fitted with hooks or with special fasteners at each end, one for attaching to the tractor and the other to hook over the cable after it has been passed around the log. One widely used type of cable choker hook, known as the Barden hook, consists of a hook with socket that slides along the cable. The choker end of the cable has an enlarged lug fastened over it so that when the cable has been passed around a log, the end can

FIG. 66. Rear view of an arch load of logs showing use of multiple chokers, on a Pacific Northwest logging operations. (*Courtesy of the Young Iron Works.*)

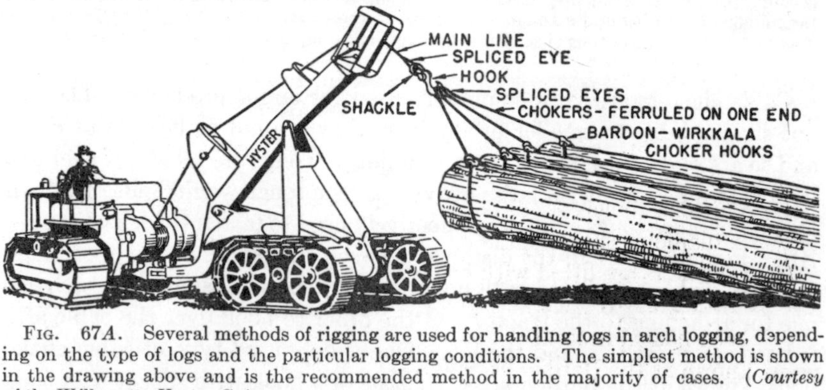

FIG. 67*A*. Several methods of rigging are used for handling logs in arch logging, depending on the type of logs and the particular logging conditions. The simplest method is shown in the drawing above and is the recommended method in the majority of cases. (*Courtesy of the Willamette Hyster Co.*)

be placed in the socket so it will not pull out. This makes a firm attachment but still permits the socket to slide along the cable so a tight noose is formed when the pull comes on the choker. This type of choker hook has the added advantage of permitting a tight noose to be formed without kinking the cable, which occurs when a simple hook is used. When the end of the cable is to be attached to a tractor drawbar, a similar end lug is often

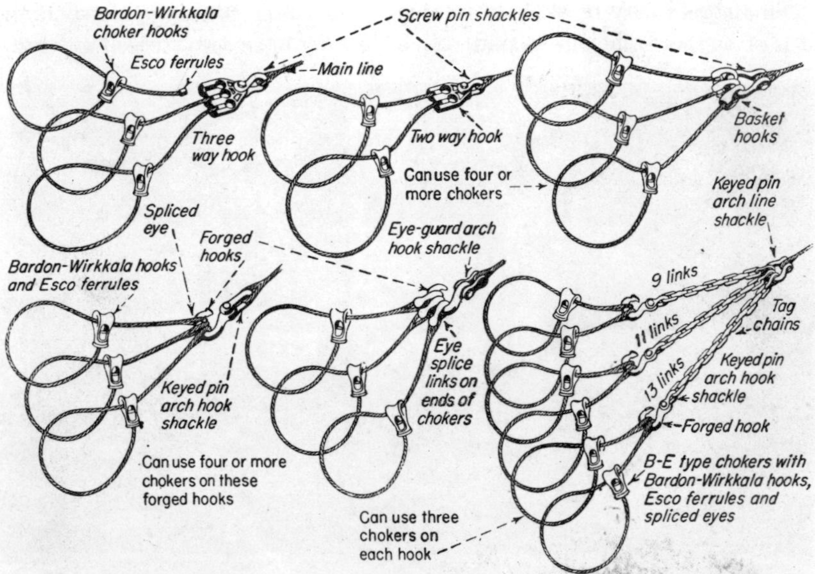

FIG. 67*B*. These drawings show several practical methods for making up rigging for arch logging. (*Courtesy of the Willamette Hyster Co.*)

used that fits into a slot making for ready attaching and unattaching. When the choker is to be attached to a cable skidding line, either tractor line or stationary skidder line, it is equipped with a hook or snap that can be fastened to a ring or with a ring that can be placed on a hook.

SKIDDING WITH TEAMS AND TRACTORS

Snaking or Ground Skidding. Teams and tractors are effectively used to skid small products by dragging them on the ground. Even larger products can be skidded in this way for short distances if the skidding unit is powerful enough to move the weight of the product and overcome the resistance of the ground. A downhill direction makes ground skidding possible for longer distances and for heavier loads than on level terrain. Up-slope skidding obviously has the reverse effect. Skidding logs, trees, poles, and other products by dragging them on the ground is called "snaking," or ground skidding.

Snaking is an efficient method for skidding distances so short that it would be too cumbersome to use some form of cart or arch or other auxiliary device. When skidding extends back from a road or railroad, some products to be skidded are usually located very near the yard or loading point. These can be snaked, while, for the more distant products, some skidding attachment can be more effectively used.

Small trees may be skidded full length and later cut into pulpwood and it is often more efficient to skid even as few as one or two trees at a "turn,"

FIG. 68. Ground skidding or snaking logs with a tractor in Arkansas. Tongs are used to attach the log shown here to the tractor which is not equipped with a winch. (*Courtesy of the Allis-Chalmers Mfg. Co.*)

or trip, to eliminate the cost of bunching larger loads. Tractors and teams have a surplus of power in such circumstances, but no effort is lost by skidding less than maximum loads when the skidding distance is so short that it would be inefficient to take time to hook on a capacity load.

Snaking is the least efficient method of skidding in the effective use of power. The product is fastened to the tractor drawbar or the team's whiffletree by means of a chain. As the log is pulled it gouges into the ground, requiring added effort to overcome the resistance. To reduce the resistance in ground skidding, logs are sometimes "nosed" by chopping off the sharp edge of the forward end with an ax. This is done more often in snaking or trailing logs down steep slopes for distances of a ¼ mile or more than for short-distance snaking.

Trailing is a method of ground skidding whereby several logs, fastened end to end with short chains and dogs driven into each log, are dragged along an earthen trough. The dogs used in fastening the logs together are simply short chains with wedge-shaped spikes at either end. One is driven into each log, thus fastening the two logs together endwise, this being a more common skidding method in the southern Appalachian region than elsewhere. Generally the gradient is down so the resistance is reduced, although there may be slight rises or level stretches at some points. Teams or single horses or mules are commonly used, as only a narrow trail is needed with but little or no improvement. Tractors may be used, but if the way

FIG. 69. Ground skidding or snaking hardwood logs in a Southern swamp. The muddy ground condition is typical of Southern river bottoms in wet weather. (*Courtesy of the Caterpillar Tractor Co.*)

is wide enough for a tractor, it is usually more economical to use one of the several types of tractor skidding attachments. As some portions of such trails are so steep that logs will slide by gravity, it is necessary to have a quick unhooking method to disconnect the single animal or team if the trail of logs starts to overrun it.

In snaking, the logs usually are not bunched as the distance to the yard is ordinarily so short that the added expense required to assemble and pick up a large load is not justified. Instead, the tractor picks up each log as it passes until all the choker sockets have been filled or all the logs have been hooked. With team-snaking the turn is so short that nothing is gained by attempting to snake more than one or possibly two logs at a time. A tractor-skidding crew usually consists of a tractor driver, two choker setters per tractor, and an unhooker at the landing. The unhooker disengages the choker cables and places them on the tractor to be returned to the choker setters. One man at the landing can unhook for several tractors. With team-snaking the driver usually does the choking or tonging and unhooks at the yard or landing to which the products are skidded.

The advantage of ground skidding with tongs is that friction is reduced measurably as compared to that of a log with a choker around it because the tongs do not extend under the log as does a choker. The disadvantage is that the tongs occasionally come loose and release the log if the pull is relaxed momentarily or if the skidding line is jerked.

FIG. 70. Skidding logs with a pan on a snow road in northern Wisconsin. (*Courtesy of the Caterpillar Tractor Co.*)

Skidding Pans. To prevent the ends of logs being snaked to a landing from gouging into the ground, skidding pans and other devices have been developed. A skidding pan consists of a sheet-metal plate curved upward at the front to permit it to slide over the ground and low obstructions. It is chained to the rear of the tractor.

Logs or tree lengths are fastened to the tractor as in the case of snaking, but the forward ends rest on the pan, which prevents them from gouging into the ground. Larger loads, therefore, can be ground-skidded for longer distances with pans than without, but usually it is not economical to pan-skid more than 800 to 1,000 ft. Skidding pans are not used with teams as they are too heavy and would tire the teams unduly. Another device, not commonly used, is a sheet-metal cone that fits over the end of a log. A hole in the point of the cone permits the cable or chain to pass through it. The purpose is to prevent the log from gouging into the ground and to prevent it from hanging up on stumps and other obstructions.

Equipment to Elevate Front Ends of Logs. In ground skidding, with or without pans or cones, the logs are dragged without any elevation of the log and the choker, chain, or cable is attached directly to the log. Thus the pull is on the log. Other skidding devices have been designed to facili-

FIG. 71. The type of pan used in skidding is shown above on a Western operation. (*Courtesy of the International Harvester Tractor Co.*)

tate skidding logs and similar products which elevate the front ends of the logs and, in some, even the entire logs.

These devices may be classified according to the method of elevating the log and the extent to which the log is elevated, as follows:

1. Elevation of the front end of a log *over* an axle or sled bunk.
2. Elevation of a log *under* an axle.
3. Elevation of the entire log over two axles or sled bunks.

Elevation of a Log over an Axle or Sled Bunk. The simplest device of this type is a "lizard." It is a Y-shaped section of a forked tree trimmed and smoothed with an ax, and with the stem of the Y hewed in a rounded fashion so it will slide readily over the ground when pulled forward. It is used where the ground is slippery or smooth as in river bottoms and is used only with teams. The forward end of a log is rolled onto the "lizard" with a peavey.

The "bummer" is another device used in team logging for elevating the front end of a log. It consists of two low wheels made of wood or steel, with treads about 12 in. wide, mounted on a heavy axle, with a wooden cap

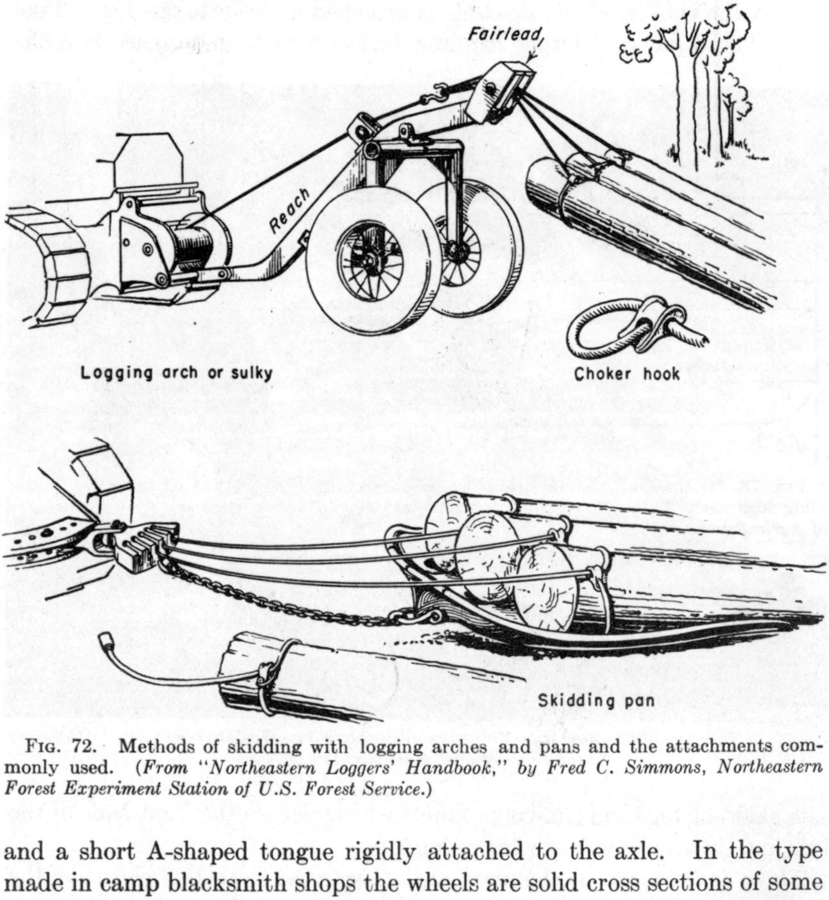

FIG. 72. Methods of skidding with logging arches and pans and the attachments commonly used. (*From "Northeastern Loggers' Handbook," by Fred C. Simmons, Northeastern Forest Experiment Station of U.S. Forest Service.*)

and a short A-shaped tongue rigidly attached to the axle. In the type made in camp blacksmith shops the wheels are solid cross sections of some cross-grained wood such as black gum. The axle is about 2½ ft long and a bunk, hollowed slightly on the upper side to accommodate a log, is mounted on the top of the axle. A steel plate with a slot in it is placed across the top side of the tongue to hold a chain attached to a pair of tongs.

In use, the bummer is backed up against the side of a log near one end and the axle is raised to a vertical position. The log is tonged on its far side near the end and the chain is placed in the slot in the steel plate across the tongue. The team then pulls on a chain attached to the tongue which

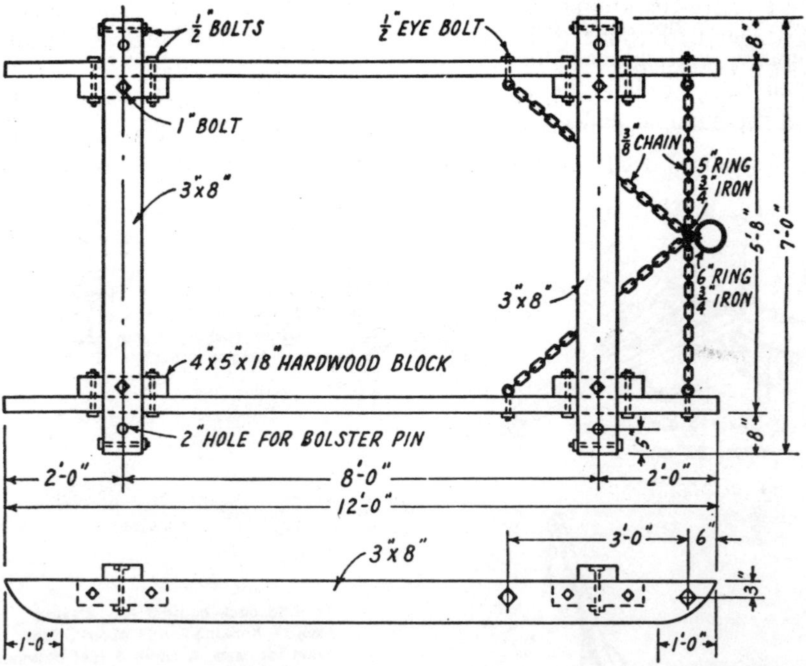

FIG. 73. A working drawing of a "log scoot" used on bare ground or snow for assembling logs short distances. (*From Farmers Bulletin* 1907, *Forest Service, U.S. Department of Agriculture.*)

FIG. 73A. A sledge sometimes used for skidding pulpwood in the South.

pulls it to a horizontal position, which in turn pulls the log up on it as the bummer is straightened up. Then the team pulls the bummer to the yard while the tongs hold the log on it. To unload it the tong is knocked loose and the bummer pulled out from beneath the log to one side.

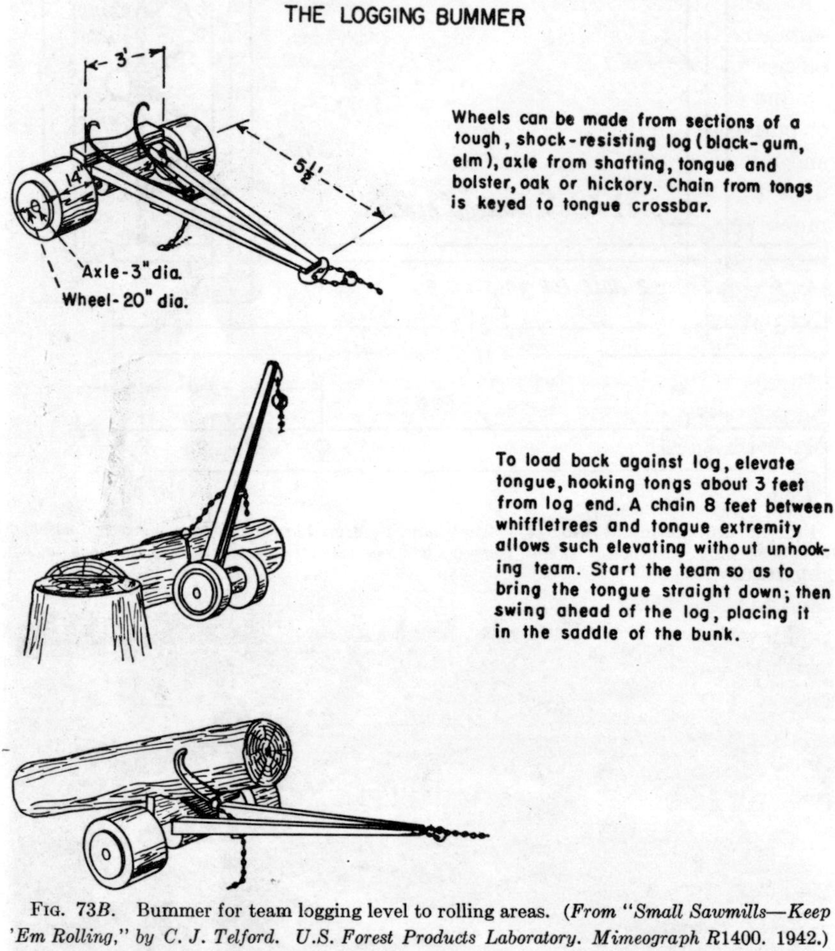

FIG. 73*B*. Bummer for team logging level to rolling areas. (*From "Small Sawmills—Keep 'Em Rolling," by C. J. Telford. U.S. Forest Products Laboratory. Mimeograph R*1400. 1942.)

Bummers are effective for skidding distances up to 500 to 700 ft if the ground is not so soft that the wheels sink in. It is not effective for small logs, as only one can be skidded at a time. Large logs, over 30 in. in diameter, are too heavy and too wide for the effective use of bummers. Bummers have been and still are more widely used in the South than other sections because of the generally level topography and because the timber is of a size suitable for bummer skidding. Bummers similar in principle but much larger and of heavier construction are used for skidding large logs with tractors. This type is constructed of steel and some of them have crawler-tread wheels to support heavy loads on soft ground.

A sledge, called a "go-devil," is commonly used in the North instead of a wheeled bummer for raising the front end of a log or logs for skidding with teams, usually in winter but in some cases on muddy ground in the summer. The go-devil is simply a low-bunk, two-runner sled that accomplishes the same object as the bummer but is adapted to snow and icy ground conditions. The go-devil is wider than a bummer and several logs or pulpwood bolts can be placed on it. Some types have two side stakes and a load of up to half a cord can be placed on the bunk. When this is done the load is fastened on by means of a binder chain functioning in much the same way as a choker. When only one to three logs are carried and the load is not chained on, the logs are held on the bunk by means of a binder chain or by sharp upright spikes driven through the bunk. The weight of a log or bolt forces the spikes far enough into the wood to prevent the dragging rear ends from pulling the logs off.

Logs or bolts are loaded on go-devils with hand tools or they are rolled on by teams, and they are unloaded the same way. With all types of skidding equipment used to raise the front end of a log or other product off the ground, the axle and wheels or bunk with sled runners must be low enough so the product can be readily rolled or placed upon it. The only purpose for using such equipment is to reduce the resistance due to friction by transferring a part of the weight of the products to wheels or runners and to prevent the front ends from gouging into the ground.

Elevation of a Log under an Axle. Most equipment designed for use with mobile skidding units, such as teams and tractors, is based on the principle of elevating one end of one or more logs beneath an axle, or in some types, the entire log. Such devices are always wheeled or equipped with crawler treads, not being adaptable to sled runners. This type of skidding equipment can be used for hauling longer distances than is practical with bummers and other overaxle devices, and larger loads can be carried.

Straight-axle Logging Wheels. When the axle is straight it is necessary for the wheels to be very large, as the axle is only half the height of the wheel diameter. Thus, for a 25-in. log, with an allowance for 3 to 4 in. sag after it is chained below the axle and 1 ft clearance above ground level, the wheels would have to be at least 6½ ft in diameter. When larger logs or trees are to be skidded with wheels, their diameters must be proportionately larger. When "big wheels," as this equipment is called, were used in large timber in the Lake states, South, and West, it was not uncommon to find carts with wheels up to 10 and 12 ft in diameter.

Big-wheeled carts with ox teams were used on some of the earliest logging operations, as this method provides for easy loading and unloading and gives good utilization of the power available either in teams or tractors. The most modern equipment, while in no way resembling big wheels as such, still embodies the principle of elevating logs, either wholly or in part, beneath an axle.

The simplest form of big wheels consists of a single axle with a pair of large wide-tired wheels and a short tongue fastened rigidly to the axle. The top side of the axle is built up with a wooden block to form a bunk with a

FIG. 74A. Old-time big wheels are still occasionally used to skid long logs with teams on Southern logging operations where fairly level topography prevails. South Carolina.

height of 8 to 10 in. above the axle, and provision is made for fastening one end of a chain over either a hook or a self-releasing catch. The other end of the chain, which is from 6 to 10 ft long, depending upon the size of the wheels, is permanently fastened to the axle or is looped over it and fastened to the base of the tongue.

The wheels are spotted over and near the end of the log or bunch of logs to be skidded by the team (tractors not ordinarily being used with this type of wheels) so that the wheels straddle the load. The chain from the team to the tongue is then loosened and the tongue is elevated to a perpendicular position. The loose end of the chain is then passed beneath the load and fastened to the axle. On some types, one or more chains with tongs are used to fasten onto the logs instead of passing the chain beneath them. The wheels are then blocked and, as the team pulls forward, the logs are raised off the ground by the lever action exerted by the built-up bunk on top of the axle. The distance the load is raised is only the height of the bunk above the axle, so a great deal of clearance is not obtained by this method and the load is limited, usually to one or two logs.

This general type of big-wheel cart was formerly widely used for skidding long logs in the Southern pine region, where it was customary to cut longleaf pine timber into tree-length logs up to 60 or more ft long. The cart was backed over the log with the wheels parallel to it and slightly ahead of the center of gravity so that the rear end of the log would tend to touch the ground lightly. The log was fastened with a chain or tongs, and when the tongue was pulled forward it was chained to the log to hold it horizontal. The team was then hitched to the front end of the log, which was then pulled to the yard or landing.

FIG. 74B. A pair of straight-axle logging wheels designed for use with teams. The built-up bunk is all that provides leverage for elevating logs under the axle. Georgia.

This principle of raising the load off the ground by the lever action of the top of the axle is the same used in all types of wheels not using the "fair-lead" method, to be discussed later. The slip-tongue cart applies this principle somewhat differently, in that the pair of big wheels with a high axle is equipped with a straight tongue from 15 to 25 ft long. The tongue consists of a sliding center piece that extends to the front and is attached to the team. It is so designed that when the team backs up, the tongue slips to the rear and a rear section jackknifes, permitting the axle to rotate to the rear while the long tongue remains horizontal. The chain is slipped beneath the log or logs and fastened to the bunk over the axle, or tongs on chains attached to the bunk may be used. When the wheels are blocked the team pulls ahead and the axle slips forward until the jackknifed portion reaches horizontal, where it automatically locks in place, holding the load off the ground. This type of cart formerly was widely used for skidding long logs from 40 to 60 ft in length. When the log was raised it was clear of the ground for its entire length and the balance was slightly forward of the axle. After the log was raised it was chained to the forward part of the tongue and the harness of the team kept the front end off the ground.

Arched-axle Logging Wheels. Because the capacity of straight-axle skidding wheels is limited by the height of the axle off the ground, wheels with arched axles have been designed and are now very widely used. The arched axle permits larger logs or bunches of logs to be carried, or, for the same size of load, smaller wheels can be used. Several types of arched wheels are used for both team and tractor skidding and the load is picked up in much the same manner as previously described.

A four-wheeled mule cart is common in the comparatively small-sized timber of the Atlantic coastal pine forests. The front and rear axles are readily disconnected and the rear pair used as a single set, if desired, with an arched axle and with the tongue leading from the top of the arch. Leverage for raising the log or logs off the ground is obtained without building up a bunk, as is necessary with a straight axle.

FIG. 75A. The four-wheeled mule cart still finds wide use in the Atlantic Coast states where it has long been a common method of skidding. The cart shown above is being loaded. As the tongue is pulled to a horizontal position the bunch of logs is elevated beneath the arched rear axle.

The arched axle is tipped backward by raising the tongue to a vertical position. The chain fastened to the arch is then placed beneath the log or bunch of logs, as the case may be, or tongs may be used if only one or two logs are skidded at a time. Pulling forward lowers the tongue to the horizontal and the arch to a vertical position, which lifts the load off the ground. The arch gives a greater lift because it can be higher above the hub level than a straight axle, and because of the curve of the arch, this type of wheel is well adapted to skidding small logs in bunches of six or eight at one time. When several logs are bunched, too many tongs would be needed, so the chain-binder method is always used.

When the rear arch of the mule cart is loaded, the tongue is pulled forward and hooked to the center of the front axle, which usually also has an arched axle. The load can be raised completely off the ground when the front axle is used by picking up the load slightly behind the center of gravity with the rear wheels, so that there is a tendency for the front end to rest on the ground. By chaining the load to the tongue, this down pressure is borne by the front wheels.

All arches of this type are so designed with respect to the tongue that when the tongue is horizontal in normal skidding position, the arch is tilted slightly forward. This gives a downward pressure to the tongue, which is held off the ground either by a front pair of wheels, as with the mule cart, or by the team harness, or by the drawbar fastening of the tractor.

Heavier and larger arched wheels of this type are built for use with tractors not equipped with winches. Most types of tractor skidding wheels, however, are so designed that it is not necessary to unfasten the tongue from the drawbar of the tractor to load the arch. The tongue, instead of being rigid, as in team types, is hinged. To tip the arch to the rear, a pin holding the tongue in a rigid horizontal position is removed or a catch is released. As the tractor backs, the tongue jackknifes upward and the arch tips over the load, which is then chained to the arch. Pulling forward, the hinged tongue comes to the horizontal position and locks in place. The tractor then proceeds with its load to the landing. This type of arch is known as a "breakdown" because of the hinged tongue, which "breaks" to permit the axle to rotate to the rear for loading without being detached from the tractor.

FIG. 75B. The mule cart loaded. Note that load is clear of the ground.

The Fair-lead Arch. An entirely different and much more effective method of loading arches is possible when the skidding tractor is equipped with a winch on its rear. The winch consists of a drum with 100 ft or so of cable wound on it, powered by a power take-off from the tractor motor. The cable can be unwound and fastened to logs or other products. When the power is applied the drum revolves, winding the cable on it and drawing the log or bunch to the tractor.

The most efficient use of a winch on a tractor in skidding is with a fair-lead arch. This consists of an arch with an elevated roller over which the cable is passed before it is fastened to the logs. Side rollers are also usually present, as the line pulls from both sides in assembling a load. Skidding arches are mounted on metal or rubber-tired wheels or, in the heavier types, on crawler treads. The arch is fastened to the drawbar of the tractor and this is where the pulling force in skidding is applied. The arch varies from the simple type similar to arched wheels to one designed with a projecting beam supported by the arch. Fair-lead arches are sometimes called "sulkies," especially when designed for small timber.

FIG. 76*A*. For light tractors, such as shown above, a fair-lead block and wheels operate efficiently on firm ground. This type of equipment is common throughout the Carolina Piedmont and coastal plain regions in small timber.

In all types the roller or fair-lead is supported by the arched axle and this gives an elevated point over which the winch cable or line can be led out. It permits drawing the logs or other products up under the arch with their front ends off the ground so that as the tractor pulls the arch to the yard the arch bears most of the weight of the load and the logs are elevated in front to offer a minimum of resistance. When a winch and

FIG. 76*B*. A logging arch, commonly called a "break-down," is often used with a tractor not equipped with a winch. The tongue remains attached to the tractor while the hinge permits the arched axle to be lowered to pick up a bunch of logs as with a mule cart. By eliminating the need of unhitching the tongue, however, greater speed is achieved. (*Courtesy of the Lindsay Wagon Company.*)

FIG. 77A. A pneumatic-tired fair-lead arch skidding two bunches of Southern pine logs each choked as a bunch. South Carolina.

FIG. 77B. A crawler tractor and crawler arch skidding tree-length logs in South Carolina. The tree lengths are bucked at a landing before being trucked to the mill.

FIG. 78. Skidding a Southern pine log with a fair-lead arch, or sulky, with metal rim wheels. This equipment is adapted for light- to medium-sized timber. (*Courtesy of the Pacific Car & Foundry Co.*)

FIG. 79. Crawler tractor and arch skidding a sugar pine log in California. (*Courtesy of the Allis-Chalmers Mfg. Co.*)

FIG. 80. Skidding heavy loads down steep grades in Idaho. Note the steel shield that protects the driver against flying cables and debris. (*Courtesy of the Pacific Car & Foundry Co.*)

FIG. 81. Skidding tree-length Southern pine logs in South Carolina with a pneumatic tired arch, or sulky, well adapted to level topography. (*Courtesy of the Willamette Hyster Co.*)

FIG. 82. Skidding heavy loads up steep grades in California. When short grades are too steep to be negotiated, the load is dropped while the tractor and arch move forward, and then the load is drawn to the arch by the winch. (*Courtesy of the Willamette Hyster Co.*)

FIG. 83. Skidding can be successfully done in winter with tractors and sulkys as shown in the above illustration of skidding in Ontario. (*Courtesy of the Willamette Hyster Co.*)

fair-lead arch are used it is not necessary to tip the axle backwards to obtain leverage for loading the arch as is necessary when a winch is not used. Furthermore, it is not necessary to back the arch over the log or logs to be loaded as the winch line can be taken to the log or logs and after the chokers are attached they are then pulled up under the arch.

FIG. 84. A bird's-eye view of a heavy-duty tractor and fair-lead arch skidding ponderosa pine logs in California. (*Courtesy of the Caterpillar Tractor Co.*)

Fair-lead arches for skidding small timber are lightly constructed and consist of wheels with an arched axle on which is mounted the fair-lead block. For medium-sized timber, arches with heavier wheels and pneumatic tires and a projecting beam, at the end of which is the lead block for the cable, are widely used. For skidding very large West coast logs such as redwood, sugar pine, and fir logs, the arches are of extra-heavy construction. Tubular steel framework, crawler treads, and extra bracing are used and the resemblance to an arch has practically disappeared. The principle of raising one end of logs under an axle is, however, maintained.

A variation of this principle has recently been devised by the U.S. Forest Service for skidding with tractors. On top of the tractor itself is mounted a projecting block, elevated sufficiently to permit pulling the logs with a winch line up and on top of the rear of the tractor, which is protected with a steel-plate shield. It is very maneuverable, but the tractor itself must bear the weight of the load. This machine, which has been called a "tomcat," is still in the process of being perfected.

FIG. 85. The "Tomcat" logging tractor, illustrated above, was designed and constructed by the U.S. Forest Service in Portland, Ore. The feature of this experimental machine is that the fair-lead arch is built over the tractor and is essentially a part of it and not an attachment towed behind. The "Tomcat" has proven particularly advantageous on steep slopes and muddy ground. (*Courtesy of the U.S. Forest Service.*)

Elevation of Log over Two Axles or Bunks. While not a widely used method, logs, poles, pulpwood, and other products are also loaded over two axles or bunks for skidding. Generally, they are not loaded in this manner until they are to be hauled to destination, but under some situations hauling with log wagons or sleds is done as a skidding operation. Log wagons were commonly used throughout the Southern pine and hardwood bottom areas and still are used to a limited extent. Log wagons originally were designed for team hauling but have also been designed for tractor hauling.

Both four-wheel and eight-wheel wagons are used with teams. The principal feature of the four-wheel wagon is sturdy construction and wide wheel treads, the general design being the same as that of all wagons except that all four wheels are of the same size. The stretcher is adjustable to suit different lengths. No floor boards are used, as logs are loaded directly onto the bunks over the axles.

A load consists of from 500 to 600 ft log scale and it is held on the bunks by low stakes or by adjustable chock blocks on the bunks, and by a binder chain around the logs. Four-wheeled wagons are used for skidding small-

Fig. 86. Where skidding roads must be constructed, as on many Western operations, skidding tractors are equipped with bulldozers, as shown above. (*Courtesy of the Caterpillar Tractor Co.*)

to medium-sized timber when, for reasons peculiar to the operation, it is not feasible to use wheels or arches, or the operator prefers this method. They cannot be used on soft ground and where the slope is steep, as an adverse grade would overexert the teams and a down slope would cause the wagons to run over the teams.

In Southern hardwood bottom lands for large logs and soft ground, the eight-wheel wagon is preferable. It consists of a double set of axles at each end with small, wide-tread wheels from 24 to 32 in. in diameter. All wheels are of the same size and both axles are pivoted for easy maneuvering. The bunks of eight-wheel wagons project over the wheels, making a wider bed than usual, and, as it is lower than the four-wheel type, it is easier to load and larger logs can be more readily handled because the weight is distributed over eight wheels instead of four.

FIG. 87. Skidding hardwood logs from a Louisiana swamp with a crawler-tread wagon. (*Courtesy of the Athey Truss Wheel Co.*)

FIG. 88. Skidding long logs of longleaf pine in Alabama on an eight-wheel log wagon with a tractor.

Wagons pulled by tractors are either the eight-wheel type or four-wheeled, but with all wheels of the same size. They are essentially trailers that are loaded in the woods and pulled to the concentrating point by crawler-type tractors. Wheels often are of steel and rims may be up to 6 in. wide, or they may be equipped with pneumatic tires. Tractor wagons frequently are used in hardwood bottom lands in the South where short logs are cut and where the skidding distance is a half mile or more. A

FIG. 89. Skidding logs in a South Carolina river swamp with an eight-wheel log wagon and a "4-up" mule team. The logs shown above are being hauled to a logging railroad.

sufficient volume of short logs, under such circumstances, could not be handled with an arch and tractor to make a long distance skidding haul as economically as a tractor-wagon skidding haul. In the heavier types designed for soft ground, crawler tractor treads are used instead of wheels.

Small-sized four-runner sleds sometimes are used for skidding on winter logging and pulpwood operations in the Lake states, the Northeast, and eastern Canada. They are more commonly used for skidding pulpwood than logs, as so many more pieces must be skidded to utilize the tractive capacity of even one horse. Skidding sleds are similar to the larger hauling sleds but are operated singly and hauled by either one horse or a team. They have upright stakes set in sockets at the ends of each bunk and the pulpwood bolts are loaded, usually, by hand. This is not difficult as the bolts, even when 8 ft long, are of light-weight species such as spruce and balsam, as a rule, and often have seasoned all summer after having been peeled during the spring.

Skidding sleds do not operate over improved snow or iced roads but over trails broken by the first trip of the sled. Since only a few trips are made over any one route, except where it approaches the landing, the cost of improving, icing, and maintaining skidding roads would be prohibitive.

SKIDDING WITH STATIONARY POWER UNITS AND CABLES

Skidding logs by means of stationary power units instead of mobile power units also is an important and widely used method. There are many variations in the method of applying power from stationary units to logs in the woods to bring them to the transportation system, but all apply power through cables extending into the forest from the power unit. The logs are attached to the cable, which is then pulled to the unit by winding the cable on a drum.

This method of skidding was developed along with railroad logging in the Lake states and pullboat logging in Southern cypress swamps, as has previously been noted. Before the advent of tractors in logging, cable skidding was much more widely used than it is today. Even so, cable skidding is common on the Pacific coast where heavy stands of large-sized timber in rugged topography present problems in logging best met by some application of cable skidding.

Because of the heavy investment in cable-skidding equipment and the cost of rigging cables to trees or spars, and changing them so that all parts of the area adjacent to a skidder set will be reached, cable skidding is not economical except when heavy and fairly uniform stands of timber are to be harvested. When partial cutting is practiced or where the stand per acre is light, tractor skidding is more economical because roads need to be built and tractors run only to those points from which logs are to be skidded.

Cable skidders of any type are not adaptable to partial cutting for the purpose of reserving a growing stand, as the high-speed movement of cables and logs knocks down, breaks or seriously damages all trees not cut. They are well adapted only to clear-cutting the area to be skidded. For this reason cable skidders cannot be recommended for any cutting except clear-cutting. This does not mean, however, that they cannot safely be used in logging properties under sustained-yield management because systems have been devised, as will be shown in a later chapter, that provide for leaving areas uncut to reseed those from which the timber has been cut and skidded. Briefly, this is done by means of staggered settings, *i.e.*, skidding alternate settings along the railroad or truck route, or by leaving untouched patches of young timber or small patches of mature timber on higher elevations, from which seed will be wind-sown over the skidded areas.

While formerly extensively used in the South, cable skidders have practically disappeared from this region except in swamps and some of the eastern mountain areas where stands sufficiently heavy to justify cable skidders still exist and the topography is too wet or rugged to use other methods. The growth of conservative logging, which means saving immature trees and seed trees, has eliminated power skidders from nearly all upland Southern logging operations. Tractor skidding can be done with very little or no damage to uncut trees, so this method has been generally substituted for cable skidders. Another and perhaps more compelling reason is that the cut per acre is so much less than formerly, when the virgin stands were being removed, that it is not economical to use cable skidders.

The smaller cut per acre is due to the fact that light second-growth stands have replaced the heavy virgin stands and that partial cutting at periodic intervals to maintain a high growth rate and continuous productivity has been widely adopted. Another reason for the decline of cable skidding in the South is that much of the saw timber cut now comes from small scattered tracts that can be logged only with trucks.

With the passing of cable skidders, or because of it, logging railroads in the South have been largely replaced by truck transportation. When the heavy stands of old-growth timber in the Pacific Northwest have been cut and second-growth stands are being logged, cable skidders undoubtedly will, to a considerable extent, be discarded there also in favor of tractors.

There are three generalized methods of cable skidding—ground-line skidding, high-lead skidding, and sky-line skidding. In ground skidding, the logs are dragged on the ground to the power unit. In high-lead skidding, the cable is elevated at the pulling end so the logs being dragged in are partially elevated in front, which reduces hanging against stumps. In sky-line skidding, the pulling cable is carried through a trolley block running on a taut cable stretched between two trees or spars so that the logs can, if occasion requires, be lifted entirely off the ground. A number of different ways of rigging the cables, blocks, and auxiliary equipment necessary for each of the three methods have been devised, but each accomplishes skidding by either the ground, high-lead, or sky-line principle.

Ground-line Skidding. The least complicated method of cable skidding is by the ground-line method. As mentioned previously, this involves only dragging the logs to the power unit by pulling them on the ground. The simplest ground-line skidders consist only of a power unit with a single drum on which the cable is wound to draw in the log or logs fastened to the other end. Once the cable has been wound on the drum and the log or logs unhooked, it is necessary to pull the cable back into the woods by hand, team, or tractor power. This is a slow and tedious job and ground-line skidders of this type are not widely used except on small operations using old-type skidders or for skidding small amounts of timber from swamps or other places inaccessible with other equipment.

Much faster skidding is obtained by using two drums with a power unit instead of one, the second for rehauling the main cable back into the woods.

This is accomplished by running the rehaul cable, which is much lighter and smaller than the main cable, from the drum through two or three blocks attached to stumps that lead it out to the far end of the area to be skidded. The end of the rehaul line is attached to the skidding end of the main line, and after the main line has been drawn in with a load of logs, the line can be drawn rapidly back out into the skidding area by winding the rehaul line on the rehaul drum and paying out the main line from its drum. Likewise, when the main line is drawn in, the rehaul line is permitted to unwind from its drum.

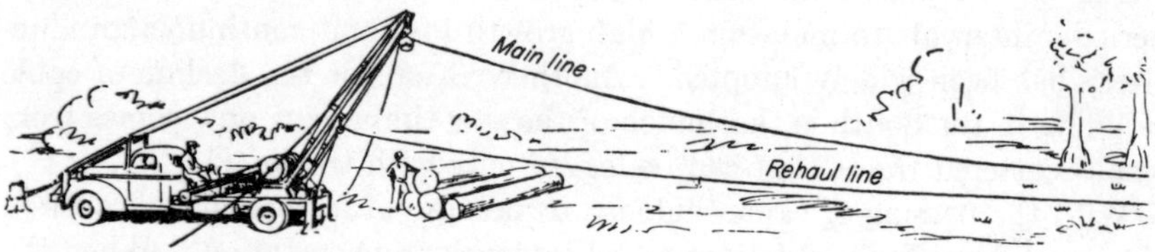

FIG. 90. A modern method of ground-lead cable skidding is illustrated by the above does not exert any lifting as logs are dragged to the skidder. (*Courtesy*

In operation, rehaul ground skidding covers only the area that can be reached by the slack of the line between the skidder and the farthest block through which the rehaul line passes. To reach other areas tributary to the same set, the rehaul blocks must be moved so the main skidding line will stretch across another segment of the circle of which the skidder is the center. In practically all cable skidding from power units which are stationary while actually skidding, the skidding line is moved around a circle by successive stages until the entire area tributary to the set is covered. When only the single skidding line is used, the line is pulled out wherever there are logs to be skidded and no relocation of blocks is necessary.

Ground-line skidders were formerly widely used in the Lake states and in the South, where level or comparatively level topography prevails. A type of ground-line skidder, known as the Clyde Skidder, is designed for skidding with railroad operations. The power unit, including boiler and engine with drums, is mounted on a platform over railroad trucks. It is moved along the railroad spur track under its own power and stops at frequent intervals to skid.

One model of the Clyde Skidder is known as the two-line skidder and another as the four-line skidder. The two-line machine has two skidding lines that lead out from separate drums through blocks on the end of a short low boom extending in front of the machine. The lines are thus slightly elevated at the machine so they will clear the area immediately adjacent to the skidder. This elevation is so slight, however, that no effect is noted in the skidding. One line is used for skidding on one side of the spur while the other is used on the opposite side of the spur, thus giving double the output for the operator, as he can handle two lines almost as easily as one. Four-line machines have this arrangement duplicated at the opposite end of the skidder, which means that each line skids from only one quarter of the area tributary to the skidder at any one setting.

Rehaul drums and lines generally are not used with four-line skidders as the area to cover is so limited for any one line that it would be hardly worth the trouble to do so. When there are no rehaul lines and since the skidder can move under its own power along the spur, it skids from more sets than those that set up for longer skidding at one point. Thus, there

drawing. The elevated block is high enough only to clear the area around the machine and *of the Taylor Machine Works, Manufacturers of the "Loggers' Dream."*)

is a much greater percentage of practically straight-line skidding from the woods to the track. All that is necessary to change setting is to haul all cables in and then move the desired distance.

This means, of course, that logs are yarded all along the spur and on both sides of it. Therefore, some type of loader that is adapted to loading all along a spur after skidding has been completed must be used instead of one that loads only from one point at a time and requires that logs be brought to it as it loads them. Since ground skidding cannot be carried on effectively for distances greater than 600 or 700 ft, railroad spurs must be closer together than with other skidding methods adapted to skidding greater distances. Ground-line cable skidding to a railroad spur is limited to heavy stands of timber and even topography that will permit regular spacing of spurs.

In the West, also, ground-cable skidding formerly was widely used but likewise has been, to a considerable extent, replaced by other methods. Such skidding in the West is commonly referred to as "donkey" skidding and the skidders are termed "donkeys." They generally operate only one line, but with a rehaul, because of the rugged terrain.

Donkeys are used to skid to either a railroad- or truck-loading point or some intermediate point from which the logs are skidded to a loading point by some other method or, in some instances, by the same method. Frequently in rugged country, railroads and truck roads cannot be located close enough so that single-stage skidding can cover all of the logging area. A ridge, also, may prevent cable skidding over it even if distance is not too great. In such cases, a donkey may be placed on the ridge to assemble logs from beyond it, after which they are skidded from that point to the transportation system. Ground-line skidders used in this way are mounted on log skids so they can pull themselves from one donkey set to the next.

In the East and South a variety of ground-cable-skidding rigs have been devised for skidding saw logs or pulp logs from small swamps and boggy creek bottoms where tractors or animals would be at a disadvantage. Such equipment consists ordinarily of a single drum and gasoline motor mounted either on a truck or on skids, so it can readily be moved. Some skidding drums of this type, when mounted on trucks, are connected with a power take-off from the truck motor instead of a separate power unit. Another type consists only of a drum mounted on the bed of a worn-out truck and powered by sprocket chains driven by the rear axle. Either the rear of the machine may be jacked up to permit the rear wheels to revolve in driving the drum or the wheels may be removed and another rear axle and wheels installed which are powered by the sprocket chains when the machine is moved.

In operation this type of ground skidder is spotted adjacent to the area to be skidded and anchored in place usually by cable or chain guy lines to trees or stumps. The skidding cable is then carried out by hand or by a mule or horse and fastened to one or more logs. As it is wound on the drum the load is pulled to the loading point. These small skidders ordinarily have a maximum skidding distance of only 200 to 300 ft. They would not be economical for handling large volumes of timber because they are slow, but they can be used to get logs to a loading point from small areas where other skidding equipment cannot operate advantageously, if at all.

Where tractors with winches are being used for skidding, the tractor is often used instead of a separate skidder as it can be spotted and anchored at the swamp or bottom edge and the winch line used for skidding short distances. Because of the heavy investment in a tractor and winch, it does not pay to use a tractor for such skidding if there is an appreciable amount to be done. A skidding drum mounted on a second-hand truck unsuited for hauling or a power unit and drum mounted on skids for pulling with a tractor can be built for only a few hundred dollars, as compared to a tractor investment of $5,000 or more.

High-lead Skidding. One of the difficulties encountered in ground skidding is the delay and breaking of lines due to logs butting against stumps and other obstacles while being pulled in. The high-lead method of elevating the skidding line at the pulling end overcomes much of this trouble because of the lifting effect of pulling from an elevated point. For this reason, high-lead skidding is widely used throughout the West.

High-lead skidding is reported to have been first practiced on pullboat operations in Southern cypress swamps. The power unit and drums were installed on a barge-like boat that was floated to the skidding set. The top of a tree adjacent to the pullboat was cut off and a block placed at the top through which the skidding line was run. Generally a rehaul cable was used due to the extreme difficulty of pulling the cable out by hand through the swamps.

In the Pacific Northwest, however, the high-lead system has reached its highest perfection and its greatest use. High-lead skidding is so adapt-

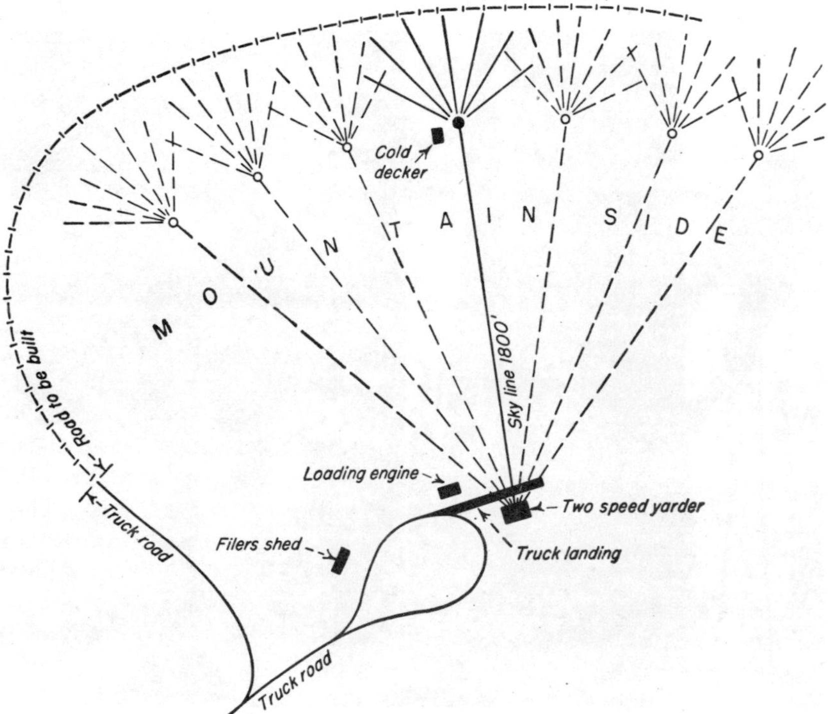

FIG. 91. The above diagram shows how cold-decking plays an important part in logging mountain slopes in Washington. (*Source: West Coast Lumberman, June,* 1937.)

able to so many situations that it is perhaps the most widely used cable-skidding method. "Cold-decking" logs at a point away from the transportation system is commonly done by a high lead, as a two-drum tractor or skid-mounted power unit can be moved to points away from logging railroads and truck roads. At such points a suitable tree is rigged for a high-lead spar and one drum of the unit is used for the main skidding line and the other for the rehaul line. Tractor winches used for such skidding are large enough to accommodate cables for standard high-lead logging.

Under Western conditions, a high-lead spar tree is selected and the top cut off. This is a specialized job requiring great skill and steady nerves. The top rigger climbs the tree by means of a light cable loop around it and climbing spikes attached to his legs, carrying a saw and ax, with which he cuts off the top. After the tree has stopped swaying he attaches a small block to the top of the spar, with a light line through it, which is used to pull up a heavy block through which the skidding cable will run.

Fig. 92. High-riggers on Western operations climb trees to be used for spars and cut out the tops. Note that the climber carries a saw and ax attached to his belt as well as a light line with which to draw up rigging blocks.

After the spar tree has been guyed with one or more sets of cables, it is ready for use in skidding. The height of a high-lead block on Western operations is from 150 to 200 ft, which provides enough elevation so that the effect of lifting on the logs as they are being skidded extends outwards from the spar tree for 400 to 600 ft. A rehaul is always used with a high-lead in the West, although on small operations in the East and South it frequently is omitted.

The high lead is not applicable for skidding farther than 800 to 1,000 ft because the effect of lifting on the logs being pulled in is lost even before these distances are reached. In practice high-lead skidding is, to all intents and purposes, simply ground-line skidding for approximately one half the distance generally skidded. As the loads are pulled towards the spar tree, the pull is practically horizontal at extreme distances due to the sag of the cable from the lead block. Closer to the tree, the lifting effect facilitates skidding by pulling up the front ends of the logs.

Overhead or Sky-line Cable-skidding Systems. The most highly developed mechanized methods of skidding are the so-called overhead, or

FIG. 93. Overhead cable skidding (left) and loading (right) in a Florida cypress swamp. Note that the skidder power unit is offset to one side of the spur and that the trestle type of logging railroad track is built on pile bents with caps serving as ties. (*Courtesy of A. Leschen & Sons.*)

sky-line, systems. Skidding by this method involves suspending a main cable between two spars, on which a block is pulled back and forth. The skidding line extends from a drum on the power unit through the block and thence to the ground. Several cables thus are required and each requires a separate drum.

After the system has been rigged to a tail tree and a head spar and the cables strung, skidding is accomplished by pulling the block out into the woods on the sky line by means of an outhaul or rehaul line. This extends from the block to the tail tree and thence to the drum on the power unit, usually through blocks anchored near the ground and off to one side. The block is spotted over the place where logs are to be attached and the skidding line is lowered through the block to the ground, where logs are attached. They are then elevated either partially or wholly off the ground and are pulled to the head spar by pulling in on the trolley block. Of course, as the block is pulled in, the rehaul or outhaul line must recede.

FIG. 94. The above illustration shows the large number of blocks and cables required to rig a head spar used for sky-line skidding and spar tree loading.

Some skidders have interlocking drums so that as one cable is wound up the other automatically unwinds at the same rate. The logs are unhooked at the landing or yard and the process is repeated. When all the logs along the lane between one pair of spar trees have been skidded, the cables are shifted to another tail spar and the logs between it and the head spar are then skidded. This process is repeated until the entire area around the skidder within reach has been covered. Then the skidder is moved to the next set and the process continued.

A number of different methods of rigging spar trees for overhead skidding have been devised but in general they operate as described above. The variations fall into two main categories. One is the tight-line method and the other the slack-line method. With the tight-line method the skidding cable extends from the drum, through a block on the head-spar tree, and thence through the trolley block to the ground. The end of the skidding line, therefore, is lowered to the ground by paying out the skidding line, and is raised by drawing in the skidding line.

With the slack-line system, a fixed length of cable is attached to the block running on the main cable. It is only long enough so that when the

FIG. 95. View showing the pattern of typical high-lead and overhead cable-skidding operations in Washington. Note the patches of uncut timber left to reseed the logged areas.

main cable is stretched normally, it clears the ground sufficiently for skidding. The logs to be skidded are attached to this length of cable. It is lowered to the ground by slacking off the main cable and is raised by tightening up the main cable. This system eliminates one cable and two drums on the skidding machine, but it is not as versatile as the tight-line method because logs cannot be raised as high as when the skidding line itself is controlled by a drum. The slack-line system is often preferred over tight-line methods when a skidding machine is moved away from the transportation system and functions as the first stage of two-stage skidding, because of the fewer cables and drums needed as compared with tight-line systems.

Most of the other variations in overhead skidding methods have to do with the method of arranging blocks, cables, and rehaul lines. All, how-

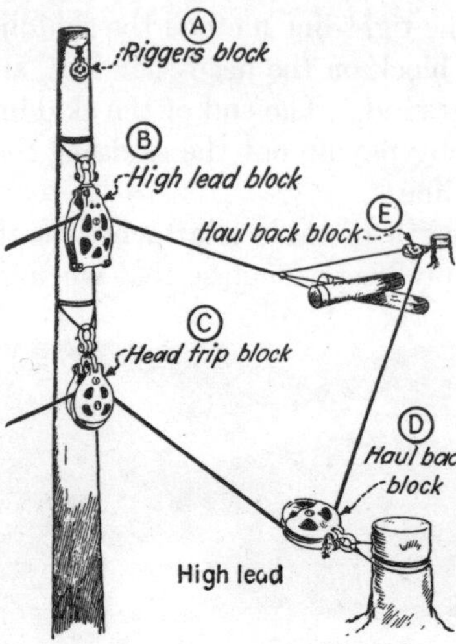

FIG. 96*A*

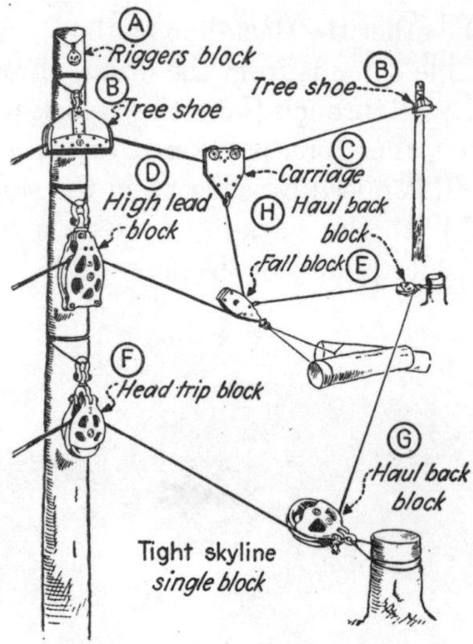

FIG. 96*B*

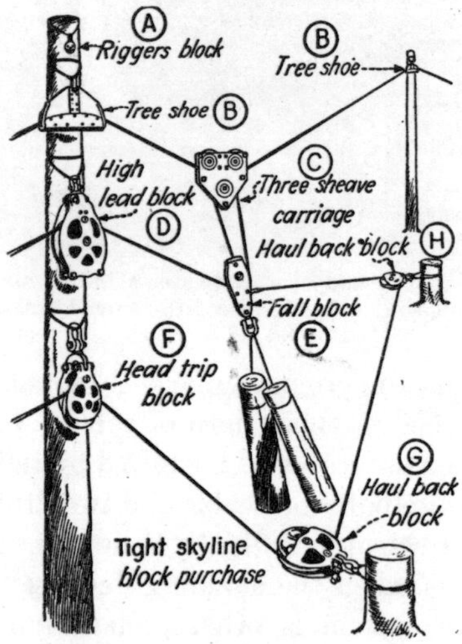

FIG. 96*C*

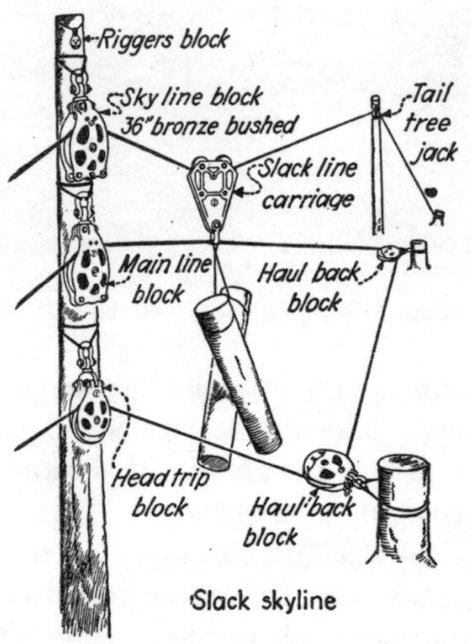

FIG. 96*D*

FIG. 96. Diagrammatic illustrations of five common types of cable skidding used on Western operations. *A*. The high lead. *B*. The tight sky line, single block. *C*. The tight sky line, block purchase. *D*. The slack sky line. *E*. The steel spar. (*Courtesy of the Young Iron Works.*)

ever, operate in much the same manner and skid logs by partially elevating them to reduce friction and to prevent hang-ups against stumps and uncut timber. Overhead systems are adaptable for skidding for distances up to 2,000 and even 4,000 ft, sometimes more. Thus, overhead systems cover distances much greater than would be possible with ground-line or high-lead skidding.

A new method of overhead skidding has been recently introduced in the West. It is revolutionary and holds much promise of success where conditions handicap more orthodox methods. The machine is known as the "Skyhook." The new feature is the method of operating the skidder itself on the cableway instead of at a fixed point on the ground.

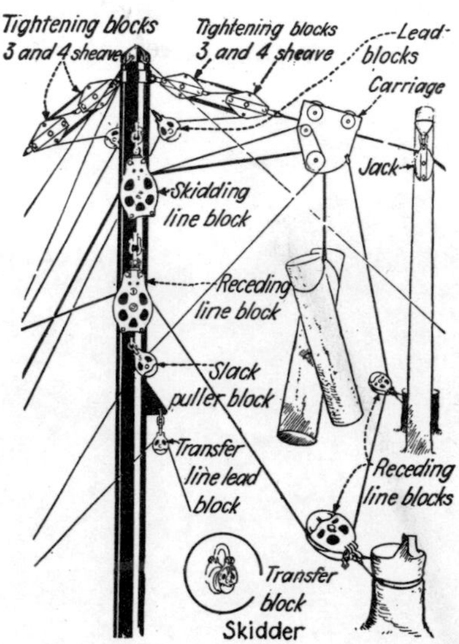

FIG. 96E. (See legend on opposite page.)

Two heavy cables about 30 in. apart are strung between spar trees and the Skyhook operates over them as a suspended carriage with engine, driving mechanism, yarding drum, and operator all contained in it. A third, lighter cable is also strung parallel to the other two, which is passed through two sets of driving pulleys so that when the wheels revolve, the carriage is propelled. The cables can be supported between the terminal-spar trees by specially designed hinged jacks attached to intermediate-spar trees in such a way that the Skyhook carriage can pass the cable supports. This feature permits the cableway to extend for distances several times the maximum that would otherwise be possible.

The advantage of the Skyhook is that the choker is taken directly to the log or to a point where the skidding line can be led out to the side, so no rehaul line is needed. Logs can be lifted clear of the ground under the Skyhook, if necessary, and carried bodily to the loading point or cold deck. The machine that rides the cables also can propel itself on the ground in moving to another location, as pneumatic-tired wheels can be substituted for the driving wheels.

POWER UNITS FOR SKIDDING

The earliest powered cable skidders for dry-land logging were mounted on railroad trucks and could operate, therefore, only on or alongside the

FIG. 97. A 200 horse-power Diesel skidder mounted on heavy log skids used in the big timber of the West. Skidders of this type can pull themselves over the ground. A cable is lead out through the blocks shown at the front end and attached to a stump or tree and then the cable is wound up on one of the drums, pulling the skidder forward. (*Courtesy of the Skagit Steel and Iron Works.*)

logging railroad and could be moved only over it. For swamp and riverbank logging, skidders were and frequently still are built on barges that can be floated from place to place for skidding. Skidders on railroad tracks are set off to one side of the track by jacking them up and moving them off, or they are placed on temporary spurs alongside the track, when it is necessary to keep the track clear for log trains. When skidding is completed before loading begins, the skidder can remain on the track, but when a spar is used to skid logs from the surrounding area to that point, it is necessary to load the logs as they are skidded to prevent too large an accumulation.

Before gasoline and Diesel power units were perfected, steam was the only source of power. Steam power units require not only quantities of bulky fuel such as coal or wood, unless oil is burned, but also a supply of suitable boiler water. Therefore, they can be operated only where it is possible and economical to keep them supplied with both fuel and water. Early power skidders were commonly fired with wood cut near the sets, but later coal and fuel oil were used extensively. Even when fuel oil is used, quantities of water, free from boiler-damaging properties, must be supplied. Tank cars often were necessary to supply suitable water and commonly are

FIG. 98. A simple type of ground-line skidder used for skidding logs from small swamps or up short slopes. (*Courtesy of the Opelika Equipment Co.*)

used today with steam power units. Steam-powered skidders, therefore, could only be used along railroads or truck roads where they could be readily supplied.

With the advent of Diesel power units suitable for the heavy and variable loads encountered in logging, the use of steam power rapidly declined. Diesel power units have many advantages over steam units. They are considerably lighter per unit of power output because no boiler is necessary and because of the compact nature of Diesel motors. Also, much less fuel oil is needed for Diesel motors than for steam boilers of equal horsepower because the oil is used for internal combustion in the cylinders of Diesel motors. Furthermore, some saving in operating labor results with Diesel power units, as a fireman is not needed. Also, there is practically no danger of starting forest fires from power units when internal-combustion motors are used. This is of considerable importance, especially in the West during dry periods, as fire hazards are great and even oil-burning steam units sometimes cause fires. Probably the most important advantage, however, is that Diesel skidders can be operated away from transportation units because enough fuel can be taken in tanks to operate them for up to 2 or 3 months without refilling the tanks.

One type of railroad-mounted skidder is equipped with a steel spar 40 to 60 ft high that constitutes the head spar. Skidding cables extend from the machine through this spar to a tail spar. The Lidgerwood skidding machine is an example of this type. When the steel spar is lacking, a suitable tree is used or, in some cases, a pole is erected and guyed for this purpose.

Skidders are also mounted on heavy beds with skids beneath so they can be pulled, by tractors in the case of lighter units, but commonly by means of their own power. This is done by leading out a cable from one of the drums and fastening it to a solid tree or stump. As the cable is wound on the drum, the skidder is moved forward to the place where the cable is anchored. By repeating this process a skidder can be moved from place to place without regard to the transportation system. Some such units have oil storage tanks sufficient for up to 2 or 3 months of operation, which greatly increases the field of usefulness of such independent units.

Small skidding units such as truck skidders consist only of one or two drums mounted on the bed of a truck and either a gasoline power unit or a power take-off from the truck motor.

Some modern Diesel-powered skidders are equipped with torque converters, which make for smooth application of power. With the sudden jerks and load changes common to skidding, this is an important improvement in internal-combustion motors. Considerable saving results from the use of torque converters with Diesel power units in that line breakage and rigging damage due to sudden jerks when logs butt against trees and stumps are reduced. Steam is highly satisfactory power because of the element of smoothness in its application to skidding and other logging steps. Without torque converters, Diesel power or any internal-combustion power is inclined to be jerky, and gear shifting is necessary with changing loads. The torque converter does away with most metal gear contacts by substituting a fluid oil.

CONDUCT OF SKIDDING OPERATIONS

Skidding operations vary greatly with the type of power unit used, the size of timber products being handled, and ground and weather conditions. With the simpler methods such as snaking, sledge, or bummer skidding, the teamster drives the team, tongs or chokes the logs, and unhooks them at the landing.

When wheels of one kind or another or sledges are used with teams or tractors, one or more men usually stay with the logs in the woods and either place chokers on the logs or assist the driver in loading. Swampers are commonly employed to cut out trails for skidding where the undergrowth is heavy enough to interfere. Some provision also must be made to feed and water animals. When a large number of them are used, this is the job for a special man or two during the noon hour. At night the animals are cared for by barn men. When tractors are used they must be greased, oiled, fueled, and overhauled at periodic intervals. This is done either at night in camp shops or garages or an extra tractor or two may be provided so that one or two can be off the job for servicing.

Cable-skidding operations are more specialized and one skidder requires several men for its operation. One man must operate the power unit and the drums. This is a skilled job as on some machines there are four or five drums, each controlling a separate cable, and their operation must be closely integrated. If steam power is used and coal or wood is the fuel, a fireman must also be provided. With Diesel power units and sometimes with oil-fired steam skidders, a fireman is not required.

Another crew remains with the logs to be skidded and fastens chokers to them so when the cable is returned to the woods no time will be lost in hooking onto the logs. The empty chokers are unhooked and the loaded ones attached to the skidding line. By the time the cable returns, another "turn" of logs will have been choked. The size of crew necessary to choke a turn of logs while one turn is being skidded varies with the distance and the speed at which the skidding line operates. Because a cable extends for considerable distances back from the machine, one or more signal men, called "whistle punks," are needed to notify the skidder operator when to stop and start the cables so the lines can be spotted at the right points and premature starting, which could cause serious accidents, is prevented.

Considerable advance work must be done at cable-skidding sets, for high-lead and overhead systems especially, because spar trees must be rigged for the cables. At each set, also, different tail trees are needed for each line or skidding road for sky-line systems and these trees must be rigged with blocks for the cables and rehaul lines. A rigging crew is, therefore, kept constantly at work ahead of the actual skidding, preparing for the next move of the cables or the machine.

To skid economically with powerful equipment representing a heavy investment, it is necessary to keep it working steadily at near maximum capacity. Delays are expensive in that skilled labor and expensive machines are not producing while costs go on. Ability of a high order, therefore, is needed in the management of timber-harvesting operations to keep all machines and crews producing and their output in balance so one stage does not overproduce or underproduce with respect to the next stage. Small harvesting operations using a minimum of equipment, on the other hand, are not so costly when idle, although the loss due to idleness must not be disregarded.

CHAPTER 13

LOADING FOREST PRODUCTS

Before forest products can be transported out of the forest they must be loaded on some kind of vehicle. Frequently, also, they must be reloaded from one type to another type of vehicle, when a transportation system of two or more stages is used, such as trucking to a railroad siding for loading on railroad cars.

Products cut in harvesting may be left at the stump for loading; or they may be bunched so several can be loaded from one point; or they may be assembled by skidding into large concentrations for loading along a transportation system. Generally, the greater the concentration of products to be loaded at one point, the more highly specialized is the loading operation. Only small products are loaded at the stump, ordinarily, as it would be too cumbersome to move a power loader with capacity for handling heavy products from piece to piece to load them.

Ties, pulpwood, and bolts of various kinds that are small enough to be lifted by one or two men are often loaded directly from where they were cut, as they can be carried short distances. Not all small products, however, are loaded without being bunched or skidded, as the most efficient method depends upon the volume removed per unit of area and the transportation system available. Heavier products, such as saw logs and poles, are either bunched or skidded, so several or more can be loaded at one point where either animal or mechanical power may be applied.

Since loading must be gauged to either cutting, bunching, or skidding and to the hauling operation as well, it is a very important step in the harvesting operation. When loading is slower than hauling, for instance, then trucks or log trains waste time waiting; and if the capacity to load is greater than hauling capacity, then the loader and crew are idle until an empty train or truck arrives. Loading is often called the keystone of a harvesting operation, as it determines the size of the cutting and bunching and skidding crews, as well as the transportation units serving the loader. Choice of loading method also is dependent upon the kind of transportation system used, as it obviously would be impossible for a log train to go into the woods from stump to stump to be loaded, which is often possible with log and pulpwood trucks.

Regardless of where loading is done and upon what kind of vehicle products are placed, it is necessary that they be elevated, except in those

TABLE 16. AVERAGE WEIGHT, POUNDS, OF LOGS PER 1,000 FT
(Doyle log-scale rule)

Species	12-in. diameter		18-in. diameter		24-in. diameter	
	Green	Dry	Green	Dry	Green	Dry
Ash, white	11,100	9,700	7,700	6,800	6,600	5,700
Aspen (popple)	10,800	6,200	7,600	4,300	6,400	3,700
Basswood	9,500	5,900	6,600	4,100	5,600	3,500
Beech	12,700	10,100	8,900	7,000	7,500	6,000
Birch, yellow	13,200	10,300	9,200	7,200	7,800	6,100
Cedar, western red	6,200	5,300	4,300	3,700	3,700	3,100
Cherry, black	10,500	8,300	7,300	5,800	6,200	4,900
Chestnut	12,600	7,000	8,800	4,900	7,500	4,100
Cottonwood	10,700	6,300	7,500	4,400	6,300	3,700
Cypress, southern	11,800	7,800	8,200	5,500	7,000	4,600
Elm:						
White	11,300	7,800	7,900	5,500	6,700	4,600
Slippery	12,600	9,200	8,800	6,400	7,400	5,500
Fir:						
Balsam	10,400	5,800	7,200	4,000	6,100	3,400
Douglas	8,700	7,700	6,100	5,400	5,200	4,600
Gum:						
Black	10,400	8,300	7,200	5,800	6,100	4,900
Red (sweet)	10,600	8,100	7,400	5,600	6,300	4,800
Hackberry	11,300	8,900	7,900	6,200	6,700	5,200
Hemlock, eastern	11,200	6,600	7,800	4,600	6,600	3,900
Hickory	14,700	11,900	10,300	8,300	8,700	7,000
Locust, black	13,400	11,300	9,300	7,900	7,900	6,700
Maple:						
Sugar	12,900	10,000	9,000	7,000	7,600	5,900
Red	11,900	8,200	8,300	5,700	7,100	4,900
Silver	10,500	7,800	7,300	5,400	6,200	4,600
Oak:						
Red	14,800	10,400	10,300	7,300	8,800	6,200
White	14,400	10,900	10,000	7,600	8,500	6,500
Pine:						
Jack	11,500	6,900	8,000	4,800	6,800	4,100
Loblolly	12,400	9,000	8,700	6,300	7,400	5,300
Longleaf	11,100	9,700	7,700	6,800	6,500	5,700
Norway (red)	9,700	7,800	6,800	5,500	5,700	4,600
Pitch	12,400	8,300	8,700	5,800	7,400	4,900
Shortleaf	10,400	8,300	7,200	5,800	6,100	4,900
Slash	12,200	10,400	8,500	7,200	7,200	6,100
Sugar	11,500	6,200	8,000	4,300	6,800	3,700
Yellow (western)	11,300	6,500	7,900	4,500	6,700	3,800
White	9,000	6,200	6,300	4,300	5,300	3,700
Poplar, yellow (tulip)	8,800	6,500	6,100	4,500	5,200	3,800
Redwood	8,900	5,900	6,200	4,100	5,200	3,500
Spruce, eastern	7,700	6,600	5,400	4,600	4,600	3,900
Sycamore	12,000	8,300	8,400	5,800	7,100	4,900
Tupelo gum (Bay poplar)	15,200	8,500	10,600	5,900	9,000	5,000
Walnut, black	11,900	8,200	8,300	5,700	7,100	4,900
Willow	11,800	6,000	8,200	4,200	7,000	3,500

Source: Report, U.S. Forest Service. (*Courtesy of Allis-Chalmers Mfg. Co.*)

rare instances when vehicles can be loaded by gravity or by horizontal movement, as by rolling. This means that power is required to elevate the products, and since green wood is a heavy substance and since many forest products are large, powerful loading machines are required on harvesting operations in big timber.

FIG. 99. Above, loading pine pulpwood bolts on a truck; below, unloading from a truck onto a flatcar by hand labor. Georgia.

The weight of freshly cut logs is given in Table 16. When such weights are to be lifted, a great deal of power, obviously, is required. Even though pulpwood is cut in short lengths from small-sized trees, the weight of an average bolt is often 100 lb and many are heavier.

The only methods by which a product can be elevated are by applying power from beneath or by applying power from above. The former is the principle of pushing, while the latter requires the application of power by lifting. As the products to be loaded are practically always on the ground or piled on others, it is never possible or, at least, practical to elevate them by pushing from beneath, except in hand loading, in which case the products are first lifted and then pushed up. Even when logs and other products floating in water are elevated by means of conveyer chains, the application of power is from above although the effect of pushing is created.

LOADING BY HAND LABOR

Hand loading, as stated, is only possible for small products such as pulpwood. Even for the smallest products, a sturdy physique is required for constant lifting and loading. The hand-loading operation is comparatively simple and is well known, as it differs in no important respect from lifting and loading any other products by hand. In loading pulpwood and ties, for instance, a man elevates one end and then either carries it in his hands or on his shoulders to the truck. If the piece does not need to be carried, it is up-ended and pushed alongside the load and placed on it. Small pulpwood bolts can be thrown on a load. Men loading by hand work either alone or in pairs, depending upon the size of the bolts being loaded. A sharp, curved hook with a handle is often used by men handling pulpwood and bolts of similar size. A blow drives the hook into the bolt, making it easier to lift it from the ground or to arrange the bolt on the truck or car.

When the haul is short, loading crews may ride pulpwood trucks, and in the Southern states they commonly do, and both load in the woods as well as unload into a car or onto a barge. Otherwise, separate loading and unloading crews are stationed at the truck-loading place and at the transfer point.

LOADING BY GRAVITY

When trucks, wagons, or sleds are loaded on sloping ground or under a road cut, logs and other round products are sometimes loaded by rolling them onto the vehicle with peaveys. The logs or pulpwood bolts are skidded to the loading point and placed on skids arranged to form a deck, and then rolled off the end of the deck onto the vehicle. This method is limited to locations where the correct degree of slope exists, as too high a drop from the deck to the vehicle would result in damage and insufficient elevation would prevent building up an adequate load. Even under favorable conditions, only small loads can be placed on the vehicle by this method.

When a considerable quantity of products are to be loaded from a deck by gravity, it is economical to build it solidly with timber cribbing, allowing a slight slope towards the loading end to facilitate rolling. In some

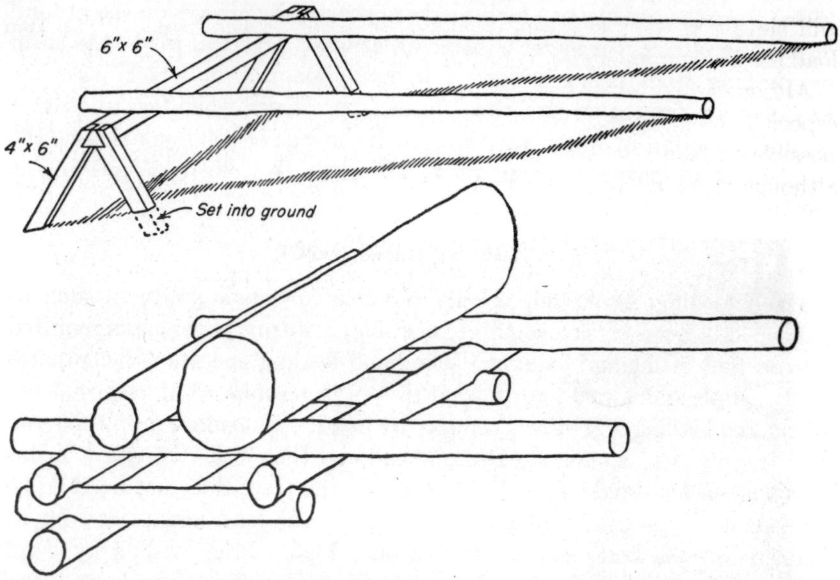

FIG. 100. Logs may be loaded by gravity with a cant hook or peavey when the slope of the ground permits the construction of simple loading docks such as are shown. (*From Farmers' Bulletin 1907, U.S. Forest Service, Department of Agriculture.*)

cases the deck is made large enough to accumulate a number of loads in advance so that loading may continue after skidding has been completed. Such decks are also known as rollways.

THE CROSSHAUL

Because primary forest products, particularly saw logs and poles, are approximately cylindrical in shape and so can be rolled readily, it is possible to elevate and load them by a very simple and convenient method known as the "crosshaul." The log or pole is, in fact, simply rolled up a pair of temporarily placed skids sloping from the ground to the bunk or bed of the truck, wagon, or car.

Power for rolling a log up the skids onto the vehicle is furnished by one or more animals or by a tractor. A forked chain attached to the vehicle near each skid is passed under the log as it lies on the ground at or near the lower ends of the skid poles. The double chain is then fastened to a single chain that passes over the log and over the vehicle to a team or tractor on the opposite side. When the chain is pulled, the log rolls up the skids to the vehicle, being cradled in the two chains attached to the bunks, which prevents it from becoming unbalanced. The operation is repeated until the bottom course of logs is completed and braced in the vehicle and then the two chains attached to the vehicle are raised, as are the skids, so they will elevate the log to the next course, or layer, of the load. When the load has been completed, it is bound with a chain and is ready to be hauled.

Although this method is comparatively slow, it has many advantages, especially on small-scale operations, because it is so readily portable. It is possible to load one log and then move the truck or wagon to another, although this is seldom done unless the logs are so large that two or three

FIG. 101. The cross-haul method of loading logs on trucks or wagons, illustrated here, is widely and successfully used when the operation does not justify a more elaborate loading machine. Cross hauling of both one log and several logs at one time is shown. Note that loading is taking place from a "bunch" of logs. Alabama.

make a load. More often, logs or poles are bunched with teams or tractors and thus assembled parallel in small groups sufficiently large to make at least one load. Logs from two or three trees commonly will make a load, so it is not necessary to move the logs very far when bunching for crosshaul loading.

It is preferable to bunch a number of single loads rather than to move the logs farther and bunch several loads at one place, unless the topography is such that it is impractical for the truck or wagon to move freely from place to place. However, when logs are moved more than ordinary bunching distance for topographic or any other reason, the operation becomes skidding instead of bunching and the size of the group is much larger.

No specialized equipment is needed for crosshaul loading. The skid poles can be cut from small trees and 50 to 60 ft of chain is ordinarily sufficient. When a wagon is used, the team can be unhitched to do the loading and even the bunching. Any farmer with a team and wagon, therefore, can load and haul logs by the crosshaul method.

One disadvantage of the crosshaul is that it is seldom practicable to load wagons or trucks with side stakes over 2 ft high. Short stakes can be placed after the first course is loaded but high stakes prevent the logs from rolling onto the vehicle. Loads loaded by the crosshaul method, therefore, are characteristically triangular shaped, being full width at the bottom and tapering to only one log at the top. When loaders that lift logs over side stakes are used, square loads can be built up, which means that larger loads can be placed on the vehicles.

It is possible to load logs and similar forest products by pulling them on the vehicle endwise, but this is never or only very seldom done. There is no advantage in this method over the crosshaul except that a truck could have a cable and drum mounted on it and thus be a complete loading and hauling unit. The method, however, is awkward and so does not find much practical application.

POWER LOADING

To apply power from above to any object requires that there be some overhead point of suspension to which a cable, chain, block, or pulley can be attached that in turn can be connected with the object to be lifted. The object then is elevated either by raising the overhead point of attachment, as in raising a boom, or by shortening the distance between it and the object, as in pulling a cable through a pulley block, or by performing both actions simultaneously.

There are many different loading methods and kinds of loading machines, also called "jammers," used in loading forest products, but all embody the principle of pulling the product up from an overhead point of suspension. Some loaders are designed for loading railroad cars and are suitable only for loading in connection with certain skidding methods; others are designed for loading trucks only. Some are self-propelled machines with booms capable of being raised, lowered, and moved in a complete circle.

While it is possible to lift an object by attaching a fixed length of cable or chain to it and to the end of a boom and then raising the boom, this method is never used in loading forest products. The method of running a cable through a pulley block and fastening the cable to the log or other product and lifting it by pulling the cable is much more rapid and versatile and is used in all cases. In more elaborate loaders, lifting is performed by pulling up the cable running through a block in the end of a boom and also by raising the boom, but the pulley block is always present.

The block through which the loading cable runs may be stationary or movable. Booms that revolve through all or a portion of a complete circle provide mobility within the reach of the boom, and with self-pro-

FIG. 102. Loading southern pine logs in Georgia with a simple tree block. The pulley is pulled away from the tree when the log is elevated which centers it over the truck being loaded. (*Courtesy of the Caterpillar Tractor Co.*)

pelled loaders with revolving booms, the block can be moved to any point to which the loader can travel. Mobility in a lineal direction only is obtained in certain types of log transfers where the block is attached to a trolley running on a cable extending between two poles or tree supports or by a combination of blocks and cables stretched between two spars. Such loading rigs are used only to transfer products from one vehicle to another or to temporary storage; they are not taken into the woods, where the primary loading operation is performed.

Stationary Pulley Blocks for Loading. A stationary overhead block for loading forest products is obtained by use of a gin pole, A-frame, or cable attached to a tree or erected pole. A gin pole is an erect pole, 30 to 40 ft tall, guyed on three sides and leaning far enough in one direction so the block attached at the top is directly over the point where the car or truck will be spotted for loading.

An A-frame gives the same effect except that two poles or sawed timbers are fabricated solidly and usually attached to a bed or base to make it readily portable. The A-frame leans outward far enough so the point of the A, to which a block is fastened, centers over the place where sleds, trucks, or other vehicles to be loaded will be spotted.

By attaching a block to a cable fastened 20 or 25 ft up a tree with the other end attached to a stump approximately 60 ft away, a stationary overhead block is obtained with room beneath to spot vehicles. A similar effect is obtained by simply attaching a block to one end of a 6- to 8-ft cable or chain and fastening the other end about 20 ft up a tree. When the loading cable running through the block is pulled, the weight of the log pulls the cable far enough away from the tree to permit spotting a vehicle beneath it for loading.

Power for raising the logs or other products being loaded with stationary overhead blocks may be obtained by the traction power of a horse or team, the pull of a tractor or truck, or by a winch operated on a tractor or powered separately. A-frame loaders for small logs and long pulpwood are called "horse jammers" when animal power is used with them to load sleds in the Lake states and the Northeast. A gasoline motor, even an old automobile motor, and a winch can be mounted in the beds of A-frame loaders for power. In the case of tree blocks, power may be by a motor and drum mounted on a truck nearby, by a team, by a tractor winch, or by the draw-bar pull of a tractor. While a winch-equipped tractor can be used very conveniently to operate such loaders, the investment in one is so great that it usually is uneconomical to do so. Adequate power for all except the heaviest products can be obtained by gasoline or Diesel power units at much less expense, and equipment that would be engaged in other productive work is not tied up doing loading.

With all types of loaders using a stationary overhead block, it is obvious that the products to be loaded must be skidded to the loading point. Furthermore, the vehicle to be loaded must be spotted directly beneath the block as that is the point where the products will be elevated. Loading can be done by stationary block methods, therefore, only when the products are skidded to the loading point from the surrounding area. When the limit of economical skidding has been reached, then the loader should be moved to load the products from the next area.

Loading cannot be done as rapidly with the use of the stationary overhead block as with the other methods, but it is adaptable to small-scale operations in that no large expensive equipment is necessary and it is comparatively easy to move such a rig to a different location. In moving an

FIG. 103. A-frame loading booms are often attached to crawler-type tractors. Right, the position of the tractor, boom, and brace when loading; left, the brace elevated (by means of the tongs) when the loader moves to another location. Alabama.

FIG. 104. A-frame loaders are also constructed on skids, so they can be readily moved from place to place in the forest. The frame is guyed to stabilize it, and the power for loading may be by a gasoline motor (shown above), by one or more animals, or by a tractor winch. The illustration above, shows logs being skidded to the loading yard and banked for quick loading by the loader. South Carolina.

A-frame, it is usually pulled on the ground if the distance is short or it can be loaded on a truck for longer moves. When such are anticipated, the frame can be built with the base of the A on swivels so when the guy lines are released it will lie flat on the bed of the frame. Nonrigid construction is advantageous also because the forward pitch of the A can be adjusted to meet varying conditions. Dismantling a tree block and a gin pole involves only releasing the cables or transporting the parts to the new location.

In loading with a stationary overhead block, the sled, truck, or other vehicle must be spotted directly beneath it and the products to be loaded must be brought near enough to be reached by the loading cable. The products may be banked ahead or brought in as loaded. The cable is then attached to the log or other product, usually by means of tongs, and the log is elevated. As it leaves the ground it swings under the block and can be lowered on the vehicle at the desired spot.

Protection must be given the vehicle to guard against the logs striking against the wheels or side of the truck bed as they are raised and swung under the block. This is provided usually by banking from one to three or more logs to form a "brow" close to the truck to receive the blow instead of the vehicle, or an earthen bank may be built up to give the same protection.

Movable Pulley Blocks for Loading. When a block used for loading is movable instead of stationary, greater speed is obtained in loading due to the greater precision that is possible in the operation. Use of a block on a moving boom permits the operator to raise the product being loaded clear of the ground and then swing it over the vehicle, where it is lowered into place. This avoids the tendency of a log to strike the side of the vehicle or its stake, as is the case with stationary-block methods.

To obtain a movable block, a boom subject to control by the operator is necessary, and this ordinarily requires some kind of a loading machine with power unit, drums, gears, cables, and control levers. Additional mobility is gained, as previously mentioned, when the loader is self-propelled and can be moved even while in operation. All loaders are portable, as they must change location at frequent intervals, but only the most modern types are completely mobile and independent of other motive power.

The self-centering boom, as used on some loaders, is not operated or controlled by the loader man but is so mounted that when the boom is pulled to one side to pick up a log, gravity pulls it back to a center position. This effect is obtained by sloping outward at the top the vertical portion of the boom, which is fabricated rigidly to the arm portion, and mounted both top and bottom in sockets so the boom can swing freely from side to side. When the loader is level, the boom is always pulled by gravity back to a position at the center. Such a loader must load from directly behind or at one end of the vehicle. Logs are then yarded or assembled at either side and can be readily elevated by means of the cable through the block.

The gravity pull back over the trucks is not so great as in the case of the stationary block, so one or two men on the ground with hand ropes can ordinarily prevent a log or other product from striking the vehicle being loaded.

Other types of loaders have powered booms that are not capable of being elevated or lowered but can be moved to either side sufficiently to permit raising a log clear of the car or truck before it is swung over the vehicle and lowered into place. These types also must load from one end of the vehicle and the logs or products are picked up on either side. There is no need to permit a log to swing into the side of the car or truck being loaded, as the boom can be held to the side until the log is clear of the car and its stakes.

Loaders with booms that swing in a complete circle are the most versatile and fastest type, as a rule. They can load a vehicle from either end or either side, depending upon the system practiced. However, most loading with movable booms also is done from one end of the vehicle, to permit yarding logs on either side. When the boom moves in a complete circle, the cab with power unit and controls all move as a unit, as this is mechanically simpler than designing one that would move the boom only. Another advantage is that it keeps the operator always facing his loading lines.

A loader that is stationary while loading seldom has a boom constructed so it can be raised or lowered, as the loader is in a fixed position with respect to the vehicle being loaded and no adjustment is necessary. Self-propelled loaders do not remain in the same position with respect to the vehicle being loaded, as they move from place to place, taking advantage of the most favorable spots; therefore, it is necessary that their booms be adjustable.

TONGS AND OTHER LOADING ATTACHMENTS

The end of the loading cable or line which is attached to the products to be loaded is equipped with one of various types of tongs or slings, or both, which are fastened to or passed beneath the products being loaded. Logs are picked up either by a pair of center tongs or end tongs, while slings are used for very large logs.

Center tongs are attached, as nearly as possible, to the center of gravity at the middle of each log. Center tongs are similar to skidding tongs and are so designed that, as the line is pulled in, they bite into the log and remain firmly embedded so long as the weight of the log is on the line. When the log is placed on the vehicle and the line is slackened, the tongs can be readily unhooked by the top loader, using his peavey when necessary.

When end tongs are used, the loading line is split into two lines at the end and one end tong is attached to each of the two ends. In practice the loading line leading from the drum over the block is equipped with a ring on the end. Two lighter cables are then attached, usually by snaps, to this ring, and the cables are long enough so that the tongs can be placed in the ends of a log. The lines should only be long enough to permit

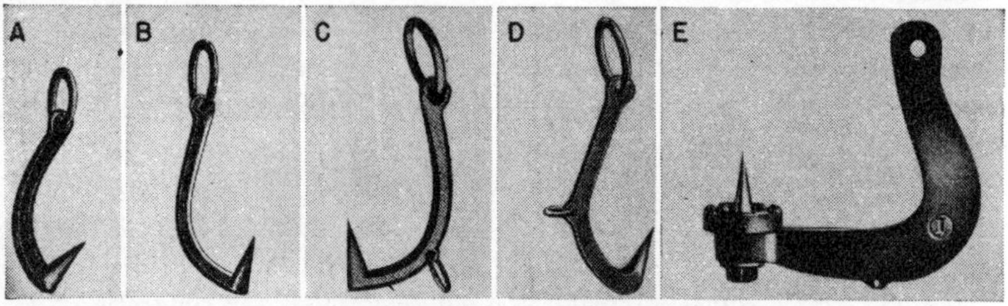

FIG. 105. Types of crotch line loading hooks or end tongs. (*A-D, Courtesy of American Logging Tool Company; E, courtesy of Young Iron Works.*)

FIG. 106. Loading with a crotch line and end tongs. Note the two ground-loader men who guide the log into position and retrieve the tongs after it has been placed. Note also the "brow" log which protects the side of the truck. California.

tonging the longest logs to be loaded, since if they are too long the log cannot be elevated sufficiently, as the doubled portion of the line cannot pass through the loading block. End tongs cannot be used very readily for loading logs longer than 20 to 24 ft, as the long cable to the end tongs would be too inconvenient to handle. Each end tong consists of a short curved steel arm with one end pointed so it will hold in the end of a log. Some styles, known as cup-point loading hooks, have a sharp ring around the point to give an increased area of contact between the tong and the wood. This also prevents the points from being forced too far into the ends by the weight of the logs, which exerts great pressure on the bearing surfaces and holds the hooks in the wood. An eye on the end of the hook opposite the point permits fastening the cable to it. In addition another

FIG. 107. Tongs used for single-line loading (known as center-tong loading because the tongs are attached to the log as near the center of gravity as it is possible to judge). This type is also used for two-tong loading with "hay-rack" booms or for heeling-boom loading. Washington.

light line, usually rope, is attached to the tong so that the end hook can be pulled out and back to the ground when the log has been placed on the load. When end tongs are used, two men remain on the ground and each handles one of the lines attached to the end tongs. They thus can control the log as it is lifted, and when it has been placed on the load they can pull the tongs free of it and place them in the next log to be loaded. A top loader, therefore, is not needed.

West coast logs frequently are so large and heavy that tongs would tear out of the wood, so slings are placed under them. Since only the largest logs need to be cradled in slings, center-type tongs are left on the loading cable and removable slings are hooked over the tong points when it is necessary to lift logs in this way. Loading booms equipped for either sling or tong loading of large logs commonly have a pair of tongs on a horizontal boom so the logs can be either tonged or placed in slings near each end. Slings are also used on some large-scale pulpwood operations to load bunches of bolts. In some cases the slings are left with the bunches to permit rapid unloading by the same method, or the slings may be pulled out of the load and used only for loading. Sling loading of pulpwood or other small products is usually done with crane-type loaders with swinging booms.

FIG. 108. Loading large logs with double tongs and slings. Washington.

When tree- or multiple-length logs are loaded, it is impossible to place the tongs at the center, as the boom would not be long enough and the near end of the log would extend behind the loader. For the same reason end tongs cannot be used. In such cases the tree-length logs or poles are yarded butt end to the loader and the center tongs are fastened about one quarter of the way towards the top end of the log. The butt end is then lifted until it strikes the under side of the boom, where it remains fixed as the continued pull on the line lifts the top end clear of the ground and of the load on which it is to be placed. Then the boom, with the long log heeled against it, is swung over the vehicle and the log lowered into place. This type of loading is known as the "heeling-boom" method. The boom must be braced, widened, and protected on the under side for this type of loading.

Skillful loading crews can load more rapidly with center tongs than by any other method when the loader is equipped with a full-swinging boom.

Some ground loaders, or "tongers," are so skillful that they seldom miss the almost exact center of gravity, and they can throw the tongs at a log and have them catch. The loaderman then maneuvers the log quickly into position with the assistance of the top loader, who must be nimble as well as skillful. The top loader also removes the tongs from the logs as soon as they have been placed on the vehicle. With end tongs no top loader is needed, as the two men on the ground, by holding the ropes attached to each end tong, can guide the log to position and then pull out the tongs and retrieve them after the line has slackened.

In some instances, long logs are loaded by first placing one end on the vehicle, then placing the next one on, and then, if necessary, adjusting it

FIG. 109. Heeling-boom spar-tree loader showing how long logs are manipulated with single tongs by raising the rear end under the armored boom. Oregon.

lengthwise. To move the log away from the loader, the tongs are placed in it nearer than the end of the boom, so when a pull is placed on it the log will slip away. To move it toward the loader, the tongs are placed in the log beyond the end of the boom.

SPECIALIZED LOG LOADERS

Since logs are the products obtained on large-scale mechanized harvesting operations, most loaders and loading-equipment accessories are designed to handle logs. They vary in design, depending upon the size and length of the logs to be handled and the kind of vehicles to be loaded, as well as the method of skidding and yarding the logs. There are certain well-defined types of log loaders and these, together with certain other ingenious types, are described in the following paragraphs. All makes and kinds are not included because of the great variety of some types.

All logs, of course, are not destined for sawmills; much timber harvested in the form of logs is used in pulp mills, veneer mills, stave mills, handle mills, dimension mills, and many other kinds of mills. Regardless of what a log is to be used for, however, it handles as a log; so in discussing equipment used in harvesting, it is not necessary to distinguish between logs to be used for sawing into lumber and logs to be refined by other utilization processes.

Harvesting operations in a forest are constantly moving from place to place, and all equipment, therefore, must be movable. Therefore, all kinds of loading equipment and loading machines are so designed that they can be readily moved. Mobility is provided in some types of loaders by mounting them on skids so they can be pulled short distances, as in the case of some A-frame loaders. Other types of loaders are designed to be carried from location to location on other vehicles such as trucks and railroad cars. Self-propelled loaders can, of course, be driven from place to place where needed and require no assistance.

Loaders designed for loading railroad cars are moved either on their own trucks or on a log car. Hence, when they are in operation loading log cars, the track must be clear so cars can be spotted. Because loading is done mostly along single-track spurs, the loader can not occupy the track while loading or only one car could be loaded, at most two, before they would have to be switched out and another one or two empties spotted. Various ways of keeping the track clear for log cars while loading have been developed and incorporated in the designs and construction of railroad log loaders.

Loaders that are stationary while loading either are off the track at one side, as in the case of the Lidgerwood combination skidder-loader, or straddle the track, permitting empties to pass beneath them. Types that move while loading, such as the Barnhart, pass over the tops of the empty cars, moving from car to car as each load is completed, but a locomotive must be in attendance to keep the loader and empties spotted where there are logs along the right of way.

Whether or not a loader can be stationary for a particular operation depends upon the skidding method. When logs are concentrated at separate points for loading, the loader can be and should be of the stationary type. Or, conversely, with the stationary type of loaders, skidding must be so conducted that the logs are concentrated at loading points at intervals along the track or road.

Logs can be skidded all along a road or railroad, instead of to yards at intervals, only when the loader can move along it to be within reach of the logs and at the same time have a vehicle spotted for loading. Self-propelled loaders and the types operating over the tops of log cars with a locomotive spotting them are adapted to this kind of loading.

Loaders that remain stationary while loading can operate only when logs or other products are skidded to them. Both skidding and loading, therefore, are carried on simultaneously. If loading were deferred until skidding had been completed, the accumulation of logs would be scattered over such a large area or the pile would be so large that the loading machines would have difficulty in functioning at all, let alone efficiently.

FIG. 110. A McGiffert loader loading logs on flatcars in a Southern swamp. Note that the train of empties has been switched beneath and behind the loader. As a car is loaded another is pulled into place forcing the loaded car ahead. (*Courtesy of Clyde Iron Works Sales Company.*)

When loaders are movable and do not of necessity remain stationary when loading, skidding can be completed before loading starts, as the logs can be loaded even if they are yarded all along a railroad or truck road, provided they are not banked or yarded out of reach of the loading cable.

The McGiffert Loader. This loader is a railroad loader only and is not adapted for use with truck logging. The machine consists of an elevated platform, or deck, on which are mounted a power unit, drums, cables, and a partially swinging boom with the usual cable block. The legs supporting the deck are wider than standard railroad gauge. The lower ends of the legs curve inward so they can rest on the ends of the ties and not interfere with the passage of log cars. Retractable railroad trucks are attached to the frame below the deck. When the loader is spotted for loading, the trucks are drawn up beneath the deck to clear the track for log cars. When the loader is to be moved, the trucks are let down so the wheels rest on the track, and power from the power unit raises the machine enough to lift the legs clear of the ties. The loader then is moved by its own power applied to the trucks, and at the next location the trucks are retracted, which permits the legs to settle again on the tie ends.

When empty log cars are switched to a McGiffert loader, they are pushed beneath it by the locomotive until the last one is in loading position. When the car has been loaded, a line is run from a drum on the loader to one of the empty cars, and as the line is drawn in the cars are moved forward. The next car is spotted and loaded, and the procedure is repeated until all are loaded.

The boom of a McGiffert loader has only a partial swing to either side, enough to place the loading block over logs yarded at either side. The logs are raised by either end tongs or center tongs, the boom swings them over the cars, and they are lowered into position on the car being loaded.

Skidder-Loader. Cable skidders of the Lidgerwood type consist of machines mounted on heavy railroad cars with spars 40 to 50 ft high. Several cables are rigged so that logs are drawn from all directions to the skidder by dragging them partially or entirely elevated on an overhead cable system. Since all logs are brought close to the spar on the machine, the logs must be loaded out as they are skidded to keep the accumulation from getting so large that skidding would be handicapped.

Therefore, a loading boom is always placed at one end of the skidder. It has a partial swing to either side for easy handling of logs. Skidders of this type, since they skid and load simultaneously, are spotted on a short switch track adjacent and parallel to the main spur. This keeps the main track clear for switching log trains, both empty and loaded, and keeps the machine near enough to the track so logs can be loaded out. Cars are spotted for loading by fastening a cable to a rear car, as in the case of the McGiffert loader and similar types of loaders.

Obviously this method of loading is possible only with cable skidding, which brings the logs to concentration points at intervals of about $\frac{1}{4}$ mile along railroad spurs or main lines. In fact, it is the only loading method reasonably possible with the type of skidder which has a spar mounted on it. Also, it is used only with railroad operations, as these skidders are always mounted on railroad trucks.

Surry-Parker Loader. This loader consists of an elevated platform, or deck, which is supported by square timbers, one at each corner of the platform, in such a manner that the platform can be raised and lowered. A jack and screw at each corner operates on the timber supporting leg so the height of the platform can be adjusted by hand operation of the jacks. The power unit, drums, controls, and boom are on the platform.

In operation the platform of the Surry-Parker loader is adjusted sufficiently high to permit the passage of empty cars beneath it, much as in the case of the McGiffert loader. It differs from the McGiffert, however, in that it must be lowered on an empty car to be moved, as it includes no features that give it independent mobility. Smaller types have been designed for loading trucks and these are lowered on empty trucks when they are to be moved. Once lowered on a car or truck, the legs of the Surry-Parker loader are raised to clear obstacles, until the machine is spotted at another loading set. Then the legs are lowered to the ground and, by means of the jacks, the platform is raised sufficiently to give the required clearance. The boom on the simpler truck loaders of this kind usually is of the self-centering type. On the larger railroad types, booms with a partial swing to either side are more common.

FIG. 111. A Surry-Parker type of log loader which operates similarly to the McGiffert but must be moved on a flatcar instead of on its own trucks. Virginia.

When loading railroad cars, the machine operates much in the same way as the McGiffert, in that the empty cars are switched beneath it to the rear and pulled in place as needed for loading. When used for loading trucks, it is not necessary to pass the trucks beneath the platform, which would, in fact, be impossible because of the height of the cabs. Instead, empty trucks back up to the loader, spotting their trailers in loading position in front of the machine. The chief advantage of the Surry-Parker for loading trucks is the ease with which it can be moved.

Decker Loader. This loading machine is suitable only for railroad operations, as it is mounted on railroad trucks. It consists of two decks. The upper one supports the loading machinery and boom; the lower deck is just above the wheels and has on it a railroad track, at each end of which are hinged lengths that can be lowered to rest on the spur track on the ground. This arrangement permits the switch engine to push a string of empty cars over the lower deck and through the machine. Empty cars can be spotted for loading by pulling in a cable attached to them in much the same manner as with McGiffert and Surry-Parker loaders. The Decker loader is self-propelled to the extent that it can be moved to the next location under its own power.

FIG. 112. Loading southern pine logs with a Barnhart-type loader using a crotch line and end tongs. Arkansas.

Barnhart Loader. The Barnhart loader is strictly a railroad-car loader. It consists of a crane-type power and control unit with a boom capable of being raised and lowered, and is so designed that it can be revolved in a complete circle, cab and all. The entire machine is mounted on wheels, of small diameter and of standard railroad gauge, which run on steel tracks on top of the log cars. The tracks may either be spiked to the floor of flatcars or to stringers on the bunks when skeleton cars are used. Hence, this loader can be used only with log cars that are equipped with tracks.

In operation, a switch engine with a string of empties and the loader are spotted on a spur where logs are banked ready for loading. The engine is at the open end of the spur, which is the end toward the mill, and the loader is on the second car toward the dead end of the track. The end car next to the locomotive is loaded and then the loader is moved to the adjacent empty car by running out a cable from a drum and fastening it to a rear car. By drawing in on the cable the machine is pulled off the car and on to the next car. The gap between cars is bridged by two lengths of U-shaped steel placed over the rails and extending between the cars. When the machine has been spotted on the car, it is held firmly in place without end play by fastening a short cable or chain from it to the car at the loading end, with tension being kept on the line used in moving it.

FIG. 113. A loader that moves over the tops of the empty flatcars to be loaded by sliding on skids. This modern machine can revolve and load from any position. (*Courtesy of American Hoist and Derrick Company.*)

A locomotive must always be in attendance to keep the loader spotted at points where there are logs to be loaded, as otherwise the loader would work itself out of position.

This method is rapid and adapted to short logs only. However, it requires a heavy investment, as the machine is expensive and a locomotive must always be kept with it while loading. Nevertheless, it is an efficient method with a large volume of logs to handle, when they have already been skidded to yards along both sides of the spurs.

American, or Rapid, Loader. This loader operates much the same as the Barnhart except that the boom does not revolve, having only a partial swing to each side, and it does not move from car to car on wheels but on skids, so tracks on the cars are not needed. They must be flatcars with solid bottoms and not skeleton cars, either of which can be used with the Barnhart loader. The American, or Rapid, loader is somewhat slower than the Barnhart but costs considerably less to keep up because of fewer moving parts.

McLean, or Hayrack Boom, Loaders. When logs are skidded to a spar tree adjacent to a logging railroad spur or truck road, the McLean boom

FIG. 114. Hayrack-type tree boom loader showing arrangement of skidding power unit and loading power unit. Oregon. (*Courtesy of the Hyster Company.*)

loading method is commonly used, particularly in the West, where large heavy logs must be handled. This method consists of a power unit, drums, and cables, and a fabricated boom heeled against the spar tree. The loading drums may be powered by the skidding power unit or the power unit may be independent of the skidder and usually is mounted on skids for ready relocation.

The boom consists of two parallel logs 20 to 30 ft long fastened together with cross members 2 to 3 ft long, so the finished boom appears somewhat like a horizontal ladder. A socket at the spar end is formed by the projecting log sides so the boom can swing freely around the spar tree. It is supported by several guy lines fastened farther up the spar in such a way that the boom is practically horizontal instead of at an angle upwards, as for most other loading booms.

Loading cables from the operating drums pass through two blocks on the boom, one near the spar and the other near the end of the boom. Tongs or slings or both are attached to each line so logs can be tonged or

FIG. 115. Crawler tractor equipped with a two-drum winch used as a power unit for loading with a tree boom. (*Courtesy of Allis-Chalmers Company.*)

swung in slings at two points, giving greater control over the logs during loading. The boom is moved laterally by power applied through a cable attached to the boom and leading to a drum through pulley blocks on trees or guy lines at appropriate places. A counterbalance, consisting of a short log section, sometimes called a "squirrel," pulls the boom back into place after a log has been loaded.

Because the boom is hung on a spar tree or pole, there is no danger of heavy logs upsetting it, as in the case of other types where only the weight

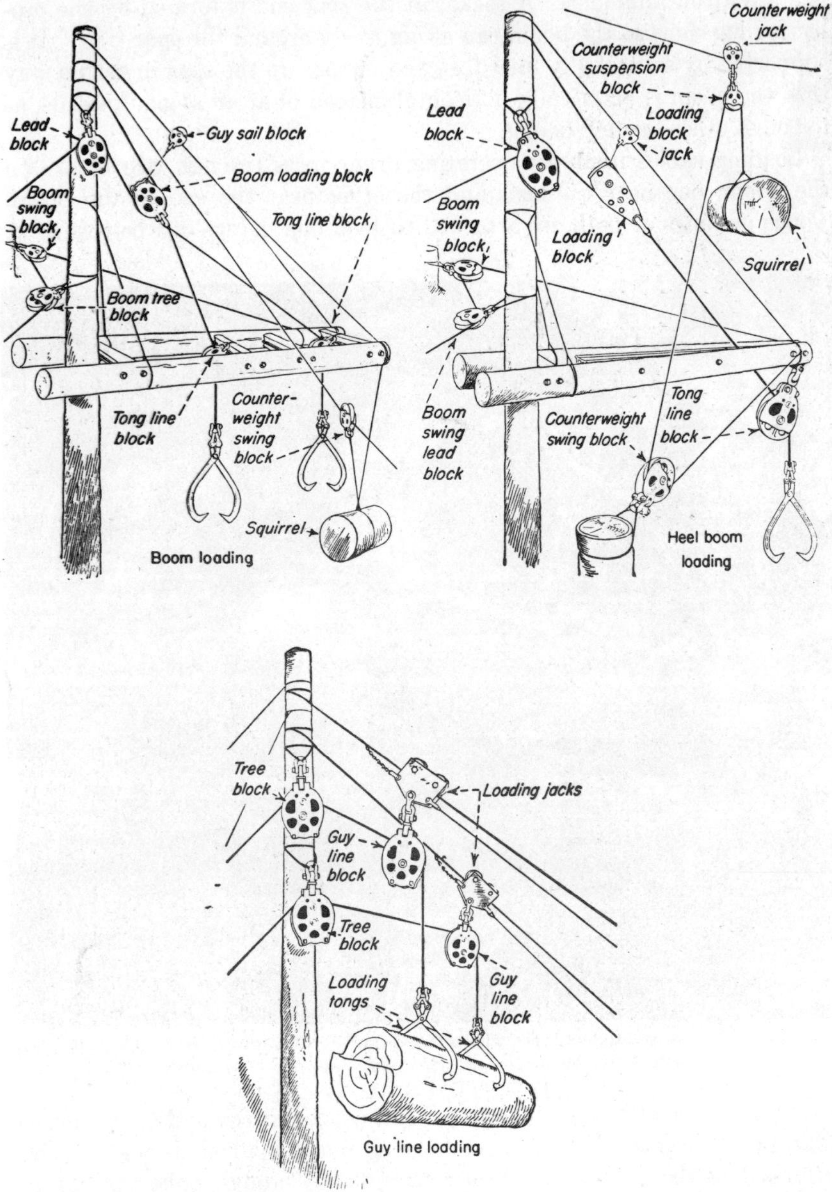

FIG. 116. Diagrammatic drawings of spar trees rigged for boom loading and guy-line loading. (*Courtesy of Young Iron Works.*)

FIG. 117. Two types of tractor-mounted loader booms used for loading trucks on Southern operations. (*Courtesy of Southern Pulpwood Conservation Association.*)

of the machine overbalances logs being lifted. Also, since there are two loading lines, there is less strain on either one.

In operation, the loading crew attaches the tongs or slings to the log to be loaded. The log is then raised under the boom and, by means of the cable controls, it is swung over the vehicle to be loaded and placed on the bed. Where slings are used they usually are looped over the tongs to avoid changing from one to the other several times a day. The power unit operating this type of loader, as with skidder power units, is usually mounted on skids and can be moved from place to place along the track or truck roads by fastening a cable ahead and pulling it along, or a crawler tractor with a double winch may be used.

TRUCK LOG LOADERS

With the rapid development of truck logging during the past 20 years, various types of truck loaders have been designed and used. They range from fully mobile crane-type loaders to simple, homemade rigs.

In medium to small-sized timber, a wide variety of homemade truck loaders are in use, which range from the A-frame type to the self-centering-boom type. Many such loaders are mounted on worn-out logging trucks and trailers that can be moved from place to place wherever log trucks can go. Some are designed to load trucks spotted broadside to them, and others are adapted to loading trucks and trailers spotted with their rears to the loader. Some consist only of a power unit, such as an automobile motor, and a winch mounted on a truck, to be used in loading with a tree block or guy-line block. Some are mounted on skids so they can be dragged from place to place. Two specialized types of truck loaders have been designed and are used rather widely, *i.e.*, tractor-mounted loaders and mobile crane-type loaders.

Tractor-mounted Loaders. Loading booms mounted on crawler tractors are suitable for loading trucks but are not adapted for loading railroad cars. A standard crawler-type tractor is equipped with a revolving boom and control seat mounted on top of it. The boom is fully mobile in that it can swing in a complete circle, move up and down, and be moved from place to place by moving the tractor.

Being mounted on a crawler tractor, the loader is capable of negotiating usual woods terrain and can load trucks in the woods and off roads. This loader can load logs or other products whether they have been bunched or skidded before loading starts, or whether they are skidded to the loading spot as loading is being carried on. In some instances, this loader also can do a certain amount of skidding when trucks are not available for loading; the loading line being pulled out by hand and fastened to logs, which are then pulled into position for easy loading.

Mobile Crane-type Loaders. Heavy-duty crane-type loaders mounted on crawler treads or pneumatic-tired wheels and fully mobile, with com-

FIG. 118. Loading pine logs in California with a mobile crane loader equipped with pneumatic-tired wheels on three axles, two of which are powered. Note that slings are used without tongs. (*Courtesy of Osgood Company.*)

FIG. 119. A log loader particularly well adapted for loading trucks since it is self-powered for mobility and has a cab and boom that can be revolved in a complete circle. Note that the boom is also adjustable. (*Courtesy of Michigan Power Shovel Co.*)

pletely revolving booms, are the latest development in truck log loaders. The largest and most versatile types have three axles, with driving power applied to two. The front end is supported by one axle with single wheels and the two rear axles have dual wheels. The wheels of the front axle and of one of the rear axles are drive wheels, which gives the loader great mobility even under adverse ground conditions.

Fig. 120. A mobile crane loader with boom armored for heeling loading, loading pulp logs on a western operation. (*Courtesy of Wickwire Rope Company.*)

These loaders are designed primarily for loading trucks, as they can move along truck logging roads and load wherever logs have been yarded. They are rapid and can handle large logs. When logs are so large that the weight of the machine alone would not give stability, side braces are extended so the side sway resulting from swinging the boom from side to side and picking up logs is eliminated.

When trucks are loaded by machines of this kind, skidding can be held to a minimum distance because it is not necessary to pull the logs to loader sets. All skidding, therefore, can be straight to the roadside instead of converging from all directions toward a series of points along the road. While self-propelled log loaders of the crane type could be used to load railroad cars, it is usually more economical to use one of the standard railroad loaders. The mobile crane type makes available to truck loading the same versatile heavy-duty loading power formerly limited to railroads.

Standard-type trucks are also equipped with power units or power take-offs and various types of booms. Some are specially designed types

FIG. 121. Loading with a specially designed loading-skidding boom and winches mounted on a standard-type truck in a Southern hardwood forest. (*Courtesy of Taylor's Machine Works.*)

with two drums that can be used for skidding as well as loading, while others are homemade units of comparatively simple design.

LOG TRANSFERS

Various kinds of overhead cable devices are in use for transferring logs from one vehicle to another or to yards for temporary storage. Generally such log transfers are used to take loads of logs off trucks and place them on railroad cars where there is a two-stage transportation system. Log transfers, therefore, are not in the same category as loaders that perform the primary loading operation in the woods. For one thing, they are stationary, so logs must be brought to them already loaded on trucks, and they may remain several months or even years in the same place.

Log transfers, when used, are located at some railroad- or truck-loading point adjacent to a body of timber all or most of which can be trucked or sleigh-hauled to the transfer point. This gives enough volume to justify the cost of installing a transfer system. Occasionally logs may be skidded to such a point by tractor, but the limited distance of economical tractor haul does not tap enough timber, in most instances. On the other hand,

FIG. 122. A twin-boom loader on a truck used in loading pulp logs on a trailer in Georgia. Note that each boom carries a block through which a cable extends, both of which operate simultaneously from a single drum. (*Courtesy of Southern Pulpwood Conservation Association.*)

trucks can haul for up to several miles and so can concentrate a large volume of logs at the transfer point.

To a limited extent log transfers are used to transfer logs from log cars on a logging railroad to cars on a common-carrier line for delivery to the mill. This may be necessary or desirable when log cars are not equipped with air brakes or side stakes that would permit them to be incorporated in freight trains, or when the logging railroad is narrow gauge.

PULPWOOD LOADERS

Mechanical methods of producing pulpwood have received much attention due to the increasing volume of pulpwood production and the desirability of developing mass-production methods. A number of so-called pulpwood mills have been developed; most of them are designed to saw and

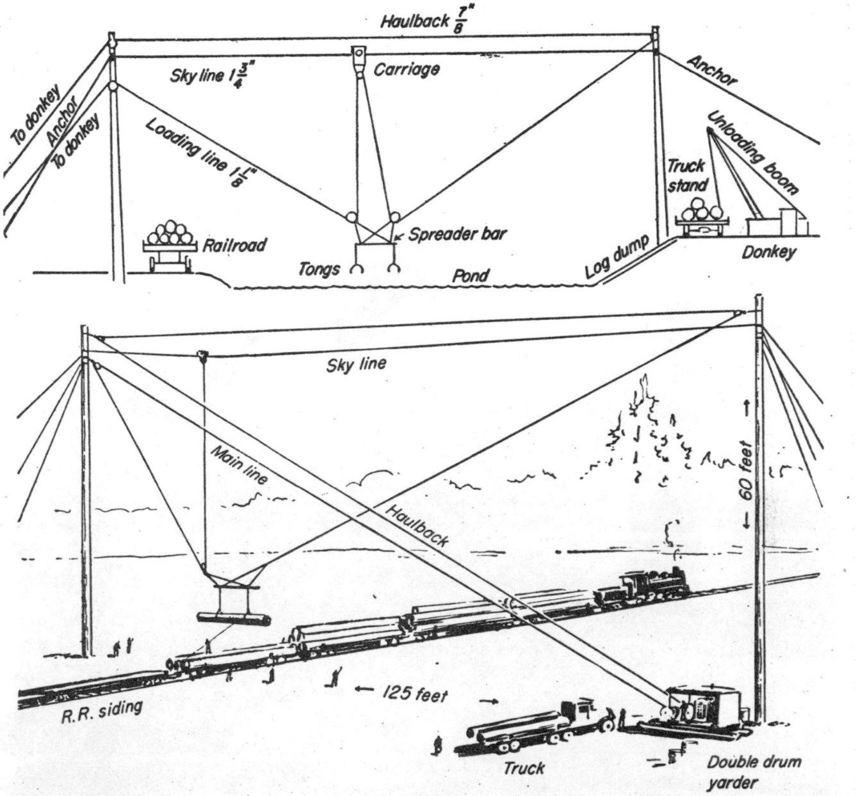

FIG. 123. Two methods of rigging log transfers and their operation. In the upper sketch logs are unloaded from trucks into a pond and then transferred to railroad cars. In the lower figure logs are transferred directly from trucks to railroad cars. (*Courtesy of the West Coast Lumberman, August*, 1938.)

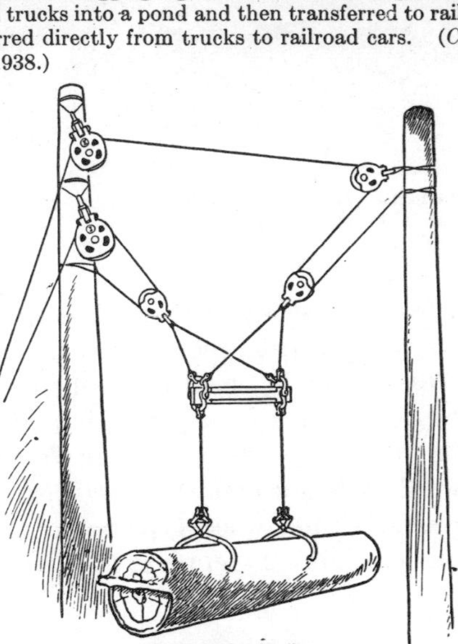

FIG. 124. Diagrammatic sketch of two spars rigged for a log transfer with a spreader bar. (*Courtesy of Young Iron Works.*)

load pulpwood on trucks in the forest. Trees are felled and skidded tree length with teams or light tractors, either of the crawler type or of the wheeled type, to a conveyer, which usually consists of dead rollers so the tree lengths can readily be pushed endwise to spot them for cross-cutting.

Bolts of pulpwood lengths are then successively sawed from the tree lengths by an operator who pulls a circular power saw of the swing type

FIG. 125. Transferring an entire truckload of logs to a logging railroad car with a spreader bar and cable system on a Western operation. (*Courtesy of Wickwire-Spencer Steel Company.*)

against it. Usually one man pushes the tree in place and another operates the saw. Sawing always begins from the butt or large end so any cutoff waste will be in the least valuable, small top section. As the bolts are severed from the tree they fall on a conveyer chain that lifts them broadside over the end of the truck, where one or two men place them on the load.

Most of the pulpwood mills are practically self-contained in that the saw and conveyer are attached to a wheel-type tractor and powered by it. This makes for easy and quick relocation when the timber at a setting has been cut and loaded. The conveyer rollers are placed on the tractor or a truck for moving and the saw and loader move with the tractor under its own power.

Successful operation of pulpwood mills of this type depends on integrating the felling, skidding, sawing, and hauling so that there are no delays. Since sawing must be discontinued when no trucks are available for loading, and since sawing and hauling must be stopped if trees are not brought rapidly enough to the mill, an unbalanced operation means lost time and

FIG. 126. The loading system used with pulpwood bucking mills. This entire mill is readily portable. (*Courtesy of Southern Pulpwood Conservation.*)

decreased production with proportionately higher costs. Pulpwood mills that saw and load in the forest have a daily capacity of up to 60 cords.

Pulpwood transfer loaders are used to a limited extent in some sections when banked or stored wood is loaded into railroad cars. Such loaders consist of sheet-metal conveyer troughs in which bolts are carried endwise from the ground into the car by endless chains with lugs at appropriate intervals. Pulpwood transfer loaders are similar to log transfers in principle, in that they do not perform the initial loading operation in the forest but move pulpwood out of storage after one stage of transportation has been completed.

CHAPTER 14

TRANSPORTATION OF PRODUCTS FROM THE FOREST

Forest crops, to have value, must be removed from the forest and delivered to manufacturing or processing plants where the wood is made into lumber, pulp, veneer, or other products or where the wood is treated with preservatives. Some few products obtained in harvesting, such as ties of durable wood, fuel, and hand-rived shingles, may be placed in the channels of trade direct from the forest without further processing or treating.

The major users of timber crops, however, need primary forest products as raw materials and a constant and approximately equal daily, or at least weekly, supply must be provided. Large sawmills, pulp mills, and similar manufacturing establishments commonly conduct harvesting operations in the forest and transport the products obtained to their mills.

Forest products are heavy and bulky and when large quantities are to be delivered regularly, some special means of transportation must be provided, unless public highways, common-carrier railroads, or waterways are available and can be used under existing state and Federal laws. When such transportation facilities are available, the movement of products from forest to mill or market is facilitated, but the actual hauling or driving still must be done.

The kind and size of transportation system needed to keep a wood-using plant or plants supplied with timber products depends upon the amount to be hauled and the distance from the forest to the mill or market where the products are delivered or sold. Obviously, the larger the mill, the more extensive must be the transportation system. Small units in many instances can dispense with a transportation system, as such, and simply operate trucks over public highways or receive the products of forest harvesting via common-carrier railroads

Because much of the forest area is in undeveloped localities or places with a minimum of transportation facilities, it is generally necessary for large users of forest products to install transportation systems, varying from truck trail roads to standard-gauge railroads, or systems of dams and stream improvements to make stream driving possible.

The kind of transportation advisable depends upon a number of factors, such as topography, distance products must be moved, volume needed each day, existing facilities, and concentration of timber.

RAILROAD TRANSPORTATION

Before motor trucks and modern highways were common, the only way large quantities of timber could be moved overland to a mill was by a logging railroad. Only in rare instances could common-carrier railroads provide this service, as their lines seldom tapped timber-harvesting areas. Therefore, logging railroads had to be constructed, owned, and operated by the companies developing the timber.

The era of railroad logging constitutes one of the most dramatic periods in American logging history. During the period of large-scale lumbering in the Lake states and the South, when the bulk of the virgin timber was removed, hundreds of log trains daily traversed the forests, hauled by locomotives burning wood. Later coal and even oil came into general use as a fuel. However, the development of trucks and the improvement in highway systems has resulted in the abandonment of railroad logging except in areas otherwise inaccessible or where existing logging railroads have been converted into main transportation arteries to which logs could be delivered by trucks.

Logging railroads are still used in Southern swamps and river bottoms where roads cannot be used throughout the year because of high water. In such cases, the track is usually built on piling, so high water does not interrupt the operation of trains, skidders, or loaders. In Eastern upland areas and in the southern Appalachian Mountains, logging railroads are still used to a limited extent for large-scale production, but truck hauling is now much more common. In the West, logging railroads are still extensively used where large bodies of timber are to be harvested over a period of years and where the haul is long and the daily volume large. Logging railroads, therefore, are still an important method of transporting products from forests to mills, primarily logs to saw mills.

Railroad Location. Because of the heavy expense involved in constructing a logging railroad, a great deal of care and thought should go into choosing the location of main lines and spurs. In rugged topography there is only limited choice, but in most cases there are several or more different possible locations. Thorough knowledge of the area to be tapped by the railroad is necessary to determine the correct location. This may be obtained by personally traversing it or by study of contour maps or both.

The entire system should be planned even if it is not to be installed completely at the start, which is usually the case, as it is uneconomical to invest money installing tracks that will not be needed for several more years. Spurs are put in only far enough in advance of use to permit the track to settle and be ready for traffic when needed. If the entire system is not planned at first, it may later be found impossible to extend tracks into certain portions of the area or unduly costly to do so.

The stand of merchantable timber seldom is uniformly dense throughout a forest area and, in addition, it may not be owned entirely by the company undertaking to harvest it and transport it to a mill. If a railroad were to be installed strictly from the standpoint of best and cheapest construction with respect to topography, it probably would be poorly situated for removing the timber owned by the operating company.

Logging railroads are planned, therefore, to permit transporting the most timber per mile of track, provided this will allow economical skidding. In locating main lines and spurs due care is taken, then, to build them where they will be as close as possible to the most timber, so long as construction cost for such a location is not prohibitive.

FIG. 127. Two kinds of logging railroad tracks in the South. Left, a spur laid on the ungraded ground on round pole ties and curved to avoid as many trees as possible; right, type of track often laid in the swamps. Another swamp type consists of pile-driven bents 4 to 6 ft apart with the rail spiked either directly to the caps or to stringers running over the caps.

Estimated construction cost and estimated cost of logging to the railroad and operating log trains over it, for several possible locations, will indicate the most economical location. Generally, logging railroads are planned to follow natural drainage systems in mountainous areas and to avoid streams in level to rolling topography. When railroads enter swamps there usually is not much choice of one location over another, except that the one which will open up the most area and the most timber to be harvested is the most desirable.

In estimating the cost of constructing a railroad such items as amount of cut and fill, type of soil or rock to be moved, grades, culverts and bridges, and curvature are involved. In estimating the cost of logging to a given railroad location such items as skidding distance, slope, stand per acre, and loading points are considered and the cost estimated. The location giving the least expensive combination of costs per unit of product to be transported naturally is the correct one for a particular operation.

The main-line location is particularly important, as it will be in use as long as the operation is conducted. Spurs leading off from the main line into the forest are used only long enough to remove the timber from the particular area they serve, after which the rail and ties are picked up and laid down again on a previously prepared spur roadbed. Thus, a railroad logging operation progresses along the main line until the entire area to be cut has been covered.

New operations commonly begin spurring at the near end of the main line to hold costs to a minimum until the operation is well under way. Those already in operation, but developing a new area, may lay all the main line and start at the far end, working toward the mill. The choice of which plan to follow depends to some extent on the financial position of the company, because if markets are good and prices high it may be advisable to make necessary future expenditures early rather than late in the operation, when the reverse may be true.

Since the main line will carry heavy traffic for the duration of the operation, it is important to hold grades to a minimum, especially "adverse" grades, or those which loaded trains must negotiate. Also, curves should be less on the main line than on spurs, as longer trains and, in many cases, large main-line locomotives do not go on spur track. Loaded cars are switched to the main line, where they are assembled into trains for the haul to the mill.

Railroad Construction. The location of main lines and spurs must, of course, be surveyed in and definitely staked out on the ground. The location is indicated by center-line stakes and, where excavating or filling is to be done, this is indicated by side stakes, as usual in highway and railroad surveying. In level topography, such as occurs in many parts of the South, the logging superintendent or foreman looks over the area ahead, sometimes on horseback, and then lines it in by means of the "three-stick" method.

This calls for placing a 6- or 7-ft pole in the ground at the starting point and then going in the chosen direction without losing sight of the starting pole. A second pole is then put in the ground, and every 100 ft or so a head-high stake is lined up with the last two or more behind it. By sighting on three sticks in this way, a careful man can carry a straight line through the woods. If it is necessary to hit a particular point because of adverse ownership or some other reason, a trial line may be run and then adjusted. Curves are also run in without the use of instruments, by measuring off tangents beginning at the point of curvature.

For more difficult topography where straight-line railroading is out of the question, railroad location should be and usually is put in by skilled railroad-location engineers or similarly trained foresters. A mistake of only a few feet or even less in mountainous localities could be very costly, not only in constructing the track, but in continual extra expense throughout the life of the operation if a grade were too steep or a curve too sharp.

FIG. 128. An incline railroad is sometimes used in steep topography to lower cars of logs downgrade. The cars are controlled by a cable with drum at the upper end. (*Courtesy of Pacific Car & Foundry Company.*)

Clearing Right of Way. After the location of a main line or spur track has been staked in the ground, clearing begins. This involves cutting and removing all trees and undergrowth that would interfere with grading or laying track. Main-line clearing is more complete than spur clearing, as the roadbed is graded and smoothed, while spur track frequently is laid on the ground without grading.

There is wide variation in the degree and method of clearing, ranging from complete removal of all trees, undergrowth, and stump to merely sawing the stumps flush with the ground. In the latter case the soil accumulated from grading and ditching covers the stumps and raises the roadbed level above them. By leaving the roots of trees intact, the ground is not disturbed and soft spots are prevented.

Merchantable trees cut from the right of way are left for the saw crews to cut up, but all other brush is burned or scattered, or sometimes, if the ground is soft, it is cross-laid on the ground where the track is to be placed to give increased support.

Crews clearing right of ways also commonly clear yards in which logs will be accumulated for loading. Yards may be continuous along both sides of a track or at intervals along the track, depending upon skidding and loading methods in use.

Grading. Once the right of way has been cleared, grading can commence. The amount and extent of grading and ditching depends on the amount of traffic that will travel over the track and the weights of the locomotives. In certain flat areas no grading is done, except of a minor nature, the track being placed directly on the ground and undulating with the topography. Whenever there is possibility of rainy weather causing soft track, the right of way is ditched, either by hand or by tractor bulldozers or grading machines. In Southern swampy locations where light locomotives are used, saplings and poles are laid lengthwise on the ground beneath the ties to give firmer support for the tracks, both main lines and spurs.

When grading has been completed, the right of way should stand for several months to settle before being used. When this has occurred, the track can be laid.

Tracklaying. The track consists of the usual kinds of ties and steel rail. Ties are commonly cut from timber adjacent to the right of way. They vary from small pole ties flattened top and bottom only to hewed or sawed squared 7 by 9 ties. Heavier, better track is needed in mountainous regions because track on grades requires sturdy ties to keep it from slipping and to support the strain of locomotives pulling loaded trains up grades. Hence, on such logging railroads full-sized well-manufactured ties, in some cases, even creosoted ties, are used when the track is to remain in use for a long enough period to justify the added expense. On spurs where tracks are laid down and then picked up to be laid on another spur, ties usually fail from mechanical wear (spiking) rather than decay. Hence there is little or no advantage in using creosoted ties that would have to be discarded while still well preserved.

Where no timber small enough for hewing into ties is found along the right of ways, the center cuts of larger trees are sawed into ties at the mill and sent back to the woods. In managed forests where future cuts are being provided for, no young trees of merchantable species are cut; only trees of little or no value are cut into ties, which helps the growth of the desirable species by removing competition.

To keep the roadbed firm in rainy weather it is necessary to provide adequate drainage in addition to ditches. Water must not be permitted to accumulate where it will soften the earth, so culverts must be put in where drainage passes beneath the track. Culverts vary from the simple type made of three logs to larger and more permanent types of corrugated steel or creosoted timbers.

Bridges vary from those over small low-banked streams to large structures bridging deep mountain gorges. Simpler types are of driven or framed pole-bents with stringers, while the more elaborate types are of multiple bents requiring a high degree of skill for their design and construction. As any type of bridging is expensive, it is avoided whenever possible by choosing the location requiring a minimum of bridge construction.

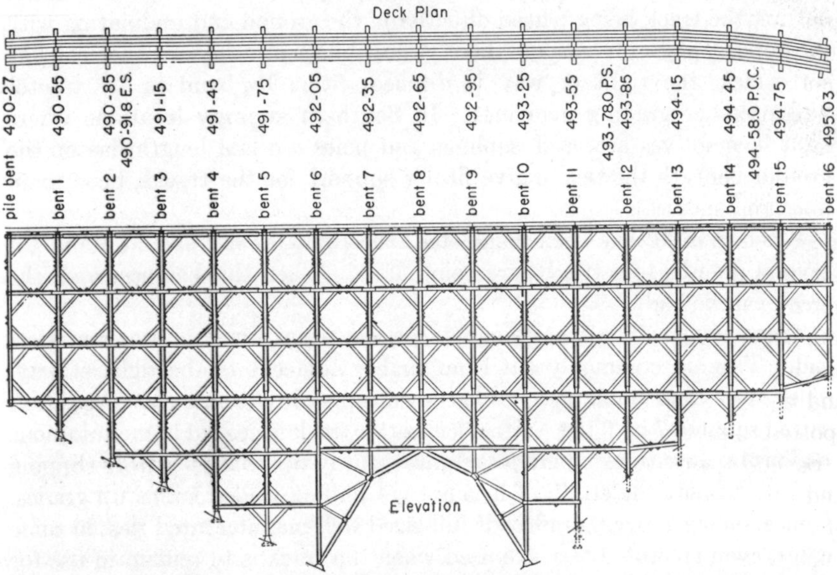

FIG. 129. Elevation and deck plan of the center or main section of the Baird Creek Bridge on the 100 line of Weyerhaeuser Timber Company, Longview, Wash. The portion shown consists of a three-hinged arch, surmounted by four 30-ft stories. The length of this section is 480 ft. It rises 235 ft above the low water level of the creek. All the material shown, with the exception of four panels of creosoted piling, is sawed timber, preframed, and pressure-creosoted. (*Courtesy of The Timberman, October*, 1940.)

Logging railroads in swamps and river bottoms subject to overflow for long periods are built entirely on pile bents instead of on the ground, with the exception of some spurs that are laid on brush on the soft ground during drier seasons. A trestle railroad bed of this type is expensive but without it the railroad would be under water for several weeks or months and not operable. In building trestle railroads, piling is driven in bents of two to three piles and then sawed off level and capped. The distance between bents varies with the weight of the rail, as the rail is spiked directly onto the caps, which serve as ties, and heavy rail with light locomotives will hold without sagging under a load if supports are not over 6 ft apart.

The pile driver moves along the bents on skids and behind it the track-laying crew lays the rail.

Rail used for logging railroads varies greatly in weight and size. Rail is measured by the number of pounds it weighs per yard; thus 40-lb rail weighs 40 lbs per yd of length. Rail for small locomotives and light service may be as light as 35 to 40 lb, but on main lines and on spurs over which heavy locomotives, skidders, and log cars must travel, it should be from 60- to 90-lb rail or, in some cases, even heavier. On spurs frequently only every other tie is fastened with one or two spikes, and no tie plates are used. Western logging railroads in many instances are as well constructed and ballasted as main-line common-carrier railroads. The logging railroad, then, varies as to kind of track with the kind of duty it is to perform, which ranges from light track on temporary spurs with no grading except leveling, to first-class track such as is found on common-carrier railroad systems.

On large railroad logging operations where considerable tracklaying has to be done, specially designed machines are used for this operation. It is not unusual for as much as ¼ mile of spur track to be picked up and laid down at another location per working day, so it is evident that this is an important job in large-scale logging. On small operations less elaborate machines are needed, sometimes consisting only of a flatcar on which rails and ties are stacked, with a small locomotive in attendance to keep the car spotted as near the end of the track as possible.

Several types of tracklayers are used, but all are so designed that ties and rail, together with spikes and fasteners, can be delivered to the end of the track being laid or loaded at the end of the track being picked up. Such tracklayers require the constant attendance of a locomotive to move the tracklayer forward or back, as the case may be.

In general, tracklayers are elongated flatcars with an apron at the receiving, or discharging, end with an endless chain that moves ties toward the working end when track is being laid and away from it when track is being picked up. The chains are operated by a small motor, and two men stack or unstack the ties on the car as they are received or are being unloaded. Men on the ground then place the ties on the roadbed and lay rail, which is also conveniently carried, under each side of the layer. As soon as the rail is laid down, it is gauged and partially spiked. The locomotive then pushes the layer to the end of the new section and the operation is repeated. When track is being picked up, the reverse procedure is followed, the locomotive pulling the layer back as rail and ties are picked up and loaded on it.

Light-weight track, including narrow gauge, is sometimes picked up in sections including ties by a small crane or elevator on specially designed tracklayers and placed behind it on a space provided for track sections. Laying track reverses this procedure. For this method obviously the rail lengths must be matched in each section, whereas with other methods, rail length can alternate, resulting in less undulating of the track with passing trains.

Locomotives. Practically all logging railroads are privately owned and are operated solely for the purpose of transporting logs from the forest to the mill, supplies and equipment into the forest, and harvesting crews to and from their places of work. The principal traffic, therefore, is log trains going empty to the woods and returning loaded from the woods.

FIG. 130. A rod locomotive headed for the mill with a train of cypress logs harvested in a Florida swamp. Note the well-constructed track.

Steam locomotives of the standard switch and freight types are used extensively. Also, on some operations where grades are steep and curves sharp, geared locomotives are used to obtain maximum power at a sacrifice in speed.

Steam locomotives are classified, first, as to the method of applying power, as rod or geared locomotives. The rod type has the pistons from the two cylinders, one on each side at the front, directly connected to drive wheels. Geared locomotives are of two types: one, the Shay type, applying power to all or most of the wheels along one side of the locomotive; the other type, the Heisler design, applies power to the axles of the drive wheels by means of a lengthwise shaft beneath the locomotive.

The cylinders of the Shay type are vertical and all are together on one side just in front of the cab. Their pistons are connected to a crank shaft that runs alongside all the wheels on that side of the locomotive. The shaft is geared to mesh with ring gears on each wheel, and as it revolves, it causes the wheels to turn, driving the locomotive. Although exhausts are rapid on geared locomotives, their forward or back motion is slow because of the sacrifice of speed to gain power. With the Heisler type, cylinders are located one on each side of the locomotive but at an angle, as the piston rods are attached to a center drive shaft. This shaft is connected to the axles of the locomotive by gears on both the axles and the shaft, so when it revolves the wheels are turned.

FIG. 131. A geared logging locomotive of the Shay type, showing the three cylinders on the side just ahead of the cab and the drive shaft driving all axles of the locomotive and tender by gears. Note the spark arrester on the stack to prevent starting forest fires. (*Courtesy of Lima Locomotive Works, Inc.*)

The advantage geared locomotives have over rod types is greater power for the same weight. Traction is good even with light weight because all wheels, even those under the tender, are drive wheels. Geared locomotives are widely used for hauling log cars in and away from spurs, which permits using steeper grades and sharper curves than would be possible with rod locomotives. The drive shafts have one or more universal joints, and wheels are pivoted so that a geared locomotive can "bend" in going around sharp curves, which is not the case with rod types.

Diesel locomotives are now coming into use. They are light in proportion to their power. Also, in some very limited instances, electric types are used. For most logging operations, however, steam locomotives are entirely satisfactory.

Comparison of a Shay Geared and a Direct Locomotive of Equal Tractive Power

Characteristic	Shay locomotive	Direct rod locomotive
Total weight	158,000 lb	262,800 lb
Weight on drivers	158,000 lb	130,000 lb
Dead weights	No useless weight; all weight on driving wheels	132,800 lb
Average weight per pair of drivers	26,330 lb	32,500 lb
Rated tractive power	30,350 lb	30,350 lb
Drawbar pull on grades:		
2% grade	26,560 lb	22,880 lb
6% grade	20,240 lb	12,370 lb
10% grade	13,920 lb	1,850 lb
Total wheel base	40 ft 2 in.	60 ft 0 in.
Rigid driving wheel base	4 ft 4 in.	14 ft 0 in.
Sharpest curve engine can take	100-ft radius	200-ft radius
Lightest rail advised	45 lb per yd	60 lb per yd

Courtesy Lima Locomotive Works, Inc.

Logging locomotives formerly used wood for fuel, but modern ones use coal or oil. The great danger of fire from logging locomotives due to sparks makes it necessary to screen the stacks if coal or wood is used. Wood-burning locomotives are especially hazardous, as large burning embers are scattered promiscuously, especially along grades where the draft is forced.

Locomotives are used, in addition to hauling log trains, for spotting loaders, tracklayers, pile drivers, and skidders, if the operation is one using cable skidders on rail tracks.

Log Cars. Log cars are of several types, varying from flatcars with platform bottoms to skeleton trucks for long logs. The usual type has standard railroad trucks with a timbered bunk over each truck. Trucks are spaced with timbered reaches to suit the log lengths of the particular operation on which they are used. The usual automatic couplers are commonly provided, but in some cases a simple link-and-pin coupling is used. Cars are not ordinarily equipped with air brakes, so a moving train is braked solely by the locomotive.

Logs are placed on the cars by loaders and are held by stakes at the sides, either inserted in sockets or of a specially designed type that lock in an upright position but can be tripped from the opposite side of the car or from one end. This is important in unloading, as there is considerable danger to men removing stakes to permit logs to roll off into a millpond or other concentrating places.

The capacity of log cars depends on the size and length of the logs carried. Loads of logs of standard length (12, 14, and 16 ft) run from 2,000 to 3,000 ft per car, but longer logs, of course, run proportionately more. In unusually large timber such as is found on the Pacific coast, logs are commonly cut up to 40 ft in length and because of this, and their large diameter, loads may average up to 10,000 ft per car and even more.

In some instances log cars are delivered from the logging railroad to common-carrier railroads for delivery to the mill, and in such cases the log cars must conform to Interstate Commerce Commission requirements and

FIG. 132. Method of loading logs on flatcars with stakes. The stakes are wired across to provide added security against the load breaking loose in transit. Wisconsin. (*Courtesy of Wisconsin Conservation Department.*)

be equipped with adequate air brakes so they can be incorporated in trains. Also, the loads on such cars must be between rigid stakes higher than the load, with the stakes wired together across the top of the load as insurance against the logs breaking the stakes and rolling off, with great possibility of damage. In other cases log trains are permitted to operate over common-carrier track provided they are operated by personnel approved by the I.C.C. and carry a caboose. As an added precaution, a railroad trainman, usually a conductor, accompanies the train to be certain it travels under correct interpretation of orders. In the latter case the log train does not necessarily have to be equipped with brakes or an ICC-inspected and -approved locomotive.

Operation of Logging Railroads. As the principal reason for operating a logging railroad is to transport loaded log cars to the mill and to return empties to the woods, log trains have the right of way over a logging railroad. Only in rare instances are double-track systems operated; so trains must be dispatched and operated on schedule. The logging superintendent usually holds this responsibility or delegates it to some one he can check with frequently.

One trainload of logs a day is delivered to a mill, except in the case of unusually large mills. One train of 50 log cars delivers from 75,000 to

FIG. 133. A trainload of southern pine logs showing a type of skeleton log car often used. Binder chains are permanently fastened to the car bunks on the near side and fastened to a release hook on the other so that logs can be readily unloaded at the mill. Note the sawed ties and the well-graded track.

100,000 ft; more if the logs are large and long. Thus, one log train a day can supply a large mill with logs.

Empty log cars are returned to the logging operations by the main-line logging locomotive. These may be taken directly to a loader, or to a switching point from which they are taken to the loader or loaders as needed. This is done by the main-line locomotive if the operation is comparatively small, or by smaller switch engines on larger operations.

Empties are spotted at loading points once a day or sometimes twice a day. It is good practice to have an afternoon spot of more than enough cars for the rest of the day so work can commence the following morning without delay. When more than one loader operates, the loaded log cars are switched to the main line, as a rule, and made into one train for the mill. If not, they are taken directly to the mill from the loading points.

Suitable water must be provided for steam locomotives where it can be furnished to them without the necessity of traveling too far for it. A water supply may be either from wells and overhead tanks or from streams or ponds, but care must be taken to avoid using water that will injure the boilers.

FIG. 134. Skeleton log cars in use on a Western logging operation, showing method of loading logs without binder chains. Note that the log cars do not have brakes, and hence a train of log cars can be braked only by the locomotive. Note also the well-constructed track. (*Courtesy of Pacific Car and Foundry Company.*)

The operation of log trains, both empty and loaded, must be geared to the requirements of the mill, and in most cases the system must be capable of delivering more logs per day than the mill requires. While bad weather may shut down cutting and skidding operations, loading and railroading must go on as usual. Nevertheless, there can be serious and unavoidable delays to the operation of trains, so there must be a surplus of logs at the mill sufficient to run it for two or three days if trains are blocked.

Regardless of how carefully log trains are operated, there are occasional wrecks, usually derailments. Ordinarily these can be cleared up within a few hours, but some can be serious enough to take a locomotive out of action for a day or two, or longer. Also, landslides may block the track, flooded streams or fires may destroy or damage bridges, and strong winds may throw trees across track; any one of such occurrences would delay the operation of log trains.

A railroad logging operation, therefore, must be equipped and ready at all times to handle wrecks and to repair damage to the tracks caused by wrecks or by natural hazards. Derailments occur most frequently on

FIG. 135. A method of loading short logs on flatcars on a Western logging operation. Two stacks of logs are loaded on each car. Note method of passing binder chains over lower portion of each load and topping it off with more logs which hold binder chains tight. In the above illustration the logs are skidded tree length to the landing and cut to short log lengths before loading. (*Courtesy of Allis-Chalmers Mfg. Co.*)

spurs, as the tracks there are less well constructed than on the main line. When a locomotive or log car is simply derailed and has not turned over or gone into the ditch, it can be placed back on the track without the assistance of cranes or other locomotives. This is accomplished by using wrecking frogs, which lead the wheels back on the track. Serious wrecks require the use of other locomotives for pulling or cranes for lifting locomotives and log cars back on the track.

Maintenance of a logging railroad is no small part of the operation. Section crews must be provided to keep the track correctly ballasted and gauged and to replace ties when needed. They also maintain ditches to drain low stretches of track and keep culverts and bridges in good order. Switches, especially, must be kept in good order to prevent derailments caused by cars' or locomotives' "splitting a switch."

Rolling stock must also be kept in good running order and shops are provided for this purpose. From time to time locomotives must be overhauled and this takes them out of service sometimes for weeks. It is ordinarily not practical to send them away for overhauling and repairing, so a suitable shop manned by competent mechanics is a necessity. Likewise, log cars, loaders, pile drivers, and all other equipment need frequent repairs and occasional rebuilding; and as there is usually no available place to send such equipment, or it would be too expensive to keep it out of service so long, adequate provision must be made to do such work either at the logging headquarters camp or at the mill. Logging shops are also used to repair and service certain other logging equipment such as skidders and tractors.

Construction and Operating Cost. Wide variation exists in the cost of constructing and operating a logging railroad, depending upon the kind of topography and the class of construction. In level topography, construction costs are at a minimum, and they are highest in mountainous topography requiring rock excavation and bridging gorges. Trestle construction in swamps is also costly railroading as compared to construction on level dry land. Even the cheapest railroad building costs around $3,000 per mile, and more costly construction ranges up to $50,000 per mile or more.

Operating costs including maintenance of track and equipment also vary with the size of locomotives, the weight of loads, and the steepness of grades. Railroad construction and operating costs vary between operations of the same general type in the same kind of topography. Even on the same operation they vary from year to year with fluctuations in wages for labor, cost of materials, weather conditions, and topographic obstacles encountered.

TRUCK TRANSPORTATION

Railroads are used primarily for transporting logs because on large operations concentrated harvesting is conducted that produces a large volume of log tonnage from a comparatively small area. On many Western and Lake states as well as some Eastern and Southern operations, the logs are actually used in the manufacture of pulp, but in so far as the operation is concerned, it is a logging operation. Poles, piling, and similar products often are moved from the forest over railroads, but seldom, if ever, are railroads installed primarily for the removal of these products; they move out over railroads installed primarily for transporting logs.

Trucks, on the other hand, are much more versatile and are widely used for transporting logs and all other forest products. Many large logging operations are conducted with fleets of trucks operating over systems of roads as elaborate in layout as logging railroads. They also are used individually or in small fleets, as the need may require, for hauling logs to small portable mills or for hauling pulpwood, ties, veneer bolts, poles, piling, posts, and, in fact, all other forest products. Large-scale trucking operations for transporting forest products from the forest to a mill, railroad loading point, waterway, or other destinations must be organized and operated with the same degree of skill as is required for a railroad operation.

FIG. 136. Binding loads of logs on trucks. *A.* Corner binders of ⅜-in. electrically welded steel on bunk logs, run through rings and rollers on end H-beam bunks and made fast with fit locks. *B.* First wrap, 25 ft of steel ½-in. cable and 10 ft of chain, wrapped around the second logs and buried in the middle. Fastened by cable binders and fit hooks. *C.* Second wrap of steel cable and chain, 43 ft long, overall, run over the third or fourth log and buried. Fastened same way. *D.* Top wrap of ⅜-in. cable and chain fastened by binders and hooks. (*Courtesy of The West Coast Lumberman, December, 1937.*)

Private roads frequently are constructed on Western operations for the sole purpose of operating log trucks over them, all the way from the cutting locations to the mill. When private roads are used, no state laws apply as to the weight of load that may be carried or its width, height, or length. Operations using private roads are in much the same category as logging railroads in that the roads are constructed and maintained by the owner for his own use only.

When trucks haul forest products over public highways all or part of the way, they must be loaded in conformance with state highway-department regulations. Frequently it is necessary to construct roads into the body of timber to be harvested, but the haul from the forest to the mill is over public highways.

Any road building for transporting forest products must be as carefully planned and laid out as are logging railroads, and the principles discussed for planning railroads apply equally to planning truck-road systems.

Truck-road Location. The kind of road to be constructed or used in a truck logging operation depends on the kind and volume of products to be

FIG. 137. Above, a type of trailer used for hauling pulpwood in the South. The trailer is readily detachable and can be spotted for loading, while the tractor hauls a loaded trailer to the mill. Such a trailer holds 4 to 5 cords. (*Courtesy of Southern Pulpwood Conservation Association.*) Below, a log truck and trailer of the type widely used in the South and East. Note two binder chains holding the load.

transported and the ground and topographic conditions. When only a few truckloads are to be hauled from one point or cutting area, obviously it would not be economical to spend considerable sums in building roads. Hence, only trails are brushed out and only a minimum of expense is incurred.

Some logging operations are conducted in very light stands with as little as 1,000 ft per acre removed. Many pulpwood-, veneer-bolt-, and pole-cutting operations likewise remove only a few truckloads per acre, and road expense must be held to a minimum. Harvesting only small amounts per acre may be due to the fact that only scattered merchantable trees are present or because the operation intentionally is held to small volume per acre to make thinning or other improvement cuts, leaving half or more of the timber to grow.

Although only a small amount is to be removed from the forest, it is still advisable to plan carefully the locations of even such roads as are only to

FIG. 138. A heavy-duty log truck equipped with a long reach so that logs of varying lengths can be hauled. British Columbia. (*Courtesy of Hayes Mfg. Co., Ltd.*)

be brushed out for light temporary traffic, so the minimum distance will have to be covered to reach the products to be removed. Road location must, therefore, be coördinated with skidding or bunching, as in some cases trucks are loaded from the stump and in other cases loads are bunched or assembled, so fewer stops and less road construction will be needed. When light harvests are to be made at frequent intervals, say every 5 or 10 years, then more care and attention is justified, as the roads or even brushed-out trails will, over a period of time, be used to transport as much volume as though the stand had been cut heavily the first time.

As in locating logging railroads, truck roads must be so located and constructed that the most economical cost is obtained per unit of product removed. This means close coordination between bunching, skidding, and road cost, giving due regard also to the size of load and speed of travel

FIG. 139. A heavy-duty log truck and trailer on a Washington logging operation. (*Courtesy of Kenworth Trucks Company.*)

FIG. 140. A heavy-duty truck and trailer designed and used for hauling bundled pulpwood in eastern Canada. (*Courtesy of Hayes Mfg. Co., Ltd.*)

possible over the roads. Skidding and road cost may be low, but if the road is so rough or soft that only partial loads can be carried, the hauling cost would be too high and the resultant total cost could be exorbitant. In such cases an additional expenditure for road improvement would be most economical.

Care, however, must be exercised to avoid the other extreme and spend too much on roads. A point of diminishing returns is reached where the expenditure of additional money results in less and less saving until an actual loss is incurred. If an expenditure of $100, or say $1 per 1,000 ft, results in a reduction of trucking cost, through permitting larger loads, of only 76 cents per 1,000 ft, then the added road cost is unjustified, except on the basis that the road will be used again in 5 or 10 years with a saving at that time of a similar amount.

The more timber to be removed per acre, the greater are the benefits in reduced costs that will result from installing the right amount of road per acre and the correct kind of road. Generally, the heavier the timber stand,

FIG. 141. A private truck-logging road, gravel surfaced, in California. It is 24 ft wide with a 12-in. rock base and a 4-in. surface. Such roads are common throughout the Western forest areas where heavy truckloads are hauled, as state highway weight limitations do not apply to private roads.

the closer together can roads be built to shorten skidding distances. A knowledge of the cost of skidding products of various kinds and sizes for varying distances is, therefore, necessary in planning road intervals and the type of road to be built. Method of loading also has a bearing on the amount and kinds of roads that should be built. When a power loader is used, logs will be skidded to the loader and roads should be so located that this can be done at points along the roads. When hand loading is used, as in pulpwood harvesting, then it may not be necessary or advisable to skid to roads, as the trucks can negotiate quickly brushed-out woods trails and the products can be loaded directly from the place of cutting.

The equivalent of the main line in logging railroads is, in truck hauling, the public highway or, in certain cases, the main private hauling road. In

(a)

(b)

(c)

FIG. 142. (*a* and *b*) Road mats constructed of low-grade hardwood lumber used on unimproved woods roads, and (*c*) stored ready for use when the ground gets soft. Such mats are used on a number of logging operations throughout the South.

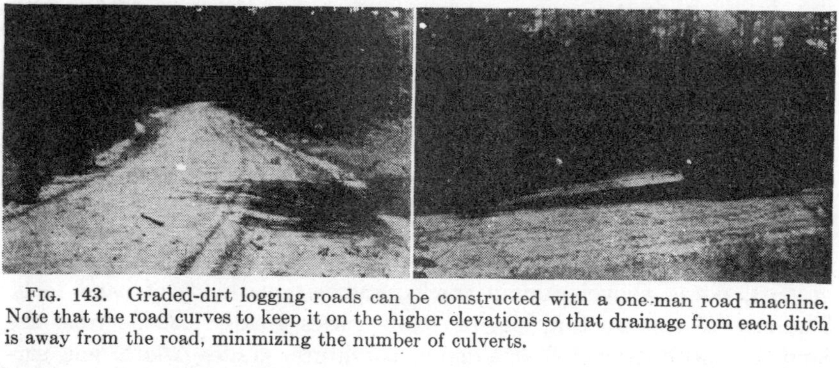

FIG. 143. Graded-dirt logging roads can be constructed with a one-man road machine. Note that the road curves to keep it on the higher elevations so that drainage from each ditch is away from the road, minimizing the number of culverts.

most areas, however, private roads are built only for access into the forest area being harvested; public roads and highways being used to transport logs, pulpwood, or other products to the mill. In the Pacific Northwest and in the redwood region, first-class (but not paved) private roads often are necessary as the heavy loads of logs hauled in many cases exceed the state highway load and weight limitations. Where it is impossible or impractical to build private roads all the distance to the mill, load size and weight must be kept within legal limits.

Construction of Truck Roads. The simplest type of truck road is a brushed-out trail, so located as to avoid stumps, trees, holes, rocks, and other obstacles. Such roads are not really cleared, as roots are not removed and no grading or other improvement work is done, except to install temporary bridges where it is necessary to cross streams.

FIG. 144. Logging roads constructed through a large private forest in North Carolina. These roads are graveled to make truck hauling independent of weather conditions. The roads are spaced about 1 mile apart and are used continually since the timber is cut selectively every few years.

When duration of use, volume to be removed, and saving in hauling justify it, more elaborate roads are built. How much can be invested in grading, bridging, and surfacing a truck logging road depends, as has been shown, on the volume of traffic, both daily and in the aggregate, that is to be moved over it and the size of each load. All types are found between the extremes of brushed-out trails and graveled private highways graded to low gradients and a minimum of curves.

The essential steps in road construction are: first, location; second, grading; third, bridging and culverts; and fourth, surfacing. These operations are performed with the type of equipment usually used on public highway-construction projects and consist of bulldozers, tractor graders, power shovels, dump trucks, and various kinds of hand tools. The cost of constructing logging roads varies greatly. Aside from the kind of road constructed (maximum curvature, grades, width, and surfacing), the kind of soil and the amount of rock to be moved, as well as the number and kind of bridges necessary, all cause wide fluctuations in cost.

It can be seen, therefore, that each project involving the location and construction of roads for logging or for hauling other forest products is a separate and distinct undertaking that must be gauged to the remainder of the operation and to the locality. No fixed set of specifications can be drawn up that will apply to even a few truck logging operations. Each must be planned and constructed quite apart from others.

It is true, however, that certain portions of a construction operation can be standardized and made uniform with others. Such items as width, degree of curvature, maximum grades, depth of surfacing, kinds of culverts, and so on can be used on all road systems of the same class of construction. The other items of cost, however, vary so greatly that uniform cost is unknown.

Equipment for Truck Hauling. Forest products can be hauled by trucks alone or in trucks and trailers of various kinds. Trucks varying all the way from standard commercial 1½-ton sizes to specially designed trucks or tractors to which trailers are attached can be used satisfactorily in transporting timber products, depending upon the kind of product. Western logging trucks nearly all are equipped with heavy-duty trailers, designed and built to withstand heavy loading because of the large size of the timber ordinarily cut.

In the East and in sections of the Rocky Mountain region where the timber is not as large, lighter equipment is preferable. For hauling pulpwood, ties, bolts of various kinds, and even small saw logs, the standard 1½-ton commercial truck is widely used and renders satisfactory service. For hauling logs the chassis should be stripped to provide a bunk for the front of the load and a trailer attached to support the rear of the log load.

Pulpwood is hauled on trucks both with and without trailers, depending upon the preference of the individual operators concerned. When pulpwood trucks leave the roads and go into the woods to be loaded, trailers seldom are practical because of the uneven and frequently soft terrain and the short turns and large number of stops that are necessary to obtain a load. On the other hand, when pulpwood is bunched along good roads, trailers can be operated efficiently, as the roads will support the heavier weights and permit normal travel speeds. In some instances, detachable trailers are used and one truck is required for two trailers; one being loaded while the other is en route. This gives greater output per unit, as but little or no waiting is required for loading.

Truck Maintenance. When large fleets of trucks are operated for hauling saw logs, pulpwood, or other products, facilities for maintenance, repair, and service of the equipment should be provided. This is advisable so that the operator is assured of prompt repairing and that it will be competently done. Also, as large-scale operations are commonly conducted away from commercial garages, an undue amount of time and expense would be incurred in taking the trucks to and from established commercial garages. Private installations usually consist of a garage where repairs can be made and tanks and pumps for storing oil and gasoline and delivering it to the vehicles as needed. Greasing and tire maintenance are important in keeping equipment in good operating condition, and this kind of service also is essential to any well-conducted truck hauling operation where the number of trucks operated justifies it.

When only from one to three or four trucks are used, as on small operations, maintaining shop and servicing facilities exclusively for them is not economical, except for minor repairs and adjustments. In such cases, local commercial garages and service stations are patronized.

When garage, repair, and servicing facilities are maintained, as is commonly the case when the number of trucks in operation makes it practical to do so, the equipment is of the usual garage kind and type of repair: overhauling, and tire-repair equipment. The usual stocks of spare parts and tires are kept on hand to keep repair time to a minimum.

Operation of Truck Transportation Systems. The principal feature of operating a fleet of trucks transporting forest products is one of maintaining a balance between loading capacity and hauling capacity and the maintenance of suitable roads. If trucks are required to wait for loads while others are being loaded, the time so lost reduces their output per day and thus increases the log-hauling cost. On the other hand if the loader has to stand idle waiting for a truck to return empty for a load, then that idle time increases log-loading cost.

Skill in maintaining a balance between loading and hauling thus is essential to economical hauling. The logging superintendent or foreman in charge of hauling any kind of forest products where fleet operations are conducted, therefore, has important responsibilities. Constantly varying hauling conditions such as length of haul, condition of roads, weather, condition of trucks, and size of product hauled all combine to make constant readjustments necessary to prevent lost time.

On well-conducted trucking operations, truck schedules are closely adhered to, and smoothly functioning fleets transporting products at maximum capacity result. When the reverse is the case, unloaded trucks will bunch up at the loading point or loaded trucks will bunch up at the unloading point, and reduced output due to such congestion results. To avoid this condition at the beginning of each day, truck dispatching sometimes is staggered, some being started early and others at later intervals, but each putting in the same number of hours per day.

When only from one to two or three trucks are operated, it usually is not necessary to plan a hauling schedule, as they are loaded by some method other than power loaders and timing is not so essential. Also, when trucks are loaded by hand, as in the case of pulpwood, ties, and other small products, there is no need to operate according to schedule. Each truck can go to a different spot for loading and no waiting time is involved. On such operations, however, it is necessary to have products cut and ready for loading to avoid lost time by the trucks due to waiting for loads.

One decided advantage of truck hauling is that operations can be conducted at several different locations at one time. With railroad transportation, the operation must be concentrated at least within the limits of the railroad system. Truck hauling, therefore, permits operators to harvest small scattered tracts that would be economically impossible to reach with a railroad. This has greatly increased the operating life of many sawmills and other wood-using plants, as large volumes of timber in the aggregate often are contained in small scattered holdings, especially in the regions east of the Rocky Mountains.

SLED HAULING

In regions with cold winters and abundant snow, hauling in the winter can be done with sleds. For successful sled transportation the topography must not be too rugged. The Lake states, the Northeast, and eastern Canada are the only regions where sled hauling is a common method of transporting logs and pulpwood during the winter.

Formerly, sled hauling of logs and pulpwood was the only way such products could be moved from the forest in the winter. Later, railroad logging operations continuing both winter and summer were conducted, and now trucks are also used for year-round transportation where suitable roads are available.

Considerable areas of forest in these regions, however, are insufficiently developed with roads to make truck hauling possible, and sled hauling still finds an important place in the transportation of saw logs and pulpwood from northern forests to streams, lakes, or railroad shipping points.

Products are seldom if ever hauled to the mill by sled because of the long slow haul that would be required and because public roads are not suitable for the wide-gauge sleds used for hauling logs and pulpwood. Sleds, however, are used to haul products to streams and lakes, where they are accumulated during the late fall and winter months. Such hauls are comparatively short, generally not over 3 to 5 miles. They are also used to haul products to railroad sidings for shipment to mills.

Sled hauling, therefore, is not a complete transportation operation, as is the case with railroad transportation and truck transportation, where the products are delivered to the mill. Sled hauling is practically always an intermediate hauling operation, in that sleds deliver products from the loading point in the forest to a stream or lakeside for driving or rafting, or to railroad sidings for loading on cars for mill delivery via common-carrier or private railroads. In some cases, sleds are used to deliver products to a roadside where trucks are loaded, and the transportation to the mill is then completed by truck over public highways.

Obviously, sled hauling must be confined to the winter months when snow and freezing weather can be depended upon. Comparatively level

Fig. 145. A typical snow road for winter hauling in Wisconsin. Such roads are frequently iced to provide a smooth, solid roadbed.

topography also is essential, or at least level routes for sled roads must be available, because of the difficulties of negotiating both upgrades and downgrades with heavily loaded sleds on iced roads. It is particularly difficult to brake a sled on a downgrade, as there are no moving parts.

Both topography and weather conditions in the Lake states, the Northeast, and eastern Canada are favorable for sled hauling, and as a consequence winter sled-hauling operations are customary. In fact, due to the abundance of lakes, streams, and muskeg swamps, it often is difficult to haul any time other than when the ground is frozen, as it then offers a solid bottom. For the same reason winter truck hauling has, in recent years, become increasingly popular.

Sled Hauling Equipment. Various types of four-runner sleds are used in hauling forest products. The principal differences between the various types have to do with the length of runners and the shape of the runner bottoms, as well as the width of bunks and the features for pulling. Because sled runners moving over ice or even snow offer a minimum of resistance, exceedingly large loads can be handled on sleds, and sleds are characteristically wide, therefore, to accommodate large loads. The bunk length, which determines the width of the load, is often as much as 16 ft, compared to about 8 ft as the maximum for trucks. Consequently the guage of the runners is various, depending upon the width of the bunks.

The wooden runners are shod with steel strips, which are of several designs. They may be flat, convex, concave, or keeled, with a raised center

FIG. 146. A train of log sleds pulled by a crawler tractor over an iced road in Saskatchewan. About 24,000 ft of logs are contained in this four-sled train. Note the wide bunks that permit large loads. (*Courtesy of Caterpillar Tractor Co.*)

running lengthwise of the runner. The patterns, other than the flat and convex, are designed to keep the runner from slipping sideways. The flat and convex types are used when the sled runners travel in prepared ruts in the roads and no such control is necessary.

Horse teams formerly were almost exclusively used to pull the sleds, but now tractors of various types do much of this work, although teams still are used on many operations. Before gasoline and Diesel tractors were developed, steam tractors were used to some extent to do the hauling. Teams handle, usually, only one sled at a time, but tractors pull up to six or eight sleds making up a train. Trucks with nonskid chains on their wheels also are extensively used to pull from one to several loaded sleds.

Tractors with bulldozers and other earth-moving attachments or teams and scoops are used to level the roadbeds during the summer months. After freezing weather sets in, snow rollers are used to pack the snow in the bottoms of the roads and snowplows are needed to keep the roads free from excessive amounts of snow. The rollers are large fabricated wooden cylinders, 6 to 8 ft in diameter, pulled by teams or tractors. After a sufficiently thick layer has been packed, the road may or may not be iced, depending upon whether or not it is to be a main road or a lateral feeder road. If it is to be iced, water is sprinkled on the snow until an ice coating has formed thick enough to support the sleds. Often only the ruts where the runners travel are iced, but in some cases the entire road is iced. For this purpose a water tank is needed. It is pulled by a team, truck, or tractor, and at frequent intervals during the hauling season the road or ruts are iced to keep a smooth surface. This operation customarily is done at night to avoid interfering with hauling.

Other equipment includes systems for braking loaded sleds down slopes too steep to be negotiated otherwise. One type of brake is the Barienger brake, consisting of drums with brakes mounted on a bed and a steel cable that attaches to the sled. The brake is anchored solidly to stumps or trees and the speed of the sled downhill is regulated by an operator who applies the brakes to the drums with a lever. Another system is of a compensating type that can be used only when there is a steep upgrade before the downgrade in the direction the loaded sleds are traveling. A sled starting down is attached by cable to a sled following and going up the grade, and the pull of the sled going down pulls the rear sled up the grade. With this system some traction power, animal or mechanical, is, of course, needed as the two would otherwise neutralize each other. Another method of braking sleds going downgrade is to spread gravel, sand, or straw over the road to increase the friction between the runners and the road.

Road rutters are used where the iced roads are to be rutted or grooved for the runners. A road rutter consists of a drag with scrapers of the width of rut desired and gauged equal to that of the sleds that will traverse the road. When this is weighted and pulled along the road, the result is a roadbed with prepared ruts in which the sled runners travel. These ruts are kept thoroughly iced, while the remainder of the road is only snow-coated. The tractor treads are gauged narrower than the sleds so the ruts are not fouled by the towing tractor. If teams or trucks are used, they travel the middle of the road, so the ruts are not disturbed regardless of what towing method is used.

Operation of Sled Hauling Systems. The first step in establishing a sled hauling operation is to locate the roads, camps, and barns or garages. As in the case of railroad and truck-road location, this is done with due regard to fitting the best road system to the topography and the timber tract to be harvested. Because of the necessity of keeping the roads as level as possible and because streams and swamps offer no particular problems with respect to soft ground during the winter months, sled hauling roads commonly follow streams and swamps when it is possible to do so and still reach the timber.

Because empty sleds and trains of sleds can be pulled up and let down much steeper grades than loaded sleds, return-trip roads over other routes are frequently used when they can be made appreciably shorter. This method has the advantage of a much quicker return trip and practically no interruption of the hauling due to the passing of loaded and empty sleds and trains, which happens with two-way traffic. Return-trip roads, when used, need not be as well constructed as the main hauling roads because of the lighter traffic they carry.

Main roads and lateral feeder roads are installed, and individual sleds are loaded at several points and assembled into a train at the main road. In layout and in operation, this system resembles railroad main lines and spurs and main truck roads and laterals.

After the roads have been located, they are cleared of vegetation and leveled. When freezing weather with snow has set in, the roads are prepared for hauling. This varies with the kind of road it is to be. Several different kinds are used, depending upon the volume of traffic. The simplest type is a snow-trough road, which consists only of a trough wide enough to accommodate the sleds, with snow sides that are built up by the snowplow pushing the snow out of the road. The bottom may or may not be iced, depending upon the preference of the operator. When roads are iced, only the portion over which the runners pass receives this attention, as a rule. The highest type of road is one with only slight grades and iced ruts. Construction of such a road involves grading and improving the roadbed during the summer or fall months, before freezing weather and snow occur. After sufficient snow has fallen, the road is rolled to pack it and a rutter is run over the road to groove it. Then the ruts are sprinkled several times to build up a firm ice coating, after which the road is ready for use.

Maintenance of sled hauling roads involves removal of excessive snow with plows and frequent icing. Main roads often are sprinkled every night to keep them in first-class condition and to avoid interfering with hauling in daytime. A water supply is, of course, necessary, but this can usually be developed near the roads where icing is done. Flowing streams or lakes are used, as the ice is much thinner over them than over shallow pools. Heated garages are needed for storing and repairing motive power equipment, mainly tractors and trucks. On animal operations, snugly built barns are essential.

The more a road is to be used, the more effort and expense can be incurred in constructing and maintaining it. Of course, when thawing weather comes in the spring, the ice and snow disappear and only the roadbed remains. Therefore, the cost of a road, other than clearing and grading costs, must be carried by the volume hauled during only a few months. Due consideration must be given to this fact in determining how much can be spent for road improvement on a particular operation.

STREAM DRIVING

Before mechanical methods of hauling were perfected, logs, pulpwood, and other forest products were commonly floated from the forest to the mills. Even with modern methods of transporting products by logging railroads and trucks, stream driving is used extensively where the correct combination of conditions makes this possible.

Stream driving means that the logs or pulpwood bolts are concentrated in streams, lakes, or other bodies of water, usually during the winter; then when spring freshets occur they are floated en masse to the mill. Even streams that have only a slight flow during the summer and fall months often have sufficient volume in the spring, when snow and ice are melting, to carry large quantities of forest products during the freshet period.

Stream driving is confined principally to those sections where snow occurs during the winter and, of course, to operations harvesting products that float. Long products such as poles and piling cannot ordinarily be stream-driven, as they would too readily tangle and jam, causing much delay and effort in freeing them, even if they did not break. Ties, although of short length and even when of lightweight wood, cannot be stream-driven as they would broom at the ends and be ruined as ties.

In the northern Rocky Mountains, the Lake states, and the Northeast, stream driving is not uncommon. Formerly, it was much more widely used, but year-round railroad and truck logging are now widely practiced instead of seasonal stream driving of logs and pulpwood.

The requirements for successful stream driving are

1. Lightweight products that will float.
2. Streams that flow from the forest to mill sites.
3. Spring freshets to flush the logs or pulpwood out of secondary streams into the main river channels.
4. Short products that will not easily jam.
5. Mill storage facilities that can accommodate a seasonal supply of wood rather than a continuous supply.
6. Streams that are not navigated.

The principal advantage of stream driving is that only a minimum of improvements is needed for transportation. Stream channels must be cleared and dams installed in tributary streams to hold back water until the drive starts. Also some booms, consisting of floating logs fastened end to end, to keep the products confined to a definite area, are needed at certain points to assemble the products. A floating camp on the larger streams often follows the drive downstream to provide quarters for the driver crew. At the mill location, if more than one company drives the stream, a sorting pen must be installed to separate the logs or wood of different owners. When two or more use the same stream, the products must be marked or branded, and the log brands are legally registered the same as are cattle brands.

Although certain improvements, such as those mentioned, must be installed, the expense is much less than would be necessary to construct a logging railroad or truck roads in undeveloped regions, which is the principal reason this method of transportation is used. It is not used in the South as the streams are sluggish, as a rule, with wide bottoms, so in periods of high water the products would strand and scatter far too widely. Also, because of the wide and periodically inundated bottoms along the channels, few mill sites are available close to the main channels and logs and pulpwood could be driven only part way to the mills.

Southern timber is much heavier than the commercial northern and Rocky Mountain species and does not float well, which would mean serious loss of products due to sinkage en route to the mill. Southern pines and hardwoods both are too heavy when green to float except, at most, for only short periods; and because decay and insect attack is so prevalent in the warm and humid Southern climate, it is impractical to season the products so they will float.

Cypress formerly was floated to mills in rafts because it grows in the overflow bottoms. A certain amount of rafting of pine logs is done in parts of the South also, as will be noted later, but special provisions have to be taken to assure that the logs will float until they reach the mill. In the case of cypress, the trees were deadened a year or more prior to cutting so they would be light enough to float when cut. For these reasons, primarily, and also because year-round land transportation is possible, stream driving never has and very probably never will be used in the South.

The disadvantages of stream driving are that it is seasonal and there is a varying degree of loss through sinking, breaking, and brooming of log and pulpwood bolts. The seasonal factor is the most serious disadvantage in that the mills dependent upon a raw-material supply from stream driving must be prepared to receive and store an entire year's supply all within a few weeks time. Furthermore, the wood when received is a year old or older, and then must be carried, some of it, for an entire year at the mill. This means a very large investment in wood inventory, which means high interest and insurance charges. Mills receiving their wood supplies, either logs or pulpwood, currently throughout the year carry an inventory only large enough for from 1 to 3 weeks' needs.

Stream Driving Operations. The conduct of a stream drive is comparatively simple and of short duration. During the brief period in the spring when the drive is on, there is, however, intense activity on the part of the crews to prevent some of the products from stranding along banks and on sand bars and to keep the main body of logs or pulpwood moving downstream. If for any reason the products do not get into the main streams and on to the mill while there is enough water to move them, they will be stranded until the following year, leaving the mill without wood.

Preliminary to the drive, as before stated, the stream channels are cleared of obstructions that would retard or block the drive and splash dams are built on tributary streams to hold back enough water to flush the drive over shallow stretches. Generally, a number of smaller streams are stocked with products during the winter and when the spring freshets occur these disgorge into the main stream. When all the wood has been assembled in the main stream, the drive begins and the crews follow it along to prevent stranding and jamming. Dams are opened when necessary to keep a full flow of water during the drive. After the spring thaws, cutting begins and production of products for the next drive the following spring gets under way.

The work of drivers is hazardous and exhausting and always is performed under uncomfortable conditions. Men leap from one floating log to another to reach threatened jams and are almost constantly wet. They wear calked boots that will not slip on the logs and are very skillful in balancing themselves on such uncertain and treacherous footing. Log "burling," or spinning logs in the water, is a sport developed by skilled drivers. Two men on the same floating log each attempt to dislodge the other by spinning the log rapidly and stopping it suddenly. The joust continues until one contestant loses his balance and falls into the water.

When jams do occur, thousands of logs or pulpwood bolts snarl and pile up many feet high. The work of dislodging the key log that is responsible for the jam is difficult and dangerous. Dynamite often is employed to blast out a log jam, but men with peaveys and pike poles ordinarily do the untangling. When the jam finally has been broken and the logs start moving, the men must get away quickly, from the churning mass forced by tremendous pressures from upstream water, to keep from being trapped and crushed or drowned.

Log and pulpwood drives are made for distances up to 200 miles or more, although the average distance is much less, usually around 50 to 75 miles. On some drives lakes must be crossed, as some streams discharge into lakes of considerable size and continue through an outlet on the other side. When winds and lake currents are not sufficient to move the drive across the lake to the outlet, it is necessary to boom the logs as they enter the lake and tow them across with power launches, making as many trips as necessary to transport them all to the far side and place them in the stream again. Narrower bodies of dead water are sometimes swept by means of a floating boom of logs stretching across it and each end is alternately moved forward by a tractor or teams, which pushes the floating products ahead of it.

BARGING

Waterways and navigable rivers are used for transporting forest products by barge when mills and shipping points are so situated that this method is feasible and economical. Both logs and pulpwood are transported by barge in considerable volume in the South. While poles, piling, ties, and other products likewise commonly are transported by barge along the Atlantic coast, this is seldom a regular part of a harvesting operation; it is a method of shipping after harvesting has been completed.

Barging is much more widely used in the South and along the middle Atlantic coast than elsewhere because many sawmills and pulp mills are located on or near navigable streams or inland waterways. Along the

Fig. 147. A flat-decked barge, fully loaded with southern pine pulpwood in a South Carolina coastal river. Three to five such barges constitute a tow for a tug.

Atlantic coast considerable timber is cut on islands without bridge connections, and pulpwood and saw logs obtained from such locations must be moved by water—and barging is the most suitable method.

For barging to be possible and practical, it is necessary that the timber being harvested be within a short haul of barge landings and that the consuming mill also be located on or within a close haul of a barge-unloading point. If it is necessary to rehandle the products from barges to trucks or railroad cars, the extra handling expense usually would make a barge operation too costly. When barging operations are carried on, the consuming pulp mill or sawmill usually owns the barges and either operates them itself or contracts for the towing.

Barges are made either of wood or steel and may be flat-topped or open, depending upon the type of loading and unloading used. Barges ordinarily hold from 150 to 300 cords of pulpwood or from 75,000 to 150,000 ft of logs or similar products. Tugboats tow barges in single line when operating in inland waterways and coastal rivers where there are no strong currents. In

FIG. 148. Above, pulpwood being loaded on barges by means of a gravity chute (*Courtesy of Southern Pulpwood Conservation Association*); below, being loaded by driving the hauling truck directly onto the barge.

the Mississippi River and similar rivers, stern-wheel river boats usually are used and the barges are lashed close together and to the boat and pushed, as better control is maintained by this method where river currents are strong and variable.

Pulpwood barges used on the south Atlantic and Gulf coasts commonly are flat-topped. The pulpwood is loaded on the barges in ricks from trucks that drive onto them. Usually two barges are spotted at a landing for loading, as this gives room enough for several trucks to unload at one time. Unloading at the mill usually is by a crane with an orange-peel or clam-shell bucket, which picks up the wood and places it on conveyers leading to the mill or to the storage yard.

Logs are too long and too large to be loaded on barges by driving trucks onto the barges. Log trucks require trailers, which are not used with pulpwood trucks loading barges, and there would not be room to spot the loads of logs. Also, loads could not be built high enough to utilize the full capacity of the barges. In the case of pulpwood, the ricks are built up by hand labor, as pulpwood bolts are small enough and light enough to be handled this way. Cranes are used, therefore, for loading barges with logs. The logs are picked up individually and placed on the barge. Trucks deliver the logs to the landing and unload on the ground. The cranes may be of the land type with crawler treads or they may be mounted on barges so they can be floated from landing to landing.

Barge landings must be available or constructed before barges can be used. For crane loading, only mooring piles in deep enough water and within reach of the crane are necessary. When trucks deliver pulpwood directly on the barges, however, a causeway of some type is required. This ordinarily consists of piling and a timbered deck wide enough to accommodate a truck safely. Because of fluctuating water levels due to tides and river variations, the barge end of the causeway must be adjustable to a range of levels within ordinary extremes of water levels. Along the lower Mississippi River some barge landing causeways are earthen. When deep water is close to a bank, it is possible and very economical to lash the barge to trees on the bank and push earth against the side with a tractor bulldozer. Trucks then can drive onto the barge over this temporary causeway.

Barge hauling is generally used in connection with other transportation methods, as the products are usually delivered to the landing by trucks. It is one link in a transportation system, and is used to avoid long truck hauls or expensive freight hauls. Only when products are loaded on barges directly from the skidding operation and are not reloaded from some other hauling vehicle is barging the complete transportation method.

On some logging operations along rivers, logs are skidded to the riverside, where they are loaded directly onto barges. Some cypress logging operations, also, are carried on from the waterside and power skidders or pullboats are mounted on barges and floated to the timber along river edges or in swamps. Channels may even be dredged to reach interior portions of

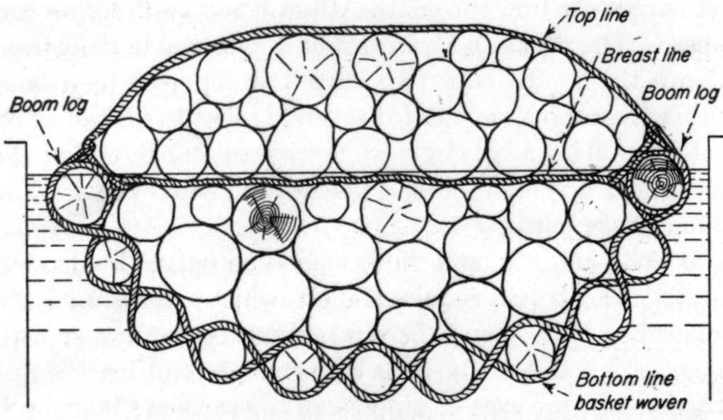

Fig. 149. Cross section of a fabricated log raft. Rafts are formed by basket weaving cables on the first row of logs and securing to boom logs. When the raft is half complete the breast lines are laid across and attached to the booms. The top lines are laid when complete. The lines are 30 ft apart and eight to a raft. (*Courtesy of The Timberman, November, 1946.*)

Fig. 150. Fabricating a Davis log raft by means of wire rope. Such rafts are used particularly in the Pacific Northwest when logs must be transported considerable distances through rough water. (*Courtesy of Wickwire Wire Rope Company.*)

swamp timber tracts. Such operations must load onto barges and, as the mills are located on water also, the barges can deliver the products directly from the harvesting operation to the mill and no other hauling method is needed.

RAFTING

Rafting forest products differs from stream driving in that the bolts or logs are not turned loose in water but are confined either in a boom or are closely bound by chains or cables. While stream driving can be done in small streams and in white water, rafting must be confined to large streams and rivers or waterways. Since products being rafted are confined closely and are under control, it is possible to transport them this way in navigable streams without obstructing shipping.

Certain highly specialized types of rafts are towed at sea from one port to another. Such seagoing rafts, however, are hardly considered as a part of harvesting, as it is a means of long-distance transportation of logs, which substitutes for cargo vessels or rail haul of the finished product.

The simplest type of raft is the floating-boom type. This consists of floating bolts or logs confined within a boom of floating logs fastened together end to end. Such rafts are towed from place to place by tugs. Very rough water would disintegrate such rafts by throwing the floating logs or bolts outside of the boom, as they are confined only by the fact that they cannot float past the boom logs in quiet water. Pulpwood is floated across Lake Superior in this way from the Canadian shore to the Wisconsin shore. Only when there is promise of quiet weather for 2 or 3 days is this attempted, due to the fact that waves would scatter the raft. Logs are also moved from place to place in the Columbia River and Puget Sound waters by towing them in booms, and the St. Lawrence River and other similar Eastern waterways are likewise used for rafting. It is an economical way to transport products that will float when suitable water routes are available.

An improvement of the open floating-boom method is a similar boom but with the added feature of a cable net suspended beneath so the products can be piled into it without danger of escaping from below. This has the advantage of greatly increasing the volume of product for the same boom area. Without the net beneath, only one layer of floating products can be towed or confined at one time. Deeper water, free of subterranean obstructions, of course, is necessary, as the net may extend several feet below the surface. Also a certain amount of products that are too heavy for open rafting can be included, as the buoyancy of the remainder keeps the entire raft floating.

In the South some rafting of logs is done but it is not common. Sawmill logs are rafted on the Mississippi River and other Southern rivers, as well

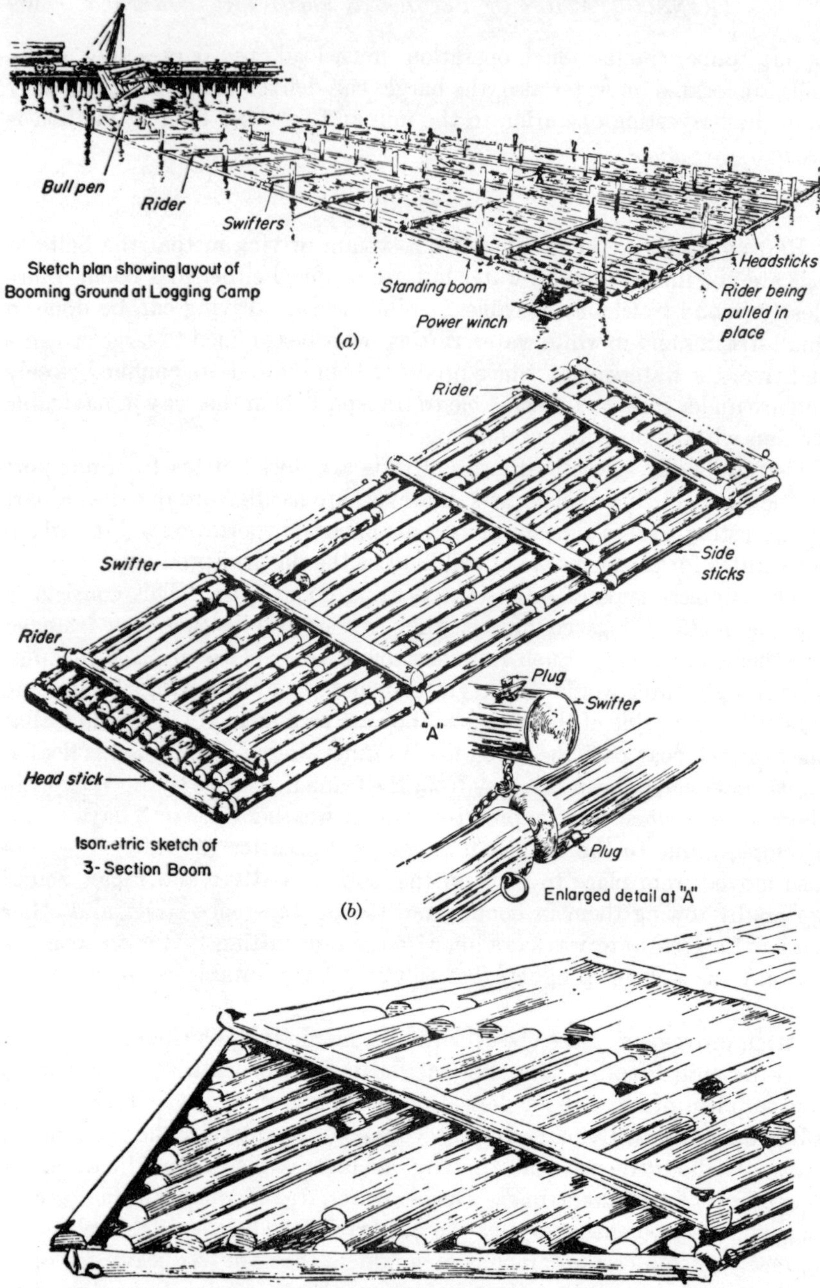

Fig. 151. Type of log boom used in British Columbia. (*a*) Sketch of a booming ground at a typical logging operation; (*b*) Isometric sketch of a three-section boom, or raft, to show members and fastenings; (*c*) Nose sticks on flat booms, or rafts. Such nose sticks are common on the Columbia and Fraser rivers but are not recommended for use in towing flat booms in rough waters. Logs in the nose have a tendency to climb. (*Courtesy of The Timberman.*)

as in eastern North Carolina waterways and rivers. Because the species are heavier, however, provision must be made to prevent sinkage. Before any rafting can be done, naturally, there must be enough buoyancy in the entire raft to support those portions that, left alone, would sink.

FIG. 152. A type of log raft used in eastern North Carolina on quiet streams.

FIG. 153. Cigar-shaped seagoing rafts of logs tightly bound with heavy chains in order to withstand heavy seas. Pacific Northwest.

Cottonwood and willow logs are rafted to some extent by fastening them together side by side with planks and spikes or looped cables. Since these species grow in the river bottoms and some mills are located on the riverbanks, this is an economical method of obtaining products from the harvesting operations.

In eastern North Carolina some pine logs are rafted in a similar way. Because second-growth pine does not float well when green, it is necessary to dry the logs out to a certain extent so there will be enough floaters to support the sinkers. This is done by felling the trees and leaving the tops on for several days. Transpiration from the tops removes an appreciable amount of moisture and the logs float readily for the towing period, which is not over 3 or 4 days. The logs are placed in the river side by side in single rows and each log at approximately the middle has a dog with a ring on it driven into it. A chain is put through the ring in each log and they are pulled close together. A tug tows several such rafts from the raft-assembling point to the mill situated where water delivery is possible.

Specialized seagoing rafts are not common. They are used on the Pacific coast to transport quantities of large logs to southern California points from the Columbia River and Puget Sound regions. Several million feet of logs are woven into a compact cigar-shaped raft by means of chains and cables. They can withstand ordinary rough seas and are towed by tugs. As previously stated, however, such rafting is not actually a part of harvesting operations but a method of shipping.

FLUMING

When harvesting operations are located at upper elevations in mountainous sections and the products are to be transported to lower elevations, it is often possible to use flumes instead of constructing roads or railroads. A flume is a trough into which water is diverted at the upper end. It is, in reality, an artificial stream. The products are then placed in the flume and they float in single file to the lower end. Frequently they travel rapidly, depending upon the gradient of the flume. The products cannot be very large, as this would require such a heavy and large flume that the cost would be prohibitive. Also, the gradient cannot be so steep that the products will bounce out of the trough.

In constructing a flume, the location is chosen with due regard to obtaining a supply of water at the upper end and a suitable gradient. As in the case of any other transportation installation, care and judgment are exercised to locate flumes so a minimum of construction will be necessary. Sharp curves must be avoided and the flume must be accessible enough to make maintenance not too difficult. Also, for long flumes, water may be needed at intervals in addition to that entering at the top, to replace water lost through leakage and splashing.

Flumes are constructed of planks on framed bents. The troughs are of several designs, ranging from a V shape to a flat bottom with sloping sides. They must be well constructed of material heavy enough to withstand the friction of rapidly floating products and prevent too much leakage. A suitable arrangement is provided at the upper end for starting the logs and bolts in the flume. This may be a pond or a banking ground from which they are rolled into the flume. At the lower end there usually are several outlets and when one discharging outlet has filled up with products, they are switched to another. The gradient at the outlet must be gradual enough so the products do not stop with so much force that they are damaged. Usually flumes discharge into lakes or rivers, which simplifies the problem of providing a suitable outlet.

Flumes are rarely used because they are costly to build and are more economical than roads or railroads only when the topography is so adverse that road and railroad construction cost would be prohibitive. Also, not a great deal of merchantable timber is found in locations where fluming would be the most economical way to transport it. Flumes are occasionally used to transport sawed products, such as ties, when harvesting operations and small mills are located at the upper elevations but the products must be moved to lower elevations for shipment to market.

CARGO VESSELS

Some mills, generally pulpmills, located on the Great Lakes, the St. Lawrence River, and tributary waterways, obtain their wood via cargo vessels from remote harvesting operations. Such transportation is, of course, costly but this is offset by the advantage of manufacturing close to markets, so a considerable amount of pulpwood moves to lakeside mills in the Lake states, the Northeast, and eastern Canada by this method.

The wood is concentrated at loading points and is loaded below decks by means of conveyers that lift it out of the water. While the wood may be stacked on the ground at loading points, it generally is put into the water and held in booms alongside the ship being loaded, as this is the easiest way to handle it.

All cargo-vessel transportation of wood is done during the ice-free seasons on the Great Lakes and tributary waterways, the wood hauled usually having been cut and sledded or trucked to the loading point during the previous winter.

COMMON-CARRIER RAILROADS

Many wood-using plants of all kinds now rely entirely or to a considerable extent on common-carrier railroads for transporting the products they require from the forest. This is the case more generally in the regions east of the Great Plains than in Western forest regions because of the many plants using wood, including sawmills, pulp mills, veneer factories, treating plants, handle factories and many others, that are not well located to develop an exclusive transportation system of their own, or who rely extensively on open-market purchases of wood from other forest owners.

Transit freight rates on raw materials have encouraged this development in many instances. Such rates give the effect of through rates from the point where the shipment originates to the final destination of outbound manufactured products, when the same railroad participates in both hauls and other requirements are met. In some instances railroads grant special commodity rates, as on pulpwood, not predicated on transit privileges but on the assumption that the outbound rail shipment from the plant will be in proportion to the amount of wood received over that railroad.

Usual shipping procedures are followed where forest products are transported by rail carriers. Producers haul by truck to railroad sidings and request suitable cars from the station agent. As the products are hauled,

FIG. 154. Specially constructed flatcars loaded with southern pine pulpwood. Two ricks of wood make up a load. These cars are standard for common carrier railroads and are equipped with air brakes and must meet I.C.C. specifications. (*Courtesy of Seaboard Air Line Railroad Company.*)

they are loaded or they may be stored on the ground and loaded later. After the cars are loaded, the shipper bills them to the mill and local freight trains pick them up.

Special cars are provided by some railroads for particular products that move regularly in sufficient volume to justify this service. An example is pulpwood flatcars, which are designed and used solely for hauling pulpwood. They hold two ricks of wood (usually cut 5 or 5.25 ft long) on each side and have an elevated outer edge so the wood in each rick slopes towards the center of the car. This prevents wood from working out and causing damage. End racks are also provided to hold ricks intact. Much pulpwood, however, is loaded in boxcars that can be used for other products also.

Poles and piling, being long, must be loaded on flatcars or in gondola cars, if they are not longer than one car length. If they are longer they must be loaded on two or more cars or an empty flatcar must be provided to take care of the overhang. Handle, veneer, excelsior, and other bolts are usually loaded in boxcars. Short logs, such as 8-ft red cedar logs produced in the South, are hauled in boxcars. Fence posts, whenever produced, also are hauled either in boxcars or flatcars.

Some large-scale logging operations conducted by means of logging railroads use common-carrier railroads for delivering the loaded log cars to the mill. Freight rates are based on the car as a unit or on carload weights. The loaded cars are delivered by the logging railroad, which either runs a special train of log cars or incorporates them in freight trains. Log cars hauled by common-carrier railroads must be in good order and have air brakes, as they must be handled according to usual safety practices. Regular log trains are operated, in some instances, over common-carrier tracks for a portion of the haul to the mill in accordance with special agreements with the railroads concerned. Such trains must be correctly dispatched by the railroad and train crews must conform to the standards of the I.C.C.

Railroads play a very important part in the final delivery of certain products harvested in the forest to the mills requiring them. Pulpwood in the South, the Lake states, and throughout the East generally is produced over such wide areas, even for individual mills, that it would be impossible for each mill to run its own railroad or to have all its wood delivered by truck or barge, except in the case of some small units using only a limited amount of nearby wood. Without the use of railroad facilities it would be practically impossible to supply large, centrally located mills with sufficient wood to run them. Some mills receive wood by rail from distances up to 300 or more miles and from several directions.

Creosoting plants treat a great many long poles and piling that are comparatively scarce in any one locality, and as these products must be concentrated from wide areas to obtain sufficient volume to treat them with preservatives economically, it is necessary to ship by rail as well as to haul by truck. Only comparatively short hauls are made by trucks. By rail these products are shipped up to 200 or more miles to the plants from the producing areas. Veneer, handle, and other bolts of high quality cannot usually be produced in quantity in any one locality or from any one stand, since only the better portions of trees of certain species can be used for these purposes. Hence, rail haul is essential to obtain an adequate volume, unless favorable trucking conditions exist.

When railroads are used to haul primary forest products to mills, such transportation is considered a part of the harvesting operation. When lumber, treated poles, ties, and other finished products are hauled, the transportation is not a part of harvesting but of delivery of finished products.

PART FOUR

ORGANIZATION AND CONTROL OF HARVESTING OPERATIONS

The several timber-harvesting steps discussed so far in this text must be closely integrated and coordinated before they can be performed economically and satisfactorily. Therefore, records of costs incurred and results accomplished must be accurately kept to maintain adequate control over operations, and much attention must be given to this phase of harvesting timber crops by those supervising or conducting them.

The chapters in Part Four, therefore, are devoted to a presentation and discussion of the organization of harvesting operations, scaling and measurement of timber products, and to the keeping of records on costs and output. Also, since the application of harvesting principles varies greatly from one forest region to another because of marked differences in terrain, size of trees, volume per acre, labor, and marketing opportunities, the concluding chapter is devoted to a presentation of regional harvesting practices by recognized authorities.

CHAPTER 15

MEASUREMENT OF FOREST PRODUCTS

Forest products are measured one or more times during the harvesting operation for correct payment in a sales transaction involving timber being harvested, for payment of labor working on a piece basis, or for cost-accounting purposes. Standing timber frequently is sold at a stated price per unit of measure on the basis of the amount actually cut; so both the buyer and the seller are very much interested in an accurate measure of timber harvested.

In many cases woods labor is paid on the basis of work actually accomplished, such as logs cut, pulpwood hauled, or ties produced; so the output of each crew must be measured to determine the amount of wages to be paid, when piecework is the form of payment. Since economical harvesting methods are the goal of all operations, measurements of output at various stages are often taken so costs of individual harvesting steps can be determined. Measuring forest products, therefore, is an essential step in any well-conducted harvesting operation.

LOGS AND BOLTS

Saw logs are universally measured in the United States by one of a number of log rules. Since it is presumed that saw logs are to be sawed into boards, dimension, or timbers, all log rules measure the volume of logs in feet. A board foot is the volume of wood in a board 1 in. thick and 12 in. square. Thus, a 1-in.-thick board, 12 in. wide and 16 ft long, has a volume of 16 bd ft. A piece of dimension 12 in. wide and 16 ft long, but 2 in. thick, contains 32 bd ft. Other sizes contain proportionately more or less board-foot volume.

Log rules were designed by their originators to express the volumes of logs in terms of their lumber content. Some rules are based on formulas, while others were constructed by drawing diagrams of logs and graphically determining the theoretical amount of lumber that logs of various sizes could yield. Still others have been constructed by measuring the amount of lumber actually cut from a representative number of logs of each size. No log rule, however, can express very closely the amount of lumber that will be sawed from the log, because so many variable factors affect the yield. The thickness of the saw that cuts a log has an important bearing on the amount of lumber produced. Some saws cut a kerf of

¼ in. each time they pass through a log. Thus, if 1-in. boards are being sawed from a log with such a saw, there is the equivalent of at least one board cut into sawdust for every four boards sawed. Most small mill circular saws are of this thickness, but large band saws are often only ⅛ in. thick and so can cut more boards from logs of the same size.

Other variations such as thickness of slabs, whether boards or timbers are sawed, waste due to crook and defect, and skill of the sawyer all affect the yield of lumber from logs. Hence, a log rule in feet log scale does not, except in rare instances, give the same volume for a given log as the actual tally of lumber or timber sawed from it. It is always advisable, therefore, to consider the feet given in a log rule or log scale as quite apart from board feet. The latter is a definite unit of lumber measure, whereas a log rule assigns fixed values in feet to logs of given diameters and lengths that are straight and sound. In referring to log-scale measurements, the expression "feet log scale" is preferable, so the expression "feet board measure" can be reserved for use in referring to measurements of sawed timber products.

Log rules assign different values for logs on the basis of their diameters in inches and lengths in feet. Such rules are tabulated when appearing in print, but for use in measuring or scaling logs or other round products, the values are transcribed on sticks so designed that they can be placed against the end of a log and the volume in feet log scale read directly from one side of the stick. A scale stick, as it is called, is shaped at one end for easy handling and has a projecting flange at the opposite end so it can be slipped over the edge of a log to hold it until the measurement is read. It is customary practice in scaling, and most log rules give consideration to this factor, to scale the smaller end of each log. This is usually scaled inside the bark; but local custom, in some cases, allows including one bark thickness. When logs are scaled to obtain a volume measurement for buying or selling, appropriate deductions are made for defect such as hollow or rotten centers, catfaces, shake, and crook. For payment of labor, such defects usually are ignored, although deductions for breakage when due to careless work are often customary.

Log scaling is done one or more times on every operation, except on some of the very small operations where the logs are skidded directly to the mill deck and only a tally of lumber sawed is kept. Felling crews are very often paid a given rate for each 1,000 ft of logs felled or felled and bucked, depending upon whether or not they perform both operations. In either event the logs cut by each crew are scaled each day, either in the woods before skidding or after they have been skidded, depending upon which is the more feasible. On some operations where skidding is by teams or tractors, it is possible for each felling crew to be assigned the job of cutting in a marked area and for the skidders to keep the logs cut by each crew separate from others.

When felling and bucking crews cannot be conveniently separated and the logs cut become mixed at the landing, the felling and bucking crews mark their numbers or initials on the scaling end of each log they cut. In recording the scale of the logs, the scaler keeps separate the scale for each crew. Scaling may be done as logs are delivered to a railroad on trucks or wagons and the results recorded separately to obtain the measure of production by each crew and also the volume hauled by each truck, when payment is on a per-1,000-ft basis.

On cable-skidding operations, it is not very convenient to scale the logs except at the point where they are felled or loaded on cars. Ordinarily there is no need to scale logs except where the trees are felled and bucked, as all labor except log makers are paid hourly wages. A rough measure of the output of cable skidders, loaders, and other heavy logging equipment is often obtained by simply counting the logs skidded or loaded or even by only counting the number of cars loaded. On some operations in the Southern pine region and when the timber is approximately uniform as to size and distribution, felling and bucking crews are paid a given amount per log or tree cut, and it is only necessary for the foreman to record the number of logs produced or the number of trees cut, as the case may be.

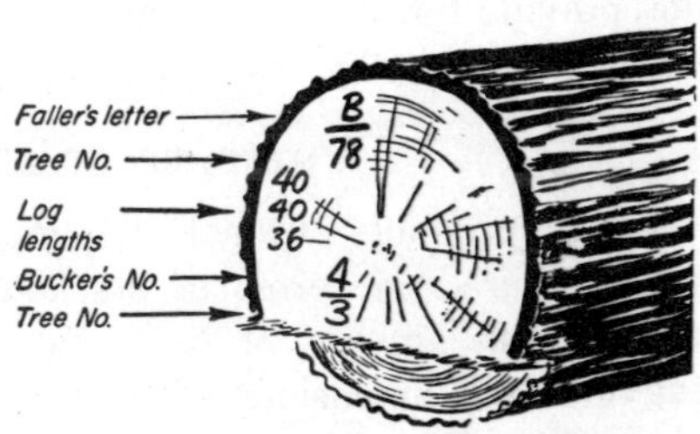

FIG. 155. Markings made on the butt of a tree to identify the crews doing the work. (*Courtesy of The Timberman.*)

Log scaling is an important job on a harvesting operation because the correctness of payment for labor, for timber, and for other steps in harvesting, when such payments are made on the basis of volume, depends upon the skill and care with which the scaling is done. Accuracy in scaling to obtain the scale of each log produced is essential, so scalers must be alert and not overlook logs in the woods, if scaling is done there, or in a ramp or pile of logs at a landing. They must also use care to scale the small end of each log and, of course, to perform their duties in conformance with company policy and accepted practices. Scalers frequently write the scale of a log on its scaling end to be certain they do not scale it a second time and to permit check scaling at the mill. In some cases the scaler simply marks the ends of logs he has scaled to avoid duplication.

In certain localities such as the Puget Sound and Columbia River sections of the Northwest, quantities of logs are scaled in the water. A large proportion of the logs sawed in that section are floated to the mills after having been logged to the rivers or sounds. Some companies engage only in logging and sell the logs after they are placed in water. Each owner's logs are branded to distinguish them and they are assembled in booms for towing to waterside mills.

Since a sales transaction is involved, a measure of the amount of logs is necessary. Expert scalers are employed by log-scaling associations or companies who perform this service at a fixed rate per 1,000 ft. A high degree of skill is required of such scalers, as they must detect defect in partially submerged logs, make due allowance for bark, and be accurate in determining log lengths correctly, as well as qualify in other respects.

Short logs or bolts such as are used in rotary veneer mills, although they are much too short for saw logs and are not to be sawed, are scaled by log rules, in most cases. There is no relation between the product and the log measure in feet log scale, as the product of veneer mills is expressed in square feet of ⅛-in. veneer. However, a log rule does give a basis of measuring round bolts irrespective of their end use and they commonly are bought or sold and their production contracted or paid for on the basis of their content in feet log scale. While the original log rules did not provide for short bolts measuring only 4 to 5 ft in length, they have been proportionally reduced to veneer-bolt lengths.

STACKED OR RICKED PRODUCTS

Forest products of small diameter and cut 8 ft or less in length often are measured in terms of stacked volume. This is practical because of the large amount of time that would be involved in measuring and tallying the dimensions of each piece when the products are small or are split. By ricking them between vertical uprights, a reasonably accurate measure of the volume of the products can be obtained.

Such products as pulpwood, fuel wood, stave bolts, chemical wood, and acid wood are measured in terms of stacked volume rather than by numerical count or measurements of individual pieces. Not all products cut in lengths of 8 ft or less, however, are measured in terms of stacked volume. Mine timbers, posts, and similar products cut 8 ft and less in length are individually valuable enough to justify and make advisable measurement by individual count. Only small products of comparatively low piece value, such as pulpwood, or of irregular shape, such as split stave bolts, are measured in terms of stacked volume.

The universally accepted standard unit of measure for wood bolts stacked together longitudinally and reasonably compact, is the cord of 128 cu ft in volume. Products measured by the cord vary in length from stovewood as short as 12 in. up to 8 ft, as is the case with some pulpwood, and they may be stacked in ricks of any convenient length and height. Only wood of one length, of course, should be piled in any one rick.

Regardless of the dimensions of a rick of wood or its shape, its cubic volume can be determined and its volume expressed in cords. The square feet of side area of a rick multiplied by the length in feet of the wood in the rick gives the cubic-foot content of the stack which, when divided by 128, gives the number of cords. In some states the standard cord of 128 cu ft is the legal unit of measure and all transactions in ricked wood must be on the cord basis.

Pulpwood is cut to different lengths in different sections of the United States, but pulpwood statistics issued by industry associations and wood records kept by the majority of mills are in cords. There is no particular reason for the standard cord to be one of 128 cu ft; it simply expresses an old custom when fuel wood and other early products were cut in 4-ft lengths and stacked the height of a stick and piled in ricks twice as long as a stick.

Throughout the Southern states most pulpwood is cut 5 ft long or 5 ft 3 in. long. This is done because pine and hardwood pulpwood bolts can be handled about as easily when of these lengths as when shorter, and because trucks and railroad cars can be loaded to capacity more readily with bolts of these lengths.

Pulpwood is piled across the truck bed for hauling; thus if the wood were shorter the capacity of the truck would be less, since it cannot be piled higher to compensate for a shorter length without causing a top-heavy load. Also, when pulpwood is loaded on open flatcars, two ricks can be placed lengthwise on each car without overhanging the sides too much. Shorter wood would mean less volume per carload, resulting in less efficiency in the use of such equipment. The effect of bolt length is less pronounced in boxcars, as the bolts are piled in ricks across the cars and the center space between the doors is filled with wood piled in the opposite direction, so the length of bolt has no particular bearing on the volume of the load. The shorter the bolts, however, the greater the number to be cut and handled for the same number of cords, so the longer bolts are more economical. In some cases the equipment used at the mills for chipping or grinding wood limits the size and length of wood that can be used. Pine pulpwood produced for one mill in the South is cut 4 ft in length because the grinders used in pulping it cannot accept longer bolts. Most pulpwood in the South, however, is chipped by being fed endwise against revolving knives; so the length of bolt is not a factor.

Ricks of pulpwood cut to lengths longer than 4 ft are referred to as units rather than cords and in many cases the unit is the basis for transactions between buyers and sellers of pulpwood. Pulpwood cut 5 ft long and ricked 4 ft high and 8 ft long is a unit containing 160 cu ft, or 25 per cent more volume than a standard cord. Wood cut 5 ft 3 in. long and ricked 4 ft high and 8 ft long is a unit containing 168 cu ft and is 1.3125 cords.

In the Lake states and the Northeast, pulpwood is cut in 4-ft lengths and also in 8-ft lengths or sometimes 50- and 100-in. lengths. The extra

2 in. for 4-ft wood and 4 in. for 8-ft wood is customary and probably originated when most wood was stream-driven, which resulted in brooming the ends, due to the bolts striking against each other and against rocks in fast water. The 2 to 4 in. of extra length made allowance for this likely loss in transit. Four-foot or 50-in. wood is measured in standard cords, the 2 in. extra in 50-in. wood generally being ignored. Eight-foot or 100-in. wood is also measured in cords, each rick of standard size (4 by 8 ft) being a double cord, or 2 cords.

Pulpwood is not always piled in the woods where cut for measurement. Cutters in some localities are paid a certain number of cents per piece, so only a count is necessary. In others, cutters are paid by the truckload or carload when the wood is hauled as soon as it is cut. In such cases there is no measurement of the wood except by hauling units, a truckload averaging about 2 cords and a carload about 15 cords. This practice is common in certain parts of the South.

When the wood is trucked directly to the mill, it is measured on the truck by the mill scalers and an accurate measure is available for payment of haulers and cutters as well as for payment for the timber. Frequently, wood is hauled to a railroad siding or barge landing and banked in ricks that can be measured daily, if desired, to obtain a record of wood production.

Pulpwood received at the mills in boxcars or on flatcars is measured before it is unloaded and mill measurement is the basis for payment to the shipper. The measurements made by producers cutting and shipping the wood are merely for the purpose of settling with labor and sometimes as the basis for paying the timber owner from whom the stumpage was purchased. In many instances, mill measure is agreed upon as the basis for paying for stumpage when there is no likelihood of mixing wood in the same car with that of other timber owners. Frequently, shippers purchase wood already cut by farmers and stacked along roads or in the woods. In such cases, the wood is measured and settlement made before the wood is hauled.

Pulpwood barges carry from 100 to 200 cords of wood and the wood usually is hauled from several tracts to load out barges more rapidly than would be possible from only one cutting area. To keep each shipper's wood separate, truckloads usually are measured as they go on the barge to unload. When wood is banked at the landing each hauler piles his wood in separate ricks to prevent mixing it. The ricks can then be measured at periodic intervals instead of continuously, as is necessary when a barge is being loaded.

When the wood is to remain where cut for some time before it is hauled, some method of measuring it is necessary to pay cutters when their wages are based on a piece rate. In such cases the wood can be ricked throughout the cutting area and readily measured each day or week, as desired. When this is done the ricks are usually less than a cord each to avoid carrying some pieces undue distances. Also, with small ricks, less sturdy end stakes are necessary and less time is required to do the ricking than would be the case with standard-sized stacks. Also, it would require careful measuring of each rick as it was piled to have it come out an even cord or unit.

A method of piling pulpwood in the South called "penning" formerly was widely used but now rapidly is being abandoned as involving unnecessary effort, and because it is an inaccurate method of measurement. The pieces are cross-piled in pairs, as described previously, until the pen reaches a height of 6 ft. Cutters are paid by the pen and producers compute roughly a given number of pens per cord of pulpwood. This varies with location, but from 4 to 5 pens per unit is the factor generally used. Actually, the number of pens per unit varies greatly with diameter of bolts. Studies have shown that, for example, for 5-ft bolts 5 in. in diameter, there are 4.88 pens per unit, and that the ratio decreases with increasing size. For pens of pulpwood made of 5-ft bolts 10 in. in diameter, only 2.40 pens are required to equal one unit.

When penning is used to measure pulpwood, the cutters customarily fell and buck trees until late afternoon and then pen their day's cut, after which the foreman or scaler counts and measures the pens. All pens need not be exactly 6 ft in height, as the height of each can be measured in inches and the total height of all divided by 72 in. to obtain the number of 6-ft pens produced by a given crew. Even when penning is used, it seldom, if ever, is a basis for measuring the wood cut for payment of the timber owner selling it because of the inaccuracy of this method of measuring pulpwood.

The solid contents of a cord vary considerably with the size and straightness of bolts. The larger the bolts in diameter, the more solid contents in a cord. When bolts average 5 in. in diameter, there are only 83.6 cu ft of solid contents in a standard cord of 128 cu ft of gross measure, the rest of the space being composed of the voids between the sticks. When bolts average 10 in. in diameter, the solid content per cord is about 92 cu ft. Perfectly round cylindrical 4-ft bolts theoretically should stack so that the solid contents would be 116.8 cu ft per cord, regardless of the size of bolt; but because no trees grow perfectly round or straight, the variations from true cylinders, which are relatively more pronounced in bolts of small diameter than large, cause the variations in solid content. Projecting limbs, knots, and other irregularities on the surface of bolts also cause looser stacking and less solid content per cord.

On pulpwood operations the wood may be measured in ricks of various sizes, shapes, and dimensions, but the volume is expressed in cords of

128 cu ft or in units which have a definite volumetric relationship to a cord. No special measuring device, such as a scale stick, is needed, as the dimensions of a rick of wood necessary to determine its cord content can be obtained with a standard rule or tape graduated in feet and inches or tenths of feet.

Judgment must be used in measuring ricks or truckloads of wood that slope in height from one end to the other or are irregular in shape. Enough measurements must be taken to obtain the average for each irregular side dimension before computing the volume.

Pulpwood is scaled, then, either in the woods where it is cut and piled or penned by the cutters, after it is loaded on trucks, or after it is loaded on railroad cars. It is not uncommon for the pulpwood to be measured more than once when a measurement is necessary to pay labor as well as for other purposes.

PIECE PRODUCTS

Forest products that are cut to specifications, such as poles, or are hewed to meet specified size classes, such as ties, are tallied by the individual piece. Such products usually are each of sufficient value to justify considering each one as a unit. While logs are scaled individually they are lumped together in summarizing their measurement, with 1,000 ft, log scale, as the unit of measure.

Poles. Trees are not cut promiscuously and then made into poles. Only trees that are of pole quality and that have the dimensions that will meet the particular classes and sizes of poles on the cutting order will be felled. Usually, there are several size classes being produced at one time, as it is impossible for even the most skilled pole producers to judge whether or not a tree will exactly meet the specifications as to length, top diameter, and freedom from defect required for one pole class only.

When trees are felled the tops sometimes break or the butt is found to be defective, necessitating cutting off the damaged portions. If no shorter poles were on order, it would be impossible to use such trees except for other products, which would sometimes be impractical and in any event would mean reduced value. Also, cutters would hesitate to fell a tree that might just meet specifications, if there were no opportunity to use it for the next lower class if it should prove to be too short or too small in the top upon measuring it exactly.

Pole sizes have been standardized to insure uniformity in them even when they are obtained by users from several different sources. The standard sizes have been set up to meet given stresses and strains. Because trees vary somewhat as to butt size with respect to top diameter for given lengths, the specifications given are minimum sizes and most poles produced exceed the minimum somewhat, although the size cannot be as large as the next size class or larger and still remain in that class.

Workers may be paid by the day or by the piece on pole-producing operations. When their pay is by the piece, the tally of poles must be so kept that the production of each man or each crew can be kept separate. This can be done by tallying the poles of each crew where they are either cut or assembled for loading. Another method is for each crew, usually two or at most three men, to put an identifying mark on each pole they produce. Poles from several crews then can be assembled before they are tallied and the sizes produced by each crew kept separate by the identifying marks on them.

Each crew cutting poles checks the length of a felled tree and the diameter at the specified points before cutting the pole from it and peeling it. The length and class is usually shown on the butt in keel with the crew number. The foreman in tallying each day's production makes the necessary check measurements to be sure the crews are cutting the poles correctly. If he finds mistakes he requires the crews to adjust them, such as cutting off the butt or top or smoothing knots. His tally of poles produced is based on his inspection, and piece-rate wage payments are made accordingly.

Poles are seldom loaded and shipped before being inspected by the buyer's representative. He may cull some poles and cut back others to the next lower class to meet the buying specifications. The poles accepted are usually branded by the inspector with a die hammer, which is the official mark of acceptance. The buyer's tally is the final measure of the products and is the basis for payment to the seller, who usually is a producer only. He in turn pays the timber owner on the basis of the buyer's tally. Some timber owners require that the producer also pay for trees that he cuts but does not use due to defect unnoticed before cutting and also for all culled poles. The producer may be required to pay either the pole rate for such trees or for the log scale they contain, depending upon the agreement between the parties.

Piles. Piles are produced in a manner similar to that for poles, as the trees must be selected to meet certain size and quality specifications. Most piles are peeled but they may or may not be treated with preservatives. Foundation piles that are to be driven all the way into the ground are not treated and may not even be peeled as they are out of sight and, being driven deep into the ground, they are largely protected against insects and decay. Thus it is not necessary to treat them with preservatives, as absence of air in a damp or saturated soil prevents destructive action by insects and fungi on wood.

Piles also are measured by the piece, but instead of each one being cut to a fixed length as in poles, they usually are cut to the maximum length possible within top-diameter specifications. Piles are priced by length classes, such as from 85- to 95-ft piling, or 75- to 85-ft piling. The price of piles is usually a given amount per foot by length classes, so piles are checked to see that they meet butt and top specifications for the class and then they are measured for length. The length is then generally marked on the butt in keel. Settlement is usually based, not on the number of piles, because of their various lengths in each class, but on the total number of lineal feet in each class.

Ties. Hewed ties are tallied by individual count by size class and are paid for accordingly. They are usually produced by the individuals selling them and are hauled to a siding or assembly point and then piled so that all surfaces can be seen. The inspector representing the buyer then examines each tie and marks on it the class into which it falls, at the same time branding it with a die hammer. Price is by size class, so the number of ties of each size is the basis for payment. The producer usually settles with the timber owner on the number of ties produced, either by size, on average price for all ties, or a certain percentage of the total value of the ties cut.

Other forest products that are cut to uniform specifications, or that must fall within certain limits of diameter and length, are measured by piece count. Mine timbers often are measured by number of pieces by size class, although they also are measured by log scale in certain instances and sometimes by the stacked cord or unit.

CONVERTING FACTORS

Frequently it is necessary or desirable to convert the measure of products from one unit to another. Pulp mills often purchase logs by log scale for pulping purposes and convert the number of thousands of feet log scale into cords of stacked measure. This is not done in connection with the harvesting operation, as a rule, but for accounting purposes in connection with relating wood consumed to pulp obtained or for similar reasons. Likewise, the measure of piece products such as poles, piling, and ties is converted into feet log scale for the purpose of crediting the stumpage account with income when records are kept in terms of feet.

Converting the measure of products from one unit to another is of concern to those conducting harvesting operations only when two or more products are harvested and records are kept, for cost accounting or settlement of timber purchases, in terms of one unit. The most common units are feet log scale and cords, so converting from one to another is done usually from other product units to cords or to feet log scale. Table 17 shows converting factors for loblolly pine logs in southeast Arkansas.

TABLE 17. CONVERTING FACTORS FOR 16-FT SHORTLEAF AND LOBLOLLY PINE LOGS IN SOUTHEAST ARKANSAS

Diameter (inside bark), small end of log, in.	Ft per log (International scale)*	Equivalent			To convert values per 1,000 ft (International scale), divide by the following figures§			Number of cu ft peeled volume per 1,000 ft		
		Doyle scale†, ft	Scribner scale‡, ft	Cu ft (inside bark)	Doyle scale	Scribner scale	Cu ft	International scale	Doyle scale	Scribner scale
6.0	19	4	12	4.0	0.2105	0.6316	0.2105	210.5	1000.0	333.3
6.5	23	6	16	4.4	0.2609	0.6956	0.1913	191.3	733.3	275.0
7.0	28	9	21	5.0	0.3214	0.7500	0.1786	178.6	555.6	238.1
7.5	33	12	25	5.6	0.3636	0.7576	0.1697	169.7	466.7	224.0
8.0	39	16	31	6.3	0.4102	0.7949	0.1615	161.5	393.8	203.2
8.5	45	20	36	7.1	0.4444	0.8000	0.1578	157.8	355.0	197.2
9.0	51	25	42	7.9	0.4902	0.8235	0.1549	154.9	316.0	188.1
9.5	58	30	48	8.8	0.5172	0.8276	0.1517	151.7	293.3	183.3
10.0	65	36	55	9.8	0.5538	0.8462	0.1508	150.8	272.2	178.2
10.5	72	42	62	10.8	0.5833	0.8611	0.1500	150.0	257.1	174.2
11.0	80	49	70	11.8	0.6125	0.8750	0.1475	147.5	240.8	168.6
11.5	88	56	78	12.9	0.6364	0.8864	0.1466	146.6	230.4	165.4
12.0	97	64	86	14.2	0.6598	0.8866	0.1464	146.4	221.9	165.1
12.5	106	72	94	15.4	0.6792	0.8868	0.1453	145.3	213.9	163.8
13.0	115	81	104	16.7	0.7043	0.9043	0.1452	145.2	206.2	160.6
13.5	125	90	113	17.9	0.7200	0.9040	0.1432	143.2	198.9	158.4
14.0	136	100	123	19.2	0.7353	0.9044	0.1412	141.2	192.0	156.1
14.5	146	110	133	20.5	0.7534	0.9110	0.1404	140.4	186.4	154.1
15.0	157	121	144	22.0	0.7707	0.9172	0.1401	140.1	181.8	152.8
15.5	169	132	155	23.5	0.7811	0.9172	0.1390	139.0	178.0	151.6
16.0	181	144	166	25.0	0.7956	0.9171	0.1381	138.1	173.6	150.6
16.5	193	156	178	26.7	0.8083	0.9223	0.1383	138.3	171.2	150.0
17.0	205	169	190	28.4	0.8244	0.9268	0.1385	138.5	168.0	149.5
17.5	219	182	203	30.2	0.8310	0.9269	0.1379	137.9	165.9	148.8
18.0	232	196	216	32.0	0.8448	0.9310	0.1379	137.9	163.3	148.1
18.5	246	210	229	33.9	0.8536	0.9309	0.1378	137.8	161.4	148.0
19.0	260	225	243	35.8	0.8654	0.9346	0.1377	137.7	159.1	147.3
19.5	275	240	257	37.8	0.8727	0.9345	0.1374	137.4	157.5	147.1
20.0	290	256	272	39.8	0.8828	0.9379	0.1372	137.2	155.5	146.3
20.5	305	272	287	41.8	0.8918	0.9410	0.1370	137.0	153.7	145.6
21.0	321	289	302	43.8	0.9003	0.9408	0.1364	136.4	151.6	145.0
21.5	337	306	318	45.8	0.9080	0.9436	0.1359	135.9	149.7	144.0
22.0	354	324	334	47.8	0.9152	0.9435	0.1350	135.0	147.5	143.1

* International ¼-in. scale is computed from the formula $V = 0.796D^2 - 1.375D - 1.230$

† Doyle scale is computed from the formula $V = \frac{(D-4)^2 \times L}{16}$

‡ Scribner scale is computed from the formula $V = 0.79D^2 - 2D - 4$

§ To convert feet in the International scale to feet Doyle scale or Scribner scale, or to cubic feet, *multiply* by the given values.

Source: Reynolds, R. R., "Factors for Converting Log and Tree Volumes or Values from One Common Unit to Another," *South. Forest Expt. Sta., Occas. Paper* 68, 1937.

Converting factors must be developed either by computation or by actually measuring a sufficient volume of product in terms of both units. This is one of the important phases of forest mensuration, which is one of the branches of forestry covered in a specialized course of study, so the methods and procedures in developing converting factors are not taken up here.

UNIFORM MEASURES

No uniform measure of wood products is used in the United States, although it would be highly desirable to have such a measure to apply to all products and to have it universally used. The cubic foot of solid wood could become such a measure and the value of wood for various products could then be expressed per cubic foot. This would permit foresters, timber owners, and operators to make accurate comparisons between products to determine the most profitable in the final analysis, after costs have been deducted. There seems to be but little prospect of establishing such a universal unit in this country, although European countries have long conducted their timber harvesting, manufacturing, and marketing on a standard unit, the cubic meter.

Weight of wood is used to a limited extent as a measure of volume, but only for the purpose of converting the measure into other units. Weight is used by some pulp mills in purchasing wood, but this is merely another way of obtaining the number of cords of wood involved without ricking and measuring it. To determine the converting factor, the weight of a number of ricks or loads of pulpwood that have been or will be measured as cords must be recorded and the weight related to the stacked volume. Weight, however, is not a stable unit of measurement unless moisture content is taken into consideration. Wood substance has definite and uniform weights per unit of volume by species, but varying amounts of moisture always found in wood cause variations in weight that make weight alone an unsatisfactory measure in most respects.

GRADING PRODUCTS

Forest products often are graded or classified at the time they are measured in connection with the harvesting operation. Saw logs commonly are segregated into quality classes based on size and freedom from defect. No standard grades are established except where a scaling association sets standards, as in the Pacific Northwest, and actually performs the scaling and grading service for buyers and sellers of logs.

Size is important because the larger the log, the more wide boards or large-sized timbers it will yield. Freedom from defect means that less loss from waste will be incurred and higher quality and hence more valuable lumber grades will be obtained. Therefore, log grades for a species always

TABLE 18. RECOMMENDATIONS FOR CLASSIFICATION OF SOUTHERN PINE LOGS

Lumber logs		Timber logs	
Lengths	Scale	Lengths	Scale
1. Regular log lengths shall be 12 to 20 ft. (Where this specification does not fit, mill will specify its requirements.) Logs cut shorter in order to prevent mill loss in crooks or woods loss in tops will be accepted. 2. Logs shall be cut in multiples of 2 ft and have at least 4 in. overlength on shortest side for mill trimming on logs 16 ft or less, and 6 in. overlength on logs 18 and 20 ft long.	1. Logs to be scaled at the small end inside the bark on one side and including the bark on the other side and along the average diameter. 2. Reduce the length of the log by 2 ft if not at least 4 in. overlength on the short side. 3. Length of log to be reduced proportionately if damaged at end by split or splinter pulls.	1. Regular lengths shall be 20 ft and over (unless shorter logs are specified). 2. Logs shall be cut in multiples of 2 ft unless otherwise specified. All logs shall have at least 4 in. trimming allowance as measured on shortest side.	1. Logs to be scaled at small end inside the bark on one side and including the bark on other side and along the average diameter. Long logs shall be scaled as two or more logs, giving proper rise in diameter for increase in accordance with the taper. 2. Reduce the length of the log by 2 ft if not at least 4 in. overlength on the short side. 3. Length of log to be reduced proportionately if damaged at end by split or splinter pulls.

Lumber logs

Log class and size	Minimum specifications	Defects allowed
A (17 in. and up)	Surface clear except one sound knot within 2 ft of end. Sweep or crook not to exceed 1 in. for each 8 ft. Uniform grain averaging 4 rings or more per in. Minimum length 12 ft	None. (Where defects as center rot, scars, crooks, and shakes occur, drop log to Class 1 if combined losses not over 25%, and to Class 2 if over 25%)
No. 1 (10-16 in.)	Same as Class A (17 in. and up)	None
(17 in. and up)	(a) Surface clear on three sides. No knot limitation on fourth side. Other specifications same as Class A.	Defects allowed on one knotty face
	(b) Surface clear, coarse-grained logs (less than four rings per inch)	None
No. 2 (10 in. and up)	(a) All other logs 10 in. and over having not more than four 3-in. sound knots—not bunched or ringed—in each 4 ft in length, if occurring on all sides or larger knots if at least two faces clear. Density same as Classes A and 1	No defects allowed in log centers. Surface defects and crook or sweep combined not to reduce volume more than 25%
(10-16 in.)	(c) Surface clear, logs having less than four rings per in.	None
No. 3 (10 in. and up)	All other logs with more and larger sound knots (and other defects) than allowed in higher classes; and logs with unsound knots not over 3 in. in size and four in number for each 4 ft of length	Combined total of all defects not to exceed 50% of contents of log
(8-9 in.)	Top logs cut under close utilization, which have sound knots and otherwise meet Class 2 specifications as to number and distribution, but not over 3 in. in size	Combined total of all defects not to exceed 25% of contents of log

Timber logs

Log class and size	Minimum specifications	Defects allowed
A (17 in. and up)	One 3-in. sound knot or equivalent in larger or smaller knots permitted for each 4 ft in length. Sweep or crook not to exceed 1 in. for each 12 ft in length. Dense-grained (four rings per in. with 50% summerwood or six rings per in. with 33⅓% summerwood beginning at 3 in. from center). (Heart content optional with each mill.)	None
No. 1 (10-16 in.)	(a) Same as Class A (17 in. and up)	None
	(b) Two 3-in. sound knots (or equivalent) permitted for each 4 ft in length. All other specifications same as Class A	None
(17 in. and up)	(c) Class A droppings	Surface defects and crook or sweep, combined effect not more than 25% of contents of log
No. 2 (10 in. and up)	Dense-grained logs containing more and larger knots or crook or surface defects than permitted in Class 1 and logs otherwise Class A or 1 but coarse grained, and/or crooked	Combined losses from defects to cause not over 25% waste of volume of log in full-length sawed product
No. 3 (10 in. and up)	All other logs containing more and larger sound knots, crook, sweep, center rot, and surface defects than in higher class, and in addition logs containing unsound knots not over 3 in. and four in number for each 4 ft in length. Log to cut out 50% in Class 2 and better	Combined total of all defects not to exceed 50% of contents of log
(8-9 in.)	Top logs produced in the course of close utilization, which have sound knots and otherwise meet Class 2 specifications as to number and distribution, but not over 3 in. in size	Combined total of all defects not to exceed 25% of contents of log

Source: *South. Pine Forestry Note* 71, *South. Pine Assoc.*, 1940.

take into consideration these two factors. Consequently the larger the log sizes and the fewer the defects, the higher the log grade. Table 18 shows recommended log grades for Southern pine logs.

When grading is practiced it is usually in connection with purchased logs, and the grade determines the price paid based on previous agreement between the parties. Log grades often are only Nos. 1 and 2, although occasionally a No. 3 grade is included. To qualify for a No. 1 grade, it is usually required that the log be of a certain diameter, say 16 in., or larger and that it be sound, straight, and free from knots. Logs smaller than this and all logs with defects are placed in the No. 2 grade. When a third grade is included, the No. 2 grade is intermediate in quality and size between the other two. When logs are sold ungraded, they are called "woods-run" logs and include all logs above minimum specifications. In areas such as the South, where timber is predominately second growth but some old growth is cut, a distinction is commonly made between virgin, sometimes also called "forest pine," or "virgin pine," and second growth, as much more heartwood is found in old-growth timber and the logs are much larger and so yield premium grades of lumber or timbers.

Hardwood logs are classified also as to species or groups of species of about equal value. Also, there may be differences in the size limits between different grades for different species or groups of species, the more valuable ones being taken to smaller sizes and including more defects. Veneer bolts, handle bolts, and other similar products are also commonly graded as to size and freedom from defects.

It is advantageous to both forest owners and wood users to grade primary forest products. If only one grade were used, there would be more waste in the woods, as portions of many trees that would not make the specifications would be left to rot on the ground unless it were possible to salvage them. When a lower grade is included, even though the price is lower, more complete utilization of timber grown and harvested is possible.

Also, when grades are recognized, the grower of quality timber is rewarded for his effort, which encourages the practice of forestry. So long as only a going price is paid for any logs, veneer bolts, or similar products that come on the market, the owner who has gone to trouble and expense in growing straight, clear trees will receive but little or no more than the timber owner whose trees just accidentally grew into merchantable size. Recognition of quality classes and the payment of premium prices for the better grades, then, encourages practical forestry in areas where such practices prevail.

CHAPTER 16

ORGANIZATION OF HARVESTING OPERATIONS

Regardless of the size of a harvesting operation or the kind of product handled, the harvesting operation must be well organized and the supervisory personnel and the crews doing the work must be competent, if an efficient operation is to be conducted. While each of the various stages of harvesting is an operation in itself, each must be coordinated with the others if costs are to be held at a reasonable minimum.

Not only must each of the steps in harvesting be gauged to the others but each must also be so handled that best advantage is taken of weather conditions and other variable factors confronting outdoor operations moving from area to area more or less continuously. If adequate thought and planning are not given to the conduct of harvesting operations, bottlenecks will develop, with output curtailed. For example, if cutting and skidding are well coordinated but loading is inadequate, then the loading operation becomes a bottleneck, and the products being skidded accumulate at loading points faster than they can be hauled.

Skidding, loading, and hauling are greatly affected by weather conditions when transportation is not by railroad or over all-weather roads, and it usually is necessary to equip and man these operations so they can handle, per working day, more than the felling crews can cut. During wet weather there are periods of no rain which permit normal felling and bucking operations, but skidding and hauling are prevented or handicapped by soft ground. These latter operations can catch up with the cutting operations when favorable conditions return if they are manned and equipped to handle more than an average day's volume. When the kind of weather that prevents certain operations occurs at more or less regular intervals and is only of short duration, it is possible to balance various operations in this way. When unfavorable weather is of a seasonal nature, then the entire operation must be gauged to it and those operations that will be handicapped or stopped must be performed when it is possible and economical to do so.

A seasonal operation is illustrated by the harvesting of logs in Southern river bottoms, using a logging railroad and a railroad loader. During late winter and spring the river bottoms are from one to several feet deep in water because of the seasonal floods that are common, and are anticipated. The bottoms are dry in summer and fall, so some operations are so planned that a 2 or 3 months' supply of timber, or at least considerably more than is currently needed, is cut, bucked, and skidded to the railroad spurs. So long as the overflow water does not cover the railroad track, it is possible to load the logs on cars by means of railroad loaders and haul them to the mill. Tracks in bottoms are often built on the highest land, and if it is necessary to traverse low spots during high water, the trestle type of construction may be used. At any rate, logs often can be loaded out when it would be impossible or impractical to cut and skid them, and a surplus of logs is produced when the ground is dry so the loading crews can regularly load out the usual amount, although cutting must be moved to other areas.

To keep harvesting operations in balance with each other and with weather, a highly skilled and experienced superintendent or manager is needed. He must not only supervise the current operations from day to day, but also plan ahead against unfavorable weather. Harvesting operations, when conducted by a lumber or pulp company on a large scale, are usually under the direct supervision of a woods-department manager or logging superintendent who has several foremen under him, each of whom is in charge of one phase of the job. The logging superintendent in turn usually reports to some official of the company in charge of raw materials.

The manager of the woods department may also, and often does, have supervision over timber lands owned by the company. When this is the case, his office handles all matters relative to protection against fire and trespass, taxation, and general administration of lands as well as operations. In some cases, actual harvesting is the sole responsibility of the logging superintendent, while a land or forestry department has charge of the lands and timber. Regardless of whether jurisdiction lies with one or two men, the harvesting and land and timber management activities must be closely coordinated to obtain logs or other products economically and yet leave the land in condition to grow another crop. Table 19 shows the organization of typical woods departments of pulp and paper mills in eastern Canada.

The larger the volume of daily output, the more specialized is the department concerned with harvesting operations. The responsibility for getting logs to a small mill or wood-using plant often rests with only one man with a helper or two, while for large companies a force of a dozen men or more, in addition to the crews actually engaged in harvesting activities, devote their entire attention to matters concerned with harvesting and land management.

The form of organization in harvesting varies greatly with the size of the output as well as with the source of raw material. Some sawmills and pulp mills obtain their raw materials from concentrated areas, while others obtain their raw materials from widely scattered locations. No two are exactly alike, however, whether large or small, because no two companies

TABLE 19. OUTLINE OF ORGANIZATION OF A WOODS DEPARTMENT

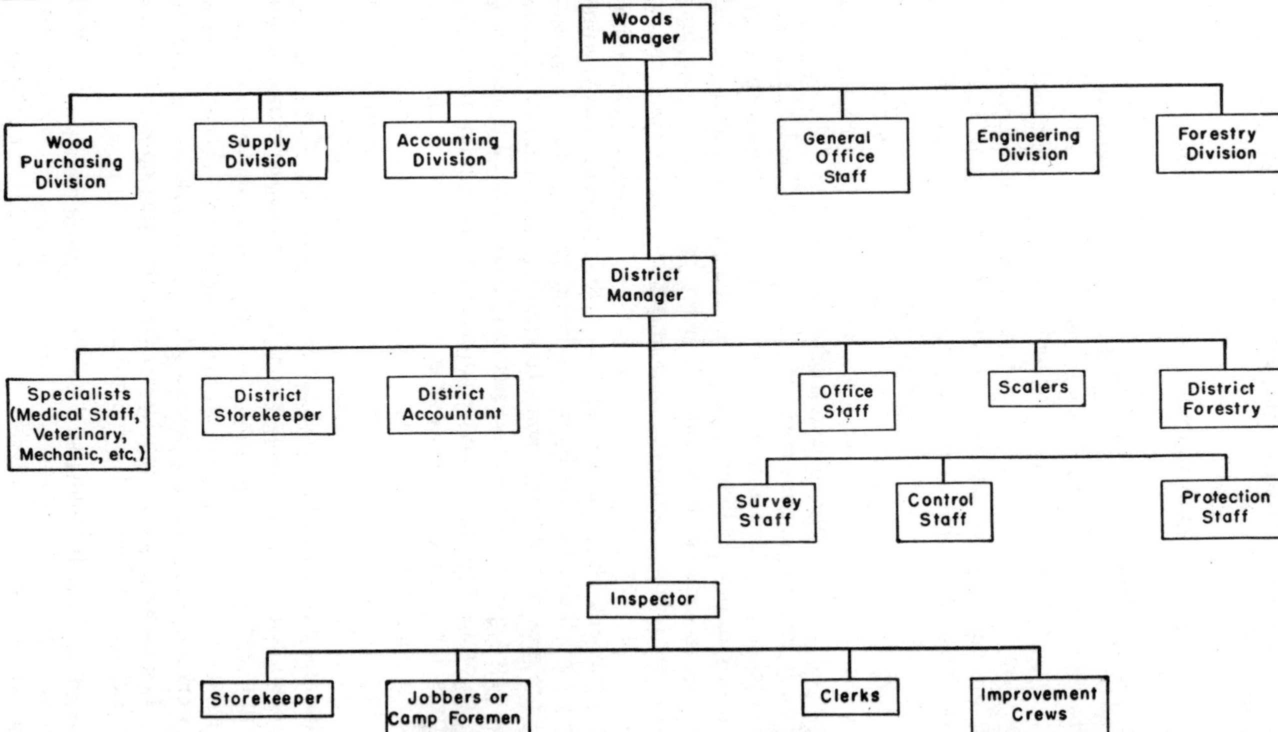

NOTE: For the sake of simplicity only one district and one sub-district are shown on the chart. Actually the district organization would be repeated in several districts and there would be several sub-districts (under inspectors) in each.

From Bentley, A.W., *et al.*, "Organization and Administration of Woods Operations," the Woodlands Section of the Canadian Pulp and Paper Association, 1938.

are confronted with the same operating conditions and each has certain features of internal organization that operate to make its own organization individualistic.

LARGE-SCALE CONCENTRATED OPERATIONS

To justify the construction of a stationary sawmill it always has been judged advisable for the owners to acquire in advance sufficient standing timber to supply the mill for at least 15 years, preferably 20 years. The timber is acquired, generally, in large contiguous blocks accessible to the mill site. As a result of this policy, logging operations supplying such sawmills are concentrated harvesting operations in a limited area, with transportation to the mills usually by means of company-owned logging railroads, truck roads, or water transport systems.

When logging can be concentrated and a large volume of timber products removed from a limited area, specialized methods can be used and costly improvements such as railroads and truck roads installed.

Large-scale logging requires crews of up to several hundred men, and since the timber being harvested is usually remote from population centers, it is necessary to provide living quarters for the crews and often for their families. Since individuals would not invest in houses for temporary occupancy near logging operations, the company conducting the operations must provide the camp facilities. A crew of several men, including carpenters, plumbers, electricians, painters, cooks, gardeners, and janitors, must be provided to keep the living and dining quarters in good repair.

Shops are provided at most camps to keep logging equipment in good condition and a crew of mechanics therefore is necessary. When a railroad is used, the camp shops are usually equipped and manned for overhauling locomotives, loaders, and other heavy equipment, as well as for repairing and maintaining trucks and tractors. Shops are maintained because it is cheaper to maintain a company garage than use commercial garages when a considerable number of machines are used. Also, too much time would be lost in taking equipment to town for repair work.

Each operation in harvesting usually is under the supervision of a foreman, who is responsible to the camp or logging superintendent, who is in turn responsible to the manager of woods operations. These crews, of course, work out of the camp, going each day to work on log trains or trucks. The crews comprising a large-scale harvesting operation getting out saw logs usually include the following:

- Camp boarding and maintenance
- Camp shop
- Cutting
- Skidding
- Loading
- Railroad (if used)
- Trucking
- Construction (in advance of logging)
- Camp office

The functions of these crews are obvious, as each conducts that operation in harvesting assigned to it. Construction in advance of logging is usually assigned to a separate crew. Maps often are needed for planning the location of roads and railroads. Also, it may be necessary to locate land lines and perform other technical duties. The foreman of this crew often is a civil engineer, who not only can locate transportation systems in advance but who can design bridges and handle the construction of improvements. When logs are floated or rafted down streams, special crews handle the necessary operations involved.

Large companies often operate two and occasionally more camps when timber is owned in different localities but is sawed at the same mill. Sometimes one operation is seasonal, depending upon weather conditions. In the Inland Empire region of western Montana and northern Idaho, for example, logging in the upper mountain areas is often done only during the summer months, while another operation is conducted either on a year-round basis or during periods when the mountain camp is closed.

Two or more operations may be carried on simultaneously when large volume is required, because if the entire amount were to come from one operation, it would mean too cumbersome a crew. Also, it is advantageous, when the volume required is sufficient to justify it, to conduct harvesting operations at separate localities so that an even flow of products to the mill can be maintained, in the event some accident such as loss of a main-line bridge, flood, or a landslide should block the transportation system.

The various activities essential in conducting a large-scale concentrated harvesting operation are commonly performed by company crews. Under this policy all of the personnel are employed by the company and receive daily or monthly wages and salaries, or in the case of certain crews directly engaged in production, payment may be on a piece rate of a given amount per 1,000 ft cut, hauled, or skidded. When harvesting operations are conducted by company crews, the company is responsible for workmen's compensation insurance, social-security pay-roll deductions and contributions, conformance to minimum-wage and hours-of-work laws, and other obligations of employers.

While company-conducted harvesting operations are common throughout the United States, it is not unusual to find operations where some, and in rarer instances all, portions of the operation are contracted. In the West, contract loggers frequently conduct the entire harvesting operation and are locally known as "gyppo loggers." They undertake to cut and deliver logs with their own crews and equipment at a given amount per 1,000 ft of logs. The contract logger usually does not own the timber but only undertakes to deliver the logs. The timber usually is owned by the company contracting for the logging.

Recent labor legislation that entails considerably more employer responsibility and obligations has caused wood-using mills to consider the advantages of contracting their logging and thereby be relieved of the extra accounting and costs which such legislation involves. While such costs are not escaped by contracting harvesting operations, large companies frequently cannot handle the accounting work as economically as smaller contractors and there has been an increasing tendency towards contracting major elements of harvesting and even the entire operation. Another factor that causes mill operators to favor contracting is the unionization of woods labor. Many companies prefer to be relieved of labor negotiations as much as possible and therefore contract their harvesting to others.

When only part of a harvesting operation is contracted, the company usually runs the camp and the transportation system. Under this arrangement log cutting may be contracted to someone who employs crews to fell and buck the logs for a specified amount per log or per 1,000 ft. Or, cutting and skidding to loading points may be contracted, in which case the contractor receives a specified sum for all logs delivered to a landing and the company loads and hauls them. It is not unusual, also, for companies to contract hauling either to the mill or to a company-owned logging railroad or to a stream, if the products are to be floated or rafted to the mill.

Large concentrated harvesting operations are usually possible only for large sawmills or pulp mills requiring a large volume of products and owning solid blocks of timber. Such cases are common in the West and to a limited extent in the South, Appalachian, and Lake states regions for sawmills. Pulp mills in the Lake states, Northeast, and eastern Canada also have timber in concentrated areas and require sufficient volume to make possible and advantageous large-scale concentrated operations such as have been discussed. Western pulp mills, and some mills in the Lake states using hemlock, obtain their wood in the form of logs, so their operations are essentially large-scale logging operations. Other rather large-scale harvesting operations are conducted for the production of western red cedar poles. Most other forest products are harvested by scattered small-scale operations that, in some cases, produce in the aggregate a large volume for a particular utilizing mill.

LARGE-SCALE SCATTERED OPERATIONS

Many wood-using plants that need large amounts of forest products each year cannot obtain their requirements from large-scale concentrated harvesting operations. Such plants include many pulp mills, veneer factories, specialty sawmills, stave factories, and wood-preserving plants.

Nearly all Southern pulp mills obtain their wood from a wide area and it is produced by a large number of small crews. A few Southern mills obtain some of their wood from large-scale concentrated operations, but this is the case only when a pulp mill and a sawmill are operated by the same owners and the pulpwood is produced by the sawmill logging operation, sometimes in the form of pulp logs.

The great majority of mills, however, are supplied with wood from a large number of producers within shipping distance of the mill, which is approximately 200 to 250 miles by rail under present practices and freight-rate structures. This is necessary because the mills do not own sufficient forest land to supply their own requirements. Even were concentrated operations possible, the abundant supply of pulp-sized timber available throughout the South and the network of transportation facilities (highways, railroads, and waterways) makes it economical to obtain wood from an extensive area.

Also, to concentrate the entire production required on company-owned or leased tracts would not be advisable for a large mill. Southern pulp mills cannot store large volumes of wood at the mill because it deteriorates too rapidly, especially during the spring and summer months, from insect and fungus activity. Therefore, production in the woods must just about keep pace with mill consumption to prevent running out of wood or building too large an inventory, either of which would result in serious losses.

Since all pulpwood is trucked from the woods at least as far as a railroad or barge landing, and, when close enough, all the way to the mill, wet weather greatly hampers production. Seasonal wet weather may be expected more or less uniformly during the winter months. During the rest of the year wet weather occurs intermittently. The advantage of widely scattered producing operations lies in the fact that wet weather seldom is so general that it affects the entire producing area. When one or more sections are too wet to permit production, the others usually can make up the deficit, at least for short periods, and thereby maintain an even flow of wood to the mill. Were all production to come from only two or three areas, wet weather could easily block all production simultaneously and cause the mill to run out of wood, or greatly increase the cost by continuing production under temporarily unfavorable conditions.

Many of the Southern pulp mills are large and some require up to 1,500 to 2,200 cords of wood per day for 7 days a week, which is in the neighborhood of 100 to 150 freight carloads daily for just one mill. At present, the light stands of timber per acre are a factor in favor of scattered small operations. On the average, the cut of southern pine probably does not exceed 5 or 6 cords per acre. Mechanized methods of production such as are possible with large-scale concentrated logging operations are not possible with such light stands, as the volume that would be produced per mile of truck road or logging railroad would be so small that the fixed cost would be too great. Also, mechanical skidding and loading are not possible, economically, unless the volume of timber per acre is large enough to provide full loads with a minimum of machine movement, and this is seldom possible when only 5 or 6 cords are cut per acre.

In the future these procurement practices may change with emphasis on large-scale concentrated operations because most mills have acquired large acreages of productive forest land, which are well restocked to young timber that will, within a decade or so, develop into heavy stands over large areas. The holdings of several companies exceed 250,000 acres and some of the larger exceed 500,000 acres. These holdings are generally concentrated in blocks of from 60,000 to 175,000 acres each, so when the stands reach harvesting size the cut per acre should be heavy enough to make mechanical harvesting methods economical.

With forests under good management for permanent production of forest products, all-weather truck roads can be built and charged off against the current cut without excessive cost per unit of product and still be available for succeeding harvests. Stands of southern pine as a rule do not now yield heavy cuts per acre because so many different forest industries draw upon them. Much pulpwood produced in the South is cut as thinnings and salvage behind sawmilling and turpentining, which leaves only the trees no longer useful for those purposes. When the stands now being grown primarily for pulpwood are ready for harvesting, yields of from 20 to 30 cords per acre may be expected from harvest cuttings.

Many Northern pulp mills find it possible and advantageous to conduct large-scale concentrated operations because the spruce, fir, and aspen timber they use grows in dense-enough stands over large-enough areas to justify the installation of camps and other production improvements. Even so, a considerable portion of the wood used by many Northern mills is supplied by small producers because of the availability of wood from scattered small tracts and suitable transportation facilities.

Other kinds of wood-using plants that obtain their supplies from many widely scattered small producers have the same gèneral problems of production. The reason veneer factories, pole- and pile-treating plants, stave mills, and specialty sawmills and similar plants have to depend on many small producers is that the kind of product they require from the forest does not occur in heavy stands per acre. Veneer-quality trees are scattered, since only trees that are clear of knots and defects and will yield bolts of 14 to 16 in. in diameter or larger can be economically cut for veneer. Likewise, only trees that have the correct dimensions and form and are free from defects meet the specifications for poles and piles. Since these specifications are rigid and require straightness and freedom from any decay, only a comparatively small percentage of the trees in most stands can be used for poles and piles. This precludes the possibility of operating stands with a large-scale or even small-scale concentrated operation for such products, except in the case of western red cedar and northern white cedar, because stands of these species often are rather uniformly of a size and of pole quality.

The organization required to maintain an adequate supply of forest products at plants relying on scattered small operations is quite different from that for a concentrated operation. A large number of producers must be given production allotments and must be paid for their shipments. The procurement organizations for most large mills of all kinds find it advantageous to deal with dealers who assume the responsibility for obtaining the needed wood supply from a given area assigned to them.

Dealers do not themselves produce pulpwood or other products but buy from producers who actually engage in hiring labor, buying equipment, acquiring timber, and doing the job of getting the products cut, loaded, and shipped to destination. Producers conduct operations varying in size from only one truck to six or eight trucks or even more. They locate and purchase stumpage and handle all the details involved in getting it cut and shipped. Small operators, each with only one or two trucks, often act as foremen and obtain timber for cutting when they can leave their supervisory jobs long enough to do so. Larger producers usually employ foremen, while they give general supervision to their operations, which may be at two or more places.

The dealers often purchase stumpage and in effect sell it to producers who cut and ship it when stumpage transactions are too large and require too much capital for the usual producer to handle. The dealer who has better financial standing can arrange for such purchases, which relieves the producer from such obligations. Producers do often handle small timber purchases from farmers but in many cases they receive a great deal of help from the dealers in the matter of obtaining stumpage for cutting. Dealers also assist many producers in the purchase of trucks and other equipment and tools, often handling the necessary financial arrangements for them.

Dealers receive, usually, a commission from the mills receiving the wood shipped by the producers, which is their main source of income. In some cases they may add to their income by negotiating advantageous timber purchases that result in lower-than-average stumpage cost. Some dealers also engage in various other businesses, such as storekeeping, cattle raising, running filling stations, operating small sawmills, and similar activities that they can manage along with their wood businesses. Dealers also find it possible, in some cases, to deal in a variety of forest products, including poles and piles, pulpwood, veneer bolts, logs, acid wood, and chemical wood.

The mills buying from dealers maintain close contact with them through field men who travel throughout the producing territory. The field men assist the dealers in arranging for cars, purchasing stumpage, obtaining trucks and other equipment, and in meeting competition from other mills that may be in the market for forest products and have dealers in the same locality. The dealers usually get the bills of lading at the end of each week for cars shipped during the week and advance the producers a substantial portion of the amount the mill will pay for the wood. The bills of lading are sent to the mill as evidence that the wood was shipped. Final measurement is made at the mill when the cars arrive and the dealers are paid the total purchase price including their commissions. The dealers then pay the producers the balance due them. Some mills deal directly with producers and maintain a larger staff of field men, as many more contacts must be maintained under such an arrangement. Generally, only the smaller mills can handle their procurement without dealers, as large mills would find the field organization necessary to do the job too large and cumbersome to be economical.

The system of dealers and producers is also widely used for practically all other wood-utilizing plants where concentrated operations are not possible, except for poles and piles. Poles and piles are produced whenever accessible timber is available that will yield the sizes on order, but company field men contact producers, as a rule, so the dealer system is not so adaptable. One reason is that the sizes of poles and piles the treating plants can accept changes from time to time in conformity with orders received at the plants. Prices also fluctuate considerably and negotiations through a third party, or dealer, would add to the likelihood of misunderstandings and dissatisfaction.

Furthermore, poles and piles are usually bought f.o.b. loading point, so a company inspector inspects the poles or piles in each shipment as they are loaded. Otherwise, the loss to the buyer could be excessive, due to rejecting culls after arrival at the plant, as freight would have to be paid on them. Therefore, since a company representative is present at each loading and, in effect, purchases the piles or poles at that point, there is no real need for dealers and they are seldom used.

Tanning-extract plants using acid wood and destructive-distillation plants using chemical wood have dealer and producer arrangements very similar to those for pulp mills. These products come from scattered small producers who deal either directly with mill representatives or dealers who maintain contacts with the buying mills. Veneer bolts, handle bolts, and, to a large extent, ties are produced under similar arrangements, as the conditions under which they are produced are such that concentrated operations are impractical.

Certain sawmills, particularly specialty sawmills cutting choice hardwoods, also obtain their logs from widely scattered producers shipping either direct or through dealers. The logs required by these mills, such as logs of clear white oak, figured red gum, large clear yellow poplar, red cedar, and walnut, do not occur in sufficient quantity even in natural stands to permit logging solely for them. Such specialty saw logs, therefore, are also produced under the dealer and producer plan of organization, along with purchases from other operators using the general run of logs and conducting their own harvesting operations.

The organization maintained by a large wood-using mill obtaining its raw materials from scattered producers consists, usually, of a department designated variously, depending upon the mill policy. A manager or chief wood buyer heads most such departments and he is responsible for keeping the mill supplied with an adequate stock of logs, pulpwood, bolts, poles, ties, or whatever the mill uses. His assistants vary in number with the area to be covered and number of producers or dealers to be contacted. When the dealer system is used, fewer company field men are needed for the same size of mill, as the dealers handle arrangements with the producers. An office force is required to keep accounts and make payments for wood purchased and to deduct for cash and equipment advances when the mill assists its producers and dealers in this way.

The mill organization has nothing to do with the actual harvesting operations when the wood is received through dealers or producers. As a matter of policy and to enhance the future supply of wood in the mill buying areas, many mills, particularly pulp mills, have from one to several foresters on their staffs, primarily to assist producers, dealers, and timber owners in cutting correctly. Frequently they mark stands for cutting according to good forestry practices to encourage timber growing by farmers and other owners and thereby maintain a perpetual timber supply. Mills also commonly finance dealers in large timber purchases when a dealer's financial resources are insufficient to handle them himself. Before advancing funds for others to purchase timber, the mills have each tract estimated to be sure the advance is not out of line with the amount of timber on the tract. To do this, the mills maintain from one to several timber cruisers who devote most of their time estimating timber on areas dealers or producers want to purchase and need financial advances from the mills to do so.

Mills owning timber may also conduct company producing operations in addition to buying products from dealers and producers. When this is done, a company crew is organized and harvesting operations are undertaken in keeping with the volume to be cut and the forest conditions. Often company crews are used to cut products where conditions are more unfavorable than usual and individual producers are unwilling to assume the risks involved in operating in such areas.

SMALL SCATTERED OPERATIONS

Many small-sized harvesting operations exist throughout forested areas in the United States, supplying small portable sawmills, box mills, stave mills, excelsior plants, dimension factories, and other small units of the wood-using industries. The most common are those operating in connection with portable sawmills.

So long as lumbering was conducted in virgin forests in undeveloped regions, large-scale operations were a practical necessity. Powerful machinery was required to handle the heavy logs and transport them to milling centers and a large volume of output was necessary to justify the heavy investment in such equipment. When populations moved into the formerly forested regions, however, and the former wilderness areas became settled communities, and second-growth forests attained merchantable size on much of the remaining forest lands, small-scale lumbering developed rapidly and now has grown to substantial proportions in all lumber-producing regions.

This occurred because extensive systems of highways and rural roads traverse most rural areas. Even sections predominantly agricultural contain a high percentage of forest land, generally in comparatively small tracts. Because the timber is of small size, because transportation routes are already installed, because trucks are now well adapted to log hauling, and because portable sawmills have been perfected, small sawmilling operations now flourish not only in the East and South, where they originated, but in the West as well, where merchantable stands have grown up on many of the areas first cut over.

Small mills, either stationary or portable, ordinarily are operated on a quick-turnover basis, and so their owners usually purchase only stumpage, whenever possible, and avoid investing capital in forest land. They purchase timber often under competitive bidding and hence feel that they must cut all the timber even to small sizes to recover the purchase price. As a result, small mill harvesting operations are often responsible for serious overcutting of the forest, unless the sellers specify methods of cutting that will keep the forest productive. Larger operations with heavy investments to protect and with large areas of owned forest land are more concerned with leaving the forest in a productive condition for a future timber supply. Because they must provide for large-scale production over a period of years, they customarily acquire timber and land in fee and find it advantageous to maintain their forests in good growing condition.

Small sawmills are either individually owned or are owned by lumber merchandisers who operate concentration yards where the rough green lumber from a number of small mills is assembled for seasoning, planing, grading, and shipment. Each mill, however, is operated in much the same way, whether individually owned or company owned. Many individual mill owners harvest timber and saw lumber on contract for concentration yards, while others run on a custom basis or saw lumber for the general market. All such mills are readily portable and are moved into the timber to be harvested. It is economical to move a small mill into a stand that will yield as little as 50,000 ft of lumber, although most little mill operators prefer to cut at least 100,000 ft at each setting, as it is called. When the logs must be moved more than ¼ mile, it is generally more economical to move the mill than to transport the logs greater distances.

Harvesting and milling are closely integrated with portable mill operations in that there is no hauling stage in the actual harvesting phases of such operations. Since the mill cuts timber only within a ¼ mile radius or less, the skidding operation delivers logs to the mill deck. Thus, the loading operation is also eliminated. In effect, the portable-mill operator intercepts the logs in the woods and saws them there. The hauling of rough green lumber away from the mill to the concentration yard is substituted for a log haul.

Felling, bucking, bunching, and skidding are done in the usual way, the actual method depending upon the extent to which the operator has mechanized his operations and the size of the timber to be harvested. Occasionally, some logs are hauled to portable mills from distances greater than ¼ mile, but this is done only when some nearby timber owner has a small amount of timber to be sawed and does the hauling himself. Such individuals often keep a portion of the lumber sawed from their logs to use in farm construction, others simply sell the logs to the mill owner.

Many small mill operators work only part time at harvesting and sawing timber, spending the remainder of their time either farming, storekeeping, or conducting various other enterprises. Such operators cannot usually afford to invest in costly equipment, such as tractors, if they use them only part of the year. As a result, their operations are less mechanized than those operating steadily and animals are much more extensively used as they, also, can be worked on a farm as well as on timber-harvesting operations. Also, much of the timber sawed at portable mills is small and can readily be handled by a lightweight team such as many farmers use.

The foreman of portable-mill operations supervises the harvesting operations in the woods as well as the sawmill operation. Some workers may live at or near the operation in temporary and usually quite inadequate quarters, but a majority commute daily between their homes and the operation, often driving their own cars under ride-sharing agreements.

HARVESTING OPERATIONS BY FOREST OWNERS

More and more forestry is being practiced as a business apart from any manufacturing affiliations. Such forest owners engage in protecting and maintaining a growing stand of timber on their lands for the sole purpose of harvesting mature trees and those that should be removed in thinning and improvement cuttings. Forest owners who are not interested in supplying themselves with any particular kind of forest product naturally seek the best markets for the timber they grow, from the standpoint of both greatest financial return and greatest benefit to their growing stands.

Forests yield a variety of products when there are markets for them, and with the increasingly diverse markets for forest products, forest owners find greater opportunity to harvest for various purposes more and more of the trees they grow. Timber crops in a managed forest should always be harvested by or under the close supervision of the owner or his personal representative. When cutting rights are sold to others, no matter how strictly the contract is drawn, the purchaser is always more interested in his own personal gain and will cut corners whenever possible to do so.

It is far better from the owner's point of view if the trees are cut and made into the products for which they are best suited under his direct supervision. A sawmill operator has no particular interest in poles and piles and will cut trees suitable for them into logs unless restrained by the owner. Even when trees are marked for certain products and each operator is limited to cutting only the trees selected and marked for him to cut, administration is difficult and mistakes are common.

The choicest products, too, occur rather rarely, and the cost to one operator in getting only one class of product, such as veneer bolts, poles, or logs of certain species, usually is high because he does not remove sufficient volume to spread the fixed costs thinly. Also, in cutting veneer bolts, for example, only the choicest lower sections of trees are suitable for these products, although the upper portions could be used for logs and pulpwood. Pole and pile cutters, if they really try to cut all such trees, occasionally make slight mistakes in judgment and after a tree has been felled it is found unsuitable for these products because of unseen decay or scantness of size at some point. Buyers must either pay for the unsuitable trees so cut, which causes them to lower bidding prices, or they will be ultraconservative and leave trees standing that would make poles or piles. In either case the owner loses. Also, if the trees are cut by one operator and not utilized, they either rot or, if salvaged, the added cost of such scattered salvage operations reduces the net value of the trees.

When it is necessary to sell products on the stump to those interested in various products, it is always advisable to require that the choicest products be cut first and that the next most valuable be taken next and so on, until the least valuable is taken last. Under such an arrangement, the trees that are felled by those cutting poles, for instance, but found not suitable can be used for the next most valuable products before they deteriorate and become valueless. Also, those portions of trees cut for such products as veneer bolts or handle bolts but not of high enough grade to be used for them can be used for logs or pulpwood. Since pulpwood is utilized by chipping or grinding it and then reducing it to fibers, the specifications for pulpwood as to form and size are the least exacting and it ordinarily is the product that should be harvested last; even the tops of many saw-log trees can be utilized for pulpwood, whereas they would otherwise be valueless except for fuel wood, the market for which is seasonal and fluctuating.

It is difficult and often impossible, however, to arrange with buyers to cut timber according to such a schedule or at least promptly enough so that each can salvage the trees or portions of trees left by previous operators. Unfortunately, there are few if any producers who handle a full range of forest products. There are many so-called dealers who handle one or two or perhaps three products, but practically no dealers to whom a forest owner could sell his entire cut and have each tree and each portion of each tree put to its best and most profitable use. Dealers in forest products usually confine themselves to buying and selling only a few bulk items that do not require too much merchandising effort to market advantageously.

A forest owner who owns sufficient timber and whose annual cut is sufficiently large to justify maintaining a harvesting crew should, by all means, do so. The foreman of that crew can be trained in the specifications and relative values of various products, and in the annual harvesting he can cut each tree and all portions of each tree into the products that will yield the greatest value and leave the least waste. By producing all the products the cutting allotment will yield, lower costs can be obtained since all come out at or about the same time and the fixed harvesting cost is spread over the total volume. A minimum of lost time in moving into one job and on to another results, and little or no duplication of effort develops when the same harvesting crew handles all products. No confusion results as to the timing of operations in correct sequence, which is difficult to obtain in the case of separate sales, and no one is prejudiced in favor of one particular product.

Such an integrated operation is entirely possible. All that is necessary is for the owner or his representative to be thoroughly familiar with the markets available to him and to know the specifications of the various products. Then all can be taken from the forest in correct sequence and at the right time, with resultant profit to the owner.

One particular advantage to forest owners who are growing timber for profit is that self-harvesting operations provide constructive work for labor crews and key foremen throughout much of the year, which keeps them available at all times for other forest work. Silvicultural work, such as noncommercial improvement cuttings, road and construction work for protection purposes, and other forest work that is intermittent or insufficient to provide steady employment, can be handled by permanent crews. They can devote most of their time to harvesting operations but can drop these jobs temporarily in case of fire or other necessary work of an urgent nature, such as the control of insect outbreaks. Growing forest crops has become an established business for many and will no doubt expand further.

This makes possible owner harvesting operations in which all products are removed, within the cutting limitation imposed by the forest management plan in effect and by market requirements, to return the highest financial yield.

Forest owners find that the same crews and much of the equipment needed for protection and other forestry operations can also be used in harvesting. There is, therefore, but little duplication of equipment and no duplication of labor in combining all forestry activities under one over-all forest operations plan, of which harvesting requires the most equipment and largest man-power supply. After all, forest owners grow tree crops for the purpose of harvesting them, so the harvesting operation should be and ordinarily is their main interest. It should not be a thing apart, however, but it should be integrated into all forest operations to make the most efficient use of forest personnel and equipment.

CHAPTER 17

COSTS AND RECORDS

Since forest harvesting is conducted for the purpose of converting tree crops into merchantable commodities and obtaining an income from their sale, it is of utmost importance that such operations be conducted as efficiently as possible. This requires an adequate knowledge of costs, and to determine the cost of harvesting operations, it is necessary to keep records of all costs incurred and of the volume of products handled.

In general, harvesting costs are of four main kinds: (1) purchase of equipment, (2) labor, (3) materials and supplies, and (4) overhead. Equipment purchases involve the expenditure of considerable sums in advance of actual operations, but the equipment obtained, such as trucks, locomotives, and loaders, has an operating life of from one to several years or longer, depending upon the particular items. Trucks may wear out in only one year's time, but a locomotive will last indefinitely. All such equipment, however, has a usable life long enough to handle the cost on a depreciation schedule rather than as a lump-sum charge against the products cut and delivered during the month in which it was purchased.

Labor costs include, of course, all wages and salaries, and such items as pay-roll taxes for social-security benefits as well as workmen's-compensation insurance borne by the employer. Materials and supplies include items that are used up or consumed in the operation, such as gasoline and oil, animal feed, small tools, repair parts to keep equipment in running order, and similar articles. Overhead costs are those that are not directly incurred by the harvesting operation but are chargeable in whole or in part against harvesting. Overhead items include such costs as property taxes on equipment, insurance of various kinds, prorata share of main-office expense, legal fees, and other items of a general nature.

Total expenditures and equipment-depreciation charges divided by the volume of products harvested give the unit cost. Thus, if the total cost of harvesting pulpwood and delivering it to the mill for a particular month is $10,000 and a total volume of 1,000 cords was handled, the unit cost would be $10 per cord. When this is compared with the costs for previous months and years, the manager can judge in a general way whether or not current costs are out of line or are satisfactory. For the purpose of fixing a selling price, when this is possible, the total cost per unit of product for harvesting is sufficient information about this phase of the business.

However, to yield information of value in competent administration of harvesting operations, much more detailed cost statements are essential.

No uniform system of harvesting accounting or record keeping is in effect. Each company or individual keeping records does so in such a way that the cost and production records furnish the information in the form most suitable for that particular operation. In general, however, harvesting costs are kept separate by operations since each operation can be readily identified and actual expenditures charged against it. Costs would be of little value if no information were available on the volume of product handled, so both costs and volume measurements must be recorded for specific units of time to provide helpful information.

Practically all accounting records for harvesting operations are concerned with costs of some kind. Income from forest products is rarely obtained until they are sold, which is after harvesting has been completed. Operations conducted to supply a mill do not show any income, as the products are manufactured in the mill before being sold. Occasionally income is obtained for the use of harvesting equipment, such as hauling shipments for others over a logging railroad. When camps are maintained, rent and board are commonly charged and the camp commissaries may show profit, so it is not unusual to deal with a few credit items in harvesting accounting. In the main, however, harvesting involves the expenditure of funds for which the return may not be received until other manufacturing steps have been completed and the finished products, such as lumber and paper, are sold.

For large harvesting operations, accurate records are kept of costs and volume of product, and generally the cost items are kept separate by operations. This permits the management at all times to be fully informed as to the cost of doing business and to control costs, in so far as is possible, if some cost items appear to be higher than is usual or necessary. Items of cost may fluctuate at times, because of costly situations that develop. Expensive bridging for a truck road or railroad, or an unusually big cut or fill, may be required, which could easily make harvesting costs for any one month strikingly above average. If the costs were not segregated by operations, it might be difficult to explain or understand why the cost for that month was excessive. Accurate costs, therefore, not only help management in conducting a sound and efficient operation but also help supervisory personnel in administering operations efficiently.

Although the degree to which costs are segregated varies greatly with different operators, costs for all are essentially of the same nature. Harvesting operations that are highly specialized, such as logging operations supplying a large permanent mill, are the ones for which costs are more generally kept in most detail. Smaller operators more often find that they have sufficient data with fewer items. Individuals conducting small-scale operations often keep their records "in their hats," as they are responsible only to themselves and can keep in mind what their costs are without detailed bookkeeping records, except such as are needed for tax purposes.

PRINCIPAL HARVESTING COSTS

Selection of Trees to Cut. Because harvesting operations seldom remove all the trees from an area, there is some selection of trees to cut, and this activity may be important enough so that its cost is charged separately. In managed forests, the selection and marking of trees for cutting is normally a part of management operations and not of harvesting, and the cost would be charged to forest-management expense. Many operating companies, however, cutting selectively or according to other forestry methods requiring the selection and designation of trees to be cut, handle the marking of trees for cutting as a part of harvesting and the cost is included as an item in harvesting cost. On many operations, however, the matter of selecting trees for cutting involves only instructing the felling crews as to the species to cut and the minimum size of tree to cut. No expense is incurred when the selection is this simple, but when someone is employed particularly to select and mark trees, a definite cost is incurred and separate cost records are desirable.

The principal items of cost in selecting and marking trees are wages, equipment, and supplies, and, if the markers drive cars for their employer, whatever travel and incidental expense incurred. When trees are marked with paint, the cost includes the paint and thinner and either brushes or spray guns. Ax blazing requires only a small ax or hatchet, which may have a die on the hammer side to use in branding trees blazed for cutting to prevent cutters from blazing trees themselves. These tools cost only small amounts and no supplies are needed, so labor expense is the principal item of cost.

Camp and Maintenance. When camp facilities are provided for a harvesting operation, a separate account is usually maintained for camp costs and income. Charges under this heading include such items as repair and maintenance of buildings and grounds, electricity, water supply, and other facilities supplied. Accounts usually are separated so that costs incurred by income-producing property, such as camp houses, store, and boarding facilities, are kept separate from those needed to conduct the harvesting operation, such as repair shops.

Commissary buildings and equipment may be leased to another party, in which case a rental is received and any costs against the property leased are shown in a separate account in which the income can also be shown. Boarding facilities may also be leased to another, and when this is the case, the only accounts kept are expenditures for maintenance and rental received. Leasing such facilities relieves the camp office of keeping detailed records of numerous minor expenditures and income items, which otherwise would be necessary.

When camp is moved, special costs are incurred in the preparation of a new site and the installation of facilities. Also, from time to time buildings may be added or old ones replaced. Expenditures of a semipermanent nature are written off currently if they occur rather regularly and are of like amount, or they are handled as investments and depreciated either on the basis of useful life or for the estimated duration of the camp. A water well, for instance, probably would flow indefinitely but it loses its value when camp is moved, and so water developments at any one site can be depreciated on the basis of the estimated number of months or years the camp will remain at that location.

When periods of heavy expense occur, as in the original construction of a camp or when one is moved, the item of camp expense may appear unduly large if the cost is not prorated over the period it will be used. In some cases camp-moving costs are written off as they are incurred, while in others they are prorated over a period. In any event, there is an item in harvesting costs for camp maintenance when such facilities are provided for employees engaged in harvesting.

Camp expense may also be considered as an overhead charge and prorated over the several actual harvesting operations. No particular advantage is gained by this, as some arbitrary basis for spreading camp cost would be necessary, and all that could result would be somewhat higher operation costs and no separate camp-cost item. To keep camp cost from becoming disproportionately large, it is much more desirable to show it as a separate item and frequently to break the camp cost into subgroups.

Felling and Bucking. The costs connected with felling and bucking trees are kept together when the same crews perform both operations. When felling and bucking are separate operations, as with many Western camps, they may be kept separate. Costs having to do with felling and bucking, especially when these operations are combined, are often designated as cutting, or log-making, costs. On some pulpwood operations, the individual cutters perform not only felling and bucking but also peeling and bunching. There is no way to keep the time of such workers separate for each step and there is no point to it, except when special time studies are made. The entire operation is often referred to as cutting and the cost of wood felled, bucked, peeled and bunched is referred to as cutting cost.

Felling and bucking costs include not only the labor cost but also tool cost and supplies, such as wedges and kerosene. When power saws are used, the investment in such equipment is so large that tool cost is usually depreciated over from 1 to 2 years. Handsaws are relatively inexpensive, so the cost of saws, axes, and similar light tools usually is treated as a supply item. With gasoline power saws, of course, the fuel and supplies essential to their operation must be included. With electric saws requiring a generator and air-driven saws requiring a compressor, the cost of the auxiliary equipment must also be included. Tool sharpening requires a skilled saw filer; he usually sharpens each day the handsaws used the previous day and he works in a shop, either in the camp or near the job. Power saws require special sharpening equipment operated by power, but the saws ordinarily do not need daily sharpening. When sharpening equipment is expensive, as for power saws, the cost is included as a part of the investment. On small operations the cutters themselves or sometimes the foremen keep tools in good condition.

Contract agreements with fallers and buckers are common whereby individuals, either singly or in small crews working as a unit, undertake to fell and buck timber at a given rate per unit of product cut. In large timber, felling and bucking contractors often own their power saws and either maintain them themselves or they are maintained in camp shops as a part of the agreement. The advantage of individuals owning the saws and cutting on contract is that the saws generally receive better care because of the direct personal interest the workers have in the equipment. Crews using hand tools, also, often work on a piece rate, sometimes furnishing their own tools and keeping them in condition and in other cases working with company-owned and -maintained tools.

With contract cutting, of course, the contract price is the cutting cost and no tools or other costs are recorded separately, as the individuals contracting the operation assume such costs. Few if any cutters working as independent contractors, however, keep books on their own costs, so the various costs generally are ignored as separate items.

Some measure of the daily work of cutting crews is usually kept, even where labor is paid by the hour. When labor works by contract, either in whole or in part, measurement is necessary in order to make correct payment for work accomplished. When this is done, a scaler measures the volume of product cut each day, either in log scale, cords, numerical count, or other units, and credits to each individual or crew the amount produced. When felling is done independently of bucking, the felling crews usually are paid a daily or hourly wage, as it is difficult to obtain the volume of a tree before it has been bucked. Felling may be several days in advance of bucking, which is a factor of safety, as bucking crews might be injured by falling trees if both operations were conducted simultaneously but by separate crews. Hence, it would involve difficulties to obtain the volume of felled trees from the log scale contained in them. On the other hand, it is relatively easy to obtain the measure of logs cut, as it only involves actual scale, so bucking crews can readily be paid according to the volume of products bucked.

Since felling and bucking crews are generally scattered for safety and because the distribution of trees requires spreading the crews over considerable areas, they are not as closely supervised as workers on more concentrated operations. Also, since the output of crews is largely dependent upon their own skill and energy, some are consistently better than others. The contract method of payment, therefore, is preferred whenever possible for the felling and bucking operation. Considerable variation exists in tree size even on one operation, and on some both pines and hardwoods are cut. When there is sufficient difference to affect materially the output of cutting crews, variable contract rates usually are adopted. A crew will work just as energetically in cutting oak as in cutting pine, but oak is much harder than pine and fewer logs are obtained for trees of equal diameter, so the contract rate established commonly is higher for hardwoods than for softwoods. Likewise, crew output with equal effort is less in log scale when small timber is being cut than when large timber is being cut. Appropriate adjustment is usually made when the average log volume for a day is below a given amount, say 100 ft.

Skidding. Skidding costs ordinarily include bunching costs when bunching is done, as the two operations are closely related and are supervised jointly. Labor cost is relatively less important in skidding, as most skidding is done by mechanized methods and machine costs make up a large part of the expense of getting forest products to loading points. Skilled labor is required to operate skidders and tractors, and close supervision is necessary to prevent costly errors in judgment, particularly with respect to the location of sets for cable skidders and roads for tractors. Wage rates are high, therefore, but a few competent men can operate machinery capable of large output.

The principal items of cost in skidding are: labor; machine repair and maintenance; rigging ahead if cable skidders are used; road construction if tractors are used; and clearing landings or yarding areas. Often the various items of expense comprising skidding cost cannot readily be separated and skidding cost, therefore, usually is a composite of several operational costs. For example, skidding tractors are often equipped with bulldozers and construct skid roads on return trips of skidding turns, and it would be difficult to keep separate the actual cost of skidding from that for road construction.

Skidding operations vary from simple team snaking to expensive overhead-cable systems costing many thousands of dollars. Cost records vary likewise, and for the more simple methods fewer cost separations are made. With team skidding, such items as labor, feed and harness, barn expense, supplies, and depreciation can be separated to give management an opportunity to analyze the cost of skidding. For mechanized operations, many more items of cost can be included, and should be, to furnish an adequate picture of the component costs of skidding. Such cost items as depreciation, fuel, repairs and maintenance, labor, supplies, wire rope, and advance rigging can be kept separate and give management a much better opportunity to know what items make up the greater part of skidding cost and where suitable economies might be effected.

It frequently is not possible to obtain a daily record of the volume of product skidded because it may not be necessary to scale the products at the loading points. When skidding is kept close behind cutting and the products are yarded separately for each cutting crew, and scaling is done after skidding to obtain a measure of work for each cutting crew, the skidding output for each team or other unit also is known and can be so recorded. Since skidding labor usually is paid by the hour or day and company-owned equipment is used, it is not necessary to obtain an exact measure of the volume handled for wage settlements. Logs are usually loaded and delivered to the mill promptly after skidding, so the mill scale of products received by months is usually used to represent the volume skidded. This is not strictly accurate, but over a period of several months the discrepancies that occur at the beginning and end of each month average out and mill scale or mill receipts for a month is considered a fair measure of volume skidded.

On seasonal operations where the products do not reach the mill until months later, as in stream driving, the measure of volume skidded is obtained by actually scaling the product at concentration points or it is obtained from the cutting scale. When logs are cold-decked or stored at a spar tree for later swinging to loading points, it is often difficult to obtain a measure of the volume in a cold deck unless records were kept of the volume cut in the area from which logs were cold-decked.

Loading. Loading costs include the depreciation on loaders, whether simple A-frame jammers, truck loaders, or railroad types, and labor, materials, supplies, and maintenance of equipment. In loading as in skidding, machine costs are more important than labor costs, as comparatively few men are needed to conduct the loading operation.

On some operations loading is closely associated with skidding, as with power skidders equipped with loading booms and drums. When this is the case, skidding and loading costs may be lumped together, since the operation is practically a continuous one, the logs being loaded on cars or trucks as soon as they are yarded. Where separate loading machines or devices are operated, the loading operation is distinct from skidding and costs can be recorded separately.

The measure of products handled by loaders is usually obtained in the same manner as for skidding, as here again there is no particular advantage in obtaining an accurate daily scale of products handled. The monthly volume delivered to the mill or the number of cars or trucks loaded per day is close enough to measure unit costs for this operation.

Delivery of Products. The designation of this account is various, not only with different companies but with respect to the method of transportation used. If logging railroads are used, it may be the railroad account, or if trucks are used, it may be called trucking cost.

Those using stream drives have a different class of costs from those using overland methods, but they are similar in that they have to do with the movement of products from loading points to the mill or other destination. One operation may have costs classified in part as sled haul, stream drive, waterway improvements, booming, and towing; while another would have main-line railroad, spurs, right of way, and track-laying cost classifications. It is obvious, however, that regardless of the actual operations involved or the designation given the accounts, they are comparable in that the expenditures are made for the purpose of delivering products harvested in the forest to consuming or other delivery points.

Some of the preliminary costs involved in opening up a timbered area may be considered as a part of the initial over-all investment. A railroad or truck road may have to be constructed for a distance of, say, 25 miles before the forest to be harvested is reached. This investment is necessary in advance of harvesting operations and often has a bearing on the price originally paid for the property. Thus, the cost of such access developments can well be capitalized along with the price paid for the tract and the legal fees and cruising expenses usually incurred in such transactions. Such installations are just as much a part of capital investment as are mills and towns constructed to manufacture the products and house employees.

After the tract to be harvested has been made accessible by transportation facilities, however, the subsequent costs having to do with transportation are logically a part of harvesting cost and should be so considered. Tracts to be harvested may already be accessible by one or more methods of transportation, such as by common-carrier railroad or public highway or both. Or, a mill may be constructed close to the forest, thus reducing the haul to a nominal distance, in which events all transportation construction and investments would be a part of harvesting cost.

Transportation costs are characterized by rather heavy fixed costs, such as for road building, railroad right-of-way clearing and grading, dam and bridge construction, and similar installations that cannot be moved from place to place but must be used only where installed. Expenditures for other harvesting operations are, in large measure, for portable equipment that does not need to be used only in place. Because of the fact that the construction of transportation facilities is comparatively costly and because the investment must be liquidated in place, a high degree of skill and good judgment are required in determining the location and kind of such improvements. A road built in the wrong place can never be moved, nor can a railroad bridge or cut, and the expense incurred is lost. On the other hand, if a loader is not skillfully placed it can be moved with but comparatively small cost.

Main-line roads and railroads must be located with due consideration not only to the immediate harvesting program but with future consideration in mind. If the future use of the facility is not analyzed, a main road or railroad might be built into a valley from which it could not be extended to the remaining area to be harvested, at least not without great expense; whereas another route, less favorable for immediate operations, would be entirely suitable for future operations, with much lower total cost.

In railroad operations, costs are usually segregated into right-of-way location and clearing, grading, tracklaying, track and rolling-stock maintenance, depreciation, and train operation. When log trains or cars are run over other railroads, the cost of such privilege is included, or when products are shipped over common carriers, freight is a cost item to be included.

Truck hauling operations include such cost items as road construction, road maintenance, truck operation, truck maintenance, and depreciation. When public highways are used road costs are not incurred, except when short tap roads are constructed. Operations transporting products by water have comparable but different items that include such costs as stream improvement, sled haul, booming, and driving. Other types of water haul would include barge or rafting expense and towing cost, and also depreciation and other usual costs.

Obviously, large investments are necessary in hauling equipment, whether for locomotives and log cars or trucks and trailers, if large volumes of products are to be transported. In fact, it is only when large volumes are to be hauled that large investments can be made, as this permits spreading the cost over enough volume to keep it from becoming excessive. Heavy investments for equipment that will last several years or longer are depreciated over either the usable life of the equipment or the period for which it can be used, whichever is shorter.

With railroad and truck operations in large blocks of timber, the mileage of main-line tracks or roads and spurs or lateral roads is about the same year by year. However, it frequently is necessary to develop an unusual mileage of transportation facilities during one calendar or fiscal year, which increases harvesting costs for that year. To indicate the reason for such an increase, it is customary to keep separate accounts for lateral roads and spurs and for main-line roads and railroads. When an unusual number of transportation improvements are made in one year, the operations for the succeeding year may require few if any expenditures for such purposes, with consequent lower harvesting costs. By keeping these accounts separately, therefore, it is always apparent why high costs or low costs occur and a true estimate of efficiency can be made.

An accurate measure is not ordinarily obtained of the volume of forest products hauled by days or months on large operations, as this would involve measuring the volume of logs or pulpwood hauled and the information would be useful only to present an accurate cost figure. The monthly mill cut of logs, which are scaled as they enter the mill, is usually taken as the volume of logs delivered to the mill, which permits determining the unit cost for products transported. When logs are delivered by truck haul and a contract rate is paid, each load must, of course, be scaled to determine the amount to be paid, and the total volume per day or per month is accurately known. Even when company-operated trucks are used, detailed operation costs and volume hauled per trip and per day are kept to permit the selecting of suitable trucks and to determine whether road improvements are justified.

Overhead. All business undertakings of any size and importance incur certain so-called overhead expenses or cost. These overhead costs are characterized by being essential to the conduct of the business but of such a nature that it is difficult or impossible to assess them accurately against each operation concerned with the undertaking. Harvesting operations are no exception to this general rule and, therefore, overhead-cost items must be considered.

No hard and fast rule exists as to what specific items are overhead and what are chargeable against particular operations. Each company or individual operator dictates his own policy with respect to such matters. Some may charge the salary and expense of a forester as overhead, particularly when his duties overlap several departments, such as fire protection, harvesting, land management, and public relations. When the forester's duties are more specifically assigned to one operation, such as marking trees for cutting, his salary and expense may be charged to that account. Likewise when equipment value is assessed as a lump sum for all types instead of by individual items, the total tax payment can logically be considered as an overhead cost.

Lumber and other forest-utilizing companies of any size must maintain main offices to transact business and keep records. Executives are required who devote their time to supervising and managing all of the operations of the particular business involved. Likewise, pay-roll and other records must be accurately kept and correspondence conducted, which requires secretarial and clerical staffs. Most, if not all, of such costs are difficult or impossible to assess accurately against the individual operations of the company. Hence, they are commonly considered overhead costs and kept separately from operational costs to present a true picture of harvesting cost for a company engaged in land ownership, harvesting operations, milling, and selling. A fair portion of the overhead cost should be charged against harvesting.

The overhead account becomes the catch-all, frequently, for miscellaneous cost items and, especially in those concerns with sufficient margins between the cost of doing business and income from sales, the overhead cost tends to become large due to what might be considered extravagances, such as contributions and good-will advertising. When the margin between income and cost narrows, executives usually look first at overhead costs with a view to lowering total cost.

Overhead costs are but little affected by the skill and ability of operating personnel, as they are incurred by the officers and executive staffs beyond their jurisdiction. They are often prorated among various operating departments, usually in proportion to the total operating costs of each. Thus, if the harvesting cost for a pulp company constituted 28 per cent of the total cost of operating the company, 28 per cent of the total overhead item would be added to harvesting cost. Many heads of harvesting departments and crew foremen dislike overhead cost being charged against their operations, as it frequently seems excessive and so makes such operations appear unduly costly. Furthermore, since they are in no position to economize on overhead costs or control them, they, therefore, resent including them in departmental cost statements. Since certain such costs must be incurred, however, in the conduct of the business, it is desirable and helpful to charge a fair portion of them against at least each major operation, of which harvesting is one.

As has been previously noted, harvesting accounting costs, such as are incurred by camp pay-roll and record-keeping offices, are actually overhead costs, although they may be called office expense rather than overhead. Office bookkeeping, record keeping, and pay-roll work cannot be avoided, although such activities do not contribute directly towards the production of forest products. Camp office overhead is, therefore, direct harvesting overhead, all of which is harvesting cost for the several operations; while overhead costs incurred by an integrated company, such as for the expenses of maintaining general offices, legal advice, and similar services, is general overhead, only a portion of which is chargeable against the harvesting operation. A concern that is solely engaged in harvesting timber crops obviously would have only harvesting overhead costs, and the functions of the general office necessary for the conduct of a harvesting business would be transferred to the office maintained solely for harvesting accounting and other record keeping.

TOTAL HARVESTING COST

The sum of the costs of the several operations involved in harvesting is, obviously, the total harvesting cost. By keeping the costs of each operation separate it is possible, as has been shown, to analyze harvesting cost records much more accurately and make appropriate adjustments where necessary and possible. In fact, this is the only reason for segregating cost items by operations, but the importance to management in efficiently conducting the operations justifies this overhead expense item unquestionably. It would be impossible to judge whether a new kind of loading machine, for example, were more economical than the one being used if the loading cost alone was unknown.

Total harvesting cost has an important bearing in forestry since, considered in connection with product values, it influences the stumpage price that can be obtained for standing timber crops in the forest, or the return that can be credited to the timber if it is harvested, manufactured, and sold by the same company. The higher the harvesting cost, the less the value of the timber crops for a given product value on the market. Thus, it is a forester's urgent concern to provide for the most economical harvesting operations consistent with the silvicultural and management policies in effect for the forest. The lower the harvesting cost, the higher the value of the crops of timber grown and the greater the profit to the forest owner in practicing forestry.

CONTROLLING HARVESTING COST

Harvesting costs can be controlled and prevented from becoming excessive by alert and capable supervisory personnel. Attention must constantly be given to labor to keep energetic and skilled workers and to eliminate inefficient and lazy men from the crews. Wages must be sufficient and living conditions good enough to attract the more desirable labor. High rates of pay do not mean high costs so long as the labor output is sufficient to give a satisfactory unit cost.

Equipment must be kept in good operating condition to avoid costly breakdowns with resultant loss of output while costs, at least for that day, continue. This requires workers skilled in the operation and maintenance of harvesting machines and adequate repair and maintenance facilities. Frequent and regular greasing and checking of tire inflation pressures, for example, result in a longer useful life for trucks and tires than when they are negligently cared for. Similarly, tractors, locomotives, loaders, and all other equipment must receive adequate maintenance service to obtain economical costs. Likewise, cutting tools such as axes and saws must be kept sharp to obtain the maximum output of labor. The same energy applied to a dull saw, for example, will result in much smaller output than if it were sharp. It is evident, then, that costs can be controlled to a considerable degree by competent supervision of labor and careful maintenance of tools and machines.

Since the trees comprising a forest vary in size and the logs, bolts, or other products vary even more because of the tapering diameter of trees from stump to top, harvesting operations must handle products of different sizes. It has been demonstrated that the cost of harvesting timber varies, in practically all cases, inversely with size of product. Thus, it costs more to cut, bunch, skid, load, and transport small logs than large logs. Of course, there is a maximum limit for economical handling when the size of product is so great that the equipment on a given operation does not have the power to handle it. For all practical purposes, however, it is universally true that small-sized products are more costly to handle than large ones in terms of cubic volume or unit volume of product.

Machines used on harvesting operations must be capable of handling the largest size of individual product. Therefore, when small products occur with the large ones, it is necessary to handle several of the smaller at one time to make full use of the power and the capacity of a machine such as a tractor. If all the logs produced on a harvesting operation supplying a sawmill were of uniform size, the machines used could be designed to handle that size and no variation in cost from one log to another would occur. Since this is not the case and never can be, those conducting harvesting operations must use equipment that will handle the largest single product that will be cut and then make the most efficient use possible of the machines in handling smaller products.

The reason for the variation in cost with the size of product is readily understood when it is considered that it takes a tractor or loader, for example, just as long to handle one log as another, irrespective of size. In loading logs, only one log can be hooked at a time, so the more volume the log contains, the more volume the loader handles in a given period of time, and hence the lower the unit cost in terms of log scale. A skidding tractor is equipped to skid several logs, when this is necessary to obtain a full load, so the actual cost of moving the load to the landing is about the same whether one log scaling 1,000 ft is skidded or four logs scaling a total of 1,000 ft are skidded. The increased cost is incurred in choking the four logs instead of one and, in some cases, in making the additional stops necessary to attach the logs to the skidding machine.

Even in felling and bucking, the cost per unit of volume is less for large than for small trees. More effort is required, it is true, to sever trees of larger diameters as compared to small diameters, but the ratio of time and effort required per unit of volume is less when large trees are cut than when small trees are cut.

In hauling, also, either with trucks or railroad cars, larger loads can be placed on the vehicles used when the logs are of large size than when they are of small size. This is true because the limiting factor in loading trucks and cars is not weight but volume capacity. When small logs, bolts, or other round products are loaded in a restricted space, more of the capacity is occupied by air space than is the case with large round products.

Thus, the larger the round products loaded on a car or truck, the more space will be occupied by solid wood volume. Since the same length of time is required to make a trip with a load of small products as with a load of large products, although the product volume is less, it obviously costs more to haul small round products than large ones. Because the weight of a load of small products is less, there is a very small saving in fuel consumption and in running time, but the amount is so small that the effect or cost is, for all practical considerations, negligible.

A high degree of skill and good judgment is essential, therefore, in selecting equipment for a harvesting operation and in organizing the operations into a balanced whole. Tools and equipment adaptable to handling the sizes of products and kinds of products that will be harvested with them is essential. Machines powered and designed for handling large-sized logs, for instance, are uneconomical for handling small products. Similarly, equipment with insufficient power to handle the products causes delays and results in costly breakdowns, which increase the cost.

Because of the variation in size of trees and products harvested in any operation, and the increasing cost of handling small as compared to large products, it is possible to control costs to a marked degree by not cutting trees that are too costly to harvest and by leaving top logs in the woods when they are too costly to remove.

Strictly on the basis of costs and returns, the minimum size that should be cut and handled in harvesting is that which costs as much to remove as it is worth. When it costs more to harvest the product than it is worth, a direct loss is incurred. Even though small-sized products yield some net return above harvesting cost, it is often desirable to leave them standing and concentrate operations on larger trees to keep costs low. This is particularly desirable when harvesting is done in a forest owned and managed for a continuous yield of products. The small, expensive-to-harvest trees should be left, first, to eliminate their excessive cost and, second, to permit them to grow so when they are harvested in the future they will have attained a size more economical to handle.

When only the timber is owned and the company doing the harvesting has no future interest in the forest, then all trees that have a positive value should, of course, be cut, except enough to reseed the area and provide for continuous productivity, if the unmerchantable stand is insufficient to accomplish this. Once a merchantable tree has been felled, it should be utilized in the top to the point where cost of handling is equal to the value of the product removed.

No log, bolt, or other product cut from trees should be left that has a finished-product value equal to or greater than the cost of handling it from that point on through the mill until the products it contains are sold. The expenditures already made or incurred in building roads or railroads, marking, and felling should not be considered because they have already been incurred and will not be influenced, whether or not the marginal bolt or log is taken or left to rot. The cost of limbing, bucking, bunching, skidding, hauling, and subsequent manufacturing costs are the only items to be considered in taking a log from a felled tree, as the other costs have already been incurred. If the top of a saw-log tree has value for other products, such as pulpwood or fuel wood, and can be utilized for them, the decision as to whether or not it should be used for lumber or for another product depends upon the comparative value of the log for such products as it can be used for.

The cost of harvesting, therefore, can be controlled to a considerable degree by limiting the range of tree sizes harvested to those that do not incur an excessive cost. Knowledge of what the cost of harvesting logs of various sizes is for any particular operation can only be determined by special cost analyses that foresters should be equipped to conduct. Many such studies have been made and much data has been published about comparative harvesting costs for trees of various sizes. Reference to this literature will often serve as a guide in setting cutting limits that are approximately correct.

Harvesting operations that can be readily moved from one area to another can be so located from time to time that a marked influence on cost results. When costs must be kept low during periods of low selling prices for the products, harvesting operations can be moved into the more accessible stands, or areas where features of the terrain, size of trees, and density of stand are such that low costs are possible. When product prices are high and higher harvesting costs can be absorbed, the operation can be located in more distant areas or in stands more expensive to harvest.

Shifting the location of harvesting operations to influence costs is only possible when the method of operation is flexible enough to permit changing locations without undue expense. When logging railroads and costly truck roads are used for transportation, the harvesting operation is tied closely to these facilities and it is economically impossible to change the location of operations at will. Likewise, when transportation is by stream drive, it is not possible to shift readily from one area to another, except within limits imposed by accessibility to the waterways being used.

The use of trucks and public highways for transporting products harvested in the forest makes the operation so flexible that it can be readily relocated to meet current conditions. Since no expensive transportation installations are necessary before harvesting can begin at a new location, it is only necessary to transport the crews and equipment to the desired location.

Truck transportation is extensively used by many sawmills, both large and small, and, especially in the Eastern and Southern forest regions, highways and roads are so numerous that changing cutting locations is not difficult. In fact on most such operations, harvesting frequently must be moved. Without such flexibility it would be impossible for many sawmills to continue in operation, as the timber available to them is not in large contiguous bodies but consists of numerous small scattered tracts.

Similar flexibility of harvesting exists for other forest-products industries offering opportunity for cost control by changing the location of harvesting. Pulpwood, poles, piles, veneer bolts, and similar products are harvested over widely scattered areas, so it is possible to locate harvesting operations for the purpose of controlling cost, within limits affected by location and variable operating conditions.

Only when harvesting operations are tied down by being dependent upon installed transportation facilities, such as logging railroads and private truck roads, or are limited to streams or other water routes, is it impossible or impracticable to control costs by changing the location of the harvesting operation. When this lack of flexibility exists, harvesting costs vary but little from year to year and fluctuating selling prices mean only greater or less margin between selling price and cost.

COST ANALYSES

For correct selection of trees to cut and for the economical conduct of harvesting operations, forest owners as well as those conducting the harvesting operations should have accurate information about the cost of harvesting trees by size and species, and by size of products cut from the felled trees. Only by making cost analyses can this information be obtained. Such analyses are conducted independently of the usual cost accounting; they are designed to show the range in harvesting cost between trees of different sizes and species and to determine the comparative efficiency of different equipment used in harvesting and different methods of conducting the operation. The value of trees and products of each size, by species, also is required, as whether or not a tree or product should be removed depends upon its value as compared to the cost of harvesting it.

Armed with this information, forest owners and managers of harvesting operations can correctly fix the size of the smallest trees to cut, by species, and the smallest log or bolt to remove in the tops; and can organize the harvesting operations on the most efficient basis. When new equipment is introduced, its efficiency can be compared to that already in use, and if a satisfactory saving in cost can be achieved by using it, the change can be made.

Making cost analyses of this kind is a specialized branch of forest harvesting. It is taken up here only to point out the importance and the usefulness of the information obtained by cost analyses in harvesting.

In conducting such analyses, it is necessary to determine, from a number of observations, the actual amount of time required to perform the operation being studied, such as felling a tree, loading a log, hauling a load of pulpwood, or skidding a log or other product. Comparable observations must be obtained for trees or products of each size being handled, or for skidding various distances, or for using equipment of different kinds, whichever is being studied. Timing operations with stop watches or standard watches is the method commonly used in obtaining the basic data for such studies. It is important to obtain enough individual records or observations to give correct averages and the data must be obtained in such a way that there is no interference with normal operations. If the study procedure creates an abnormal operation, then its purpose is defeated and its results become useless because unnatural conditions are observed.

The cost of owning and operating harvesting equipment must be determined for each type and kind of machine being used. Labor wage rates for the crew required to operate the equipment also must be known, and the total cost then is reduced to a time basis for comparison with the output of products of various kinds and sizes. The cost of operating a machine or team per minute, per hour, or per day is known as the "machine rate."

Those making cost studies should be thoroughly familiar with harvesting operations in general, as well as the particular one being studied. They must be ingenious enough to obtain the necessary data and exclude that which is unimportant and, above all, accuracy to the degree required must invariably be the rule.

In time and production analyses, costs are segregated into different categories from those previously discussed. The total cost of any operation is the sum of fixed and variable costs. Fixed costs are those that exist whether or not a machine is working or idle. A tractor may not perform any work for an hour, due to an unbalanced operation, for example, yet fixed costs continue. The crew wages, interest, taxes, and such costs do not stop when the machine stops. Fuel consumption, however, does cease if the motor is shut off, so fuel cost would be a variable cost, as it is not incurred when the output stops. Likewise, if a rainy day or a transportation breakdown occurs and the wood crews do not work, wages are not paid for that day and certain tools and equipment are not used; yet supervisory expense goes on, as such employees are paid by the month. Variable costs, in this sense, then, are those which are not incurred when no work is done, while fixed costs are those which continue regardless of output. Fixed cost is, in one respect, similar to overhead in that it goes on when production stops. It differs in that it is assignable to the operation incurring it.

Fixed and variable costs may be considered in another sense also. In skidding, for example, the cost varies with the skidding distance; the longer the turn, the more skidding costs per unit of product handled. Certain costs in skidding, however, do not vary with distance. For instance, it takes no more time to hook and unhook logs to a tractor when the haul is 500 ft than when it is 1,500 ft, so hooking and unhooking cost is considered, by some, as a fixed cost in skidding. It is true, of course, that the longer time between trips should permit the hooking-on man or crew to perform this operation for more than one tractor on the longer turns. Since skidding distance varies from zero to the maximum limit, it is difficult and impractical, however, to keep changing the crews so that they can be kept in perfect balance with the tractors.

Similarly, in hauling pulpwood with trucks, the variable time is governed by the distance it is hauled, but the time required to load and unload a truck is quite independent of the distance it travels. Hence, the cost of hauling is considered to be variable, while loading and unloading cost are considered as fixed. By excluding fixed costs from such operations in harvesting, the effect of variable skidding conditions, such as distance, slope, and kind of machine, can be better and more clearly analyzed and understood.

Correct interpretation of the data developed in cost analyses is, of course, of utmost importance. No matter how accurate the observed data may be, if the mathematical analysis is erroneous due to incorrect averaging or if unsound assumptions are made, the conclusions will be useless and may lead to costly mistakes in trying to apply the findings.

With constant improvement in harvesting equipment, the opportunity to modernize methods and make operations more economical is constantly confronting those harvesting forest crops. Likewise, changing costs and selling prices mean constantly fluctuating marginal size and quality limits of trees and portions of trees. The rapidly expanding new forest industries have created new opportunities to utilize species and portions of trees that formerly were unmerchantable. Furthermore, the growing demand for forest products gives forest owners an opportunity to weigh the value of their tree crops for one use as compared to another to determine which will return the maximum net income from a forest crop. For these reasons, it is of growing importance that adequate information on the cost and value of trees and tree products be kept up to date, if the maximum value is to be received from timber harvested for commercial use.

CHAPTER 18

REGIONAL HARVESTING PRACTICES

THE NORTHEAST

BY FRED C. SIMMONS

Northeastern Forest Experiment Station, U.S. Forest Service.

The Northeast can be divided into three main logging regions. First, at the north, comes the mountain and wilderness area of northern New York and New England. Then there is the wood-lot area of southern New England and New York, New Jersey, Delaware, and eastern Pennsylvania and Maryland. Finally there is the Appalachian mountain territory in western Pennsylvania and Maryland, which extends down through West Virginia and along the border line of Virginia, Kentucky, Tennessee, and the Carolinas, to north Georgia. These logging regions will be discussed separately.

NORTHERN NEW ENGLAND AND NEW YORK

The private lands in the Adirondack Mountains of New York state and the wilderness areas of northern New England are owned largely by pulp and paper companies and large land-holding concerns. One pulp company in Maine owns 2½ million acres. Several others own more than ½ million acres each.

The topography is generally rugged, with considerable mountain masses rounded down by glaciers, and there are great granite boulders, strewn around by the ice sheets, to contend with almost everywhere. Lakes and sizable streams are generally plentiful, and there are considerable areas of swamp, especially in eastern and northwestern Maine, northern New Hampshire, and the western Adirondacks.

The Forest. In this area the original stands were predominantly spruce and fir and the northern hardwoods—birch, beech, and maple. There was in the original stand a considerable admixture of white pine, generally giant trees towering far above the rest of the stand, survivors of previous stands or volunteers that had come in in spots where some disaster like fire or wind had struck centuries before. This white pine was generally the first species logged throughout the area. Since then the heavy demand has been for spruce and fir, first by sawmills, and now largely by pulp mills. These softwoods could be economically logged and transported out of the woods by river driving. The hardwoods have only recently become marketable. As a result, the remaining softwood stands are generally light and scattered. Much former softwood area has been taken over by inferior hardwoods. In addition there are vast areas of virgin and decadent hardwoods. Some softwood stands still run to 20,000 ft, or 30 to 40 cords per acre, but many stands are being logged that run only 10 cords to the acre. Hardwoods generally run 6,000 to 8,000 ft per acre, but high-grading operations are apt to take only 2,000 to 3,000 ft of this.

In addition to the old-growth stands and their successors of the same species group, there are second-growth stands of less tolerant species. One that is very important commercially is the white, or paper, birch. The growth of this species followed extensive fires around the turn of the century, and it has been the mainstay of the wood-turning industry. White pine second growth is also important in the northwestern part of the Adirondacks around Lake Champlain, and in southern New Hampshire and southwestern Maine. Aspen, fire cherry, gray birch, and soft maple sprout stands that are practically unmerchantable cover considerable area.

Labor. Through the years, north-woods logging jobs have been conducted mostly from logging camps. Fifty- to sixty-man camps, of rough log or lumber construction, at the end of a rough tote road, some of them 30 to 40 miles back from a highway, are still common. Much of the labor is French-Canadian, either from the families who have settled on this side of the line or of "bonded" men who have been brought in by the companies or their contractors from Canada. There is also a sprinkling of other nationalities, including old-time native loggers of English, Scotch, and Irish descent, some Indians, and immigrants or their descendants from Scandinavia and the Balkans.

The skills required for old-time methods of logging are the traditional ones with the ax, the crosscut or bow saw, the peeling spud, and the peavey. A great many horses are still used and good teamsters are in demand. The greatest skill is required on the river drive, although the general practice of driving 4-ft wood has reduced the difficulty of these drives. During the past few years considerable mechanization has been taking place, and tractor drivers, loader operators, truck drivers, mechanics, and power-saw operators are in increasing demand. These occupations are bringing more and more young men of native stock back into the woods.

Felling and Bucking. Most felling and bucking is still done with hand tools, on a piecework basis. The pulpwood cutters, particularly those of French extraction, still like stump cutting by themselves. In this method the trees are cut into 4-ft lengths at the stump. The workers can see exactly how many cords of 4-ft wood they have piled up in a day or week. A light single-bit ax and bow saw are about the only tools required. Production runs from 1 to 2 cords per man-day. Because of proven economies, and because it is less damaging to the residual stand, some pulp companies prefer to handle yarded wood that is felled and limbed in tree lengths and then skidded to a yard and bucked into the short lengths there. Production is ¼ to ½ cord greater per man-day. Saw logs are generally also cut on a piecework basis, but by crews of two or three men equipped with heavier single- or double-bit axes, a crosscut saw, and a peavey. Production runs 1,000 to 2,000 ft per man-day, depending on the species and type being logged and the skill of the men.

Recently felling and bucking have been mechanized on a great many operations. The most common mechanization in pulpwood cutting is the installation of a slasher saw at the yard for bucking. In many cases this is a gasoline- or electric-powered chain saw, but even more frequently it is a circular saw, equipped with rollers, a moving belt or chain conveyor, or a buggy to carry the tree lengths under the saw. Production with these mechanized setups at the landing varies tremendously with the efficiency of the setup and the rapidity with which the wood is supplied. A four-man crew may buck from 10 to 80 cords a day.

On saw-log jobs, mechanization has been more successful back in the woods. The gasoline-powered (and to a lesser extent the electric-powered) chain saw have proven to be great timesavers in felling as well as in bucking. On operation after operation, both in hardwoods and softwoods, it has been found that production per man-day can be about doubled by use of a chain saw instead of a crosscut, when the saw crew is properly selected, organized, and trained.

Bunching and Skidding. In this glaciated region, where the soil is full of highly abrasive particles such as quartz and granite, ground skidding has never been popular. This is more important for saw and veneer logs than it is for pulpwood, but even the latter is kept up off the ground as much as possible. Consequently scoots and sleds have been the most common means of transporting wood from the stump to the yard, in summer as well as in winter. Stump-cut pulpwood is piled on drays in 4-ft lengths, and logs are usually handled on scoots. With a reasonable grade, a good team can handle 1 to 1½ cords on a dray in the summertime and 2 to 2½ cords on well-broken-out roads in the winter. On most operations skidding ¼ to ½ mile to a yard, from six to eight turns can be handled per day. With a scoot a team can handle 300 to 500 ft of logs in the summer, and about twice as much in the winter. Tractors of 35 hp commonly handle loads about twice the size of those handled by teams, and make their turns about twice as fast. On pulpwood jobs two sleds are frequently used for each tractor; one is left in the woods to be loaded each time the machine takes the other to the landing. For logs the tractor is frequently equipped with a winch and boom, so that the machine can bunch and load its own logs.

Arches or sulkies are beginning to be used on some north-woods jobs, particularly for skidding pulpwood in tree lengths. Two and one-half cords of tree-length hardwood pulpwood is a common load for a 35-hp machine. On one job in New Hampshire such a rig is keeping a slasher saw busy sawing 30 to 50 cords a day of hardwood pulpwood, depending on the difficulty and length of haul. A similar machine can handle 1,000 to 1,500 ft of long logs per trip.

Cable skidding, other than for short distances in bunching loads, is not widely practiced in the north woods, although it has been tried in a number of places, sometimes with considerable success. Recently the development of light fast rigs for relogging on the west coast, and the fact that many north-woods operators are getting into tough logging chances on high ledgy slopes or in swamps, has led to a revival of interest in cable skidding, and several outfits are once again experimenting with it. One difficulty is the scarcity of good riggers and donkey operators.

Loading. Loading of pulpwood is done mostly by hand. Four-foot bolts are usually handled with the aid of the D-handled pulp hook. Even so it takes 15 to 20 min for a man to handle 1 cord of wood under the best of conditions. If there is a considerable walk involved from the pile to the vehicle, or, as in loading a boxcar by hand, the time can easily be doubled.

During the past few years a great many ingenious mechanical loaders, both stationary and portable, have been developed by operators handling 4-ft wood. One of the best of these is a two-part chain conveyer, mounted on, and powered by, a small wheeled tractor. One part carries the bolts from the ground up over the tractor body to about a 6-ft height. The other part, the snout, carries them on over the truck bed to a point just behind the cab. When the first tier of wood is loaded, the tractor is moved back to position for loading the second tier. With this rig three men have been timed loading 3 cords under good conditions in 15 min. They said that they have done it in 9 min.

Four-foot wood is also handled in various places by cranes and derricks, either loose or with grapples or tongs. One hydraulically operated grapple loader, mounted on a truck, shows exceptional promise. It has loaded three cars, averaging about 22 cords each, in a little over 5 hr with a crew of three men.

Long logs have been traditionally loaded from skidways, or brows, by hand with a cant hook or peavey. Suitable slopes can usually be found in the north woods so that the logs can be skidded in and then rolled directly onto these brows. Then, with the aid of short skids to the truck bunks, they can be rolled onto the load. Double-deck skidways are frequently constructed so that the upper tiers of the load can be rolled on also, without too much of an uphill struggle. Depending on the size and configuration of the logs, one or two men can usually load 1,000 ft in 10 to 20 min.

Mechanical devices are being used more and more for log loading. Most frequently used is the crane, or jammer. This may take the form of an A-frame jammer, mounted on a sled and powered by horse, tractor, or motor. Or it may be a truck- or tractor-mounted stationary or revolving crane. Such loaders frequently cut loading time by a half or a third, thus enabling the truck to spend more of its day in productive hauling, and they also frequently result in a bigger and better assembled load.

Transportation. Independent logging railroads are practically a thing of the past in the north woods. They never were very common. However, a great deal of pulpwood and a considerable number of logs, particularly of the higher grades, are hauled by common-carrier railroads. During the past few years much pulpwood has been hauled in boxcars because this was the only type of car available. They are difficult and costly to load and unload, but they do have some advantages in keeping the load clean and obviating the need for special precautions to keep the load from shifting or falling off in transit. Some old boxcars have had their tops removed to facilitate loading and unloading by crane. Recently several Northeastern pulp mills have designed, and are having built for them, special pulpwood cars with open tops and hinged sides to facilitate unloading.

The bulk of the forest products of the north woods are hauled by truck. Because of the loading time for small logs and bolts, and the condition of the public roads over which they travel, these trucks are generally small, ranging from 1½ to 3 tons capacity. Most operators, however, are installing tandem-powered rear axles, which enable the trucks to carry double or more the rated load. Trailers are not common, but there is a trend toward tree-length hauling as well as tree-length skidding, which is increasing their use. Loads of logs frequently run between 2,000 and 3,000 ft for softwoods and 1,500 to 2,000 ft for hardwoods. Round-trip speeds of 30 to 40 mph, when most of the haul is on public roads, are not uncommon.

In winter considerable hauling is done on sleds over snow or iced roads. This is the most economical method of moving forest products out of many swampy areas or across lakes. Sliding resistance is less than rolling resistance, and bigger loads can be hauled. The roads themselves are cheaper to construct, particularly in adverse situations. Motive power may be a team, a crawler tractor, or a truck. On a good iced road, with a steady downgrade not exceeding 10 per cent, 60- to 80-hp tractors frequently haul trains of 15 sleds, each loaded with about 5½ cords of pulpwood. With a Diesel-powered log-hauler, an average speed of 6 to

8 mph for the round trip can be maintained. A 2-ton truck can haul three sleds at a round-trip speed of 20 to 30 mph. Adverse grades or level stretches cut down on loads and speeds.

Stream driving is still a major means of forest transport, particularly for pulpwood. Most of the major streams in Maine, the upper Connecticut River in New Hampshire, and the Hudson and Black rivers in New York state, as well as many of the smaller Adirondack streams, still have their annual drives on the spring freshets. Long-log drives are conducted on the Machias and St. John rivers in Maine and the Moose River in New York, but practically all of the other drives are of 4-ft wood. River driving is still the cheapest method of transport. On well-developed streams pulpwood is frequently driven over 100 miles for less than 50 cents a cord.

Organization of Operations. The traditional pulpwood operation is seasonal. It begins in the spring, with felling and peeling of the tree-length pulpwood. Then, in midsummer, when the wood no longer peels easily, bucking and yarding are done. Hauling to the streamside is done in the winter on snow or ice roads. The operation ends with the river drive in the early spring, when the ice goes out. Most of the woods operations are conducted from remote camps of fifty to sixty men, with a camp superintendent and a scaler. Practically all the work is done by small jobbers or contractors. Frequently the whole camp is a contract operation, with subcontractors doing the cutting, skidding, and hauling. A "walking boss" will represent the company for five or six camps.

Some softwood saw-log operations are conducted on much the same seasonal basis, but more and more saw-log cutting is a year-round proposition, with small ten- to twelve-man camps run by a jobber, delivering the logs to a landing at the roadside for a stated price. Felling and bucking, skidding, and hauling are frequently subcontracted. A company representative will scale the logs, and check on work in the woods from time to time. Portable sawmills, located back in the woods, are not common in the north country, but when they are found, the logging and sawmilling crew are usually integrated, and they may be the same men.

Nature of Cutting Operations. Very little really destructive cutting is done in the north woods. On the other hand, practically no real high-order forestry is practiced. Cutting is usually done to what are considered to be the minimum economic limits. On softwood pulpwood jobs, this is usually stated in terms of total height of the tree; 35 ft is a common minimum. On saw-log jobs, particularly in hardwoods, it is a minimum stump diameter; 15 in. is a commonly used limit, but many trees larger than this, because of form or condition, are not cut. Not infrequently less desired species like beech are not cut at all. The minimum diameter for pulpwood is usually specified as 4 in., inside bark; for softwood saw logs, 8 in., and for hardwood saw logs, 10 in.

As a result of this recurrent high-grading, many of the north-woods stands are steadily deteriorating, in both species and condition. The softwood component is becoming reduced, year after year, and the hardwoods are becoming poorer and poorer in quality.

SOUTHERN NEW ENGLAND AND EASTERN PORTION, MIDDLE ATLANTIC STATES

This area is characterized in general by small wood lots, interspersed with agricultural land, much of it devoted to dairy farming. Logging is an adjunct to farming, and most logging is done when work on the farms is at low ebb. There are very few commercial logging operators of large size.

The topography is generally rolling, with few real mountains, and little really rough logging. The glaciated country practically all lies north of the New York-Pennsylvania boundary, but rock outcrops are not uncommon in the Pennsylvania-Maryland sector. Swamps are common in southern New Jersey, the Del-Mar-Va Peninsula, and in southern Maryland.

The Forest. The timber types of this area vary a great deal. In the north are the spruce and fir and northern hardwoods of the north woods, merging toward the south into second-growth white pine, soft maple, and gray birch, and increasing numbers of the oaks. A little further south, in eastern Massachusetts, the Hudson valley, southern New Jersey, and portions of eastern Pennsylvania, the oaks and pitch pine become predominant, with the northern hardwoods confined to the uplands. In southeastern Pennsylvania, eastern Maryland, and in Delaware the first of the southern pines appear, first Virginia pine, and then loblolly. Volumes per acre vary tremendously. On good sites saw-log stands that have been relatively undisturbed or well managed frequently run 15,000 and 20,000 ft to the acre, but the average saw-log stand is probably in the 3,000 to 5,000 ft class. Similarly good second-growth cordwood may run up to 50 cords per acre, but 5 to 10 cords is commonly considered merchantable.

Labor. Practically all logging is done by local residents. Logging camps are practically unknown, although sometimes a teamster will build a horse hovel and quarters for himself on the job. The skills required are generally the traditional ones with the loggers' hand tools, although more and more mechanical skills are called for as time goes on. Skill with loggers' hand tools is generally not high among farmer-loggers. Frequently farmers will hire themselves and their teams, tractors, or trucks out to a logging operator during their slack seasons. The nationality stocks engaged in logging are as varied as those that populate this area. English,

Scotch, and Irish probably predominate, although there are many Polish, German, French, Italian, and other Southern European nationalities represented in considerable numbers.

Felling and Bucking. Felling and bucking is still done mostly by hand. Most pulpwood, even in the pine area, is cut with the narrow-bladed crosscut saw, and two- to three-man crews are used. Saw logs, mine props, and other hardwood products are generally cut with a wider crosscut, and the three-man saw crew is common. Most logs and wood are cut to length in the woods, but tree-length logging is becoming increasingly common, especially in north Jersey and the pole and piling country in Maryland. Production per man-day probably averages a little less than it does in the north woods, about a cord, or 1,000 ft, per man-day being common.

The chain saw is used to a greater extent in this area than it is in the north woods. Transporting it around and to and from the job is easier. Service is more easily obtained in the nearby towns. Consequently not only logs are cut with the power saw, but such saws are frequently found on cordwood jobs. Many of the saws cutting smaller timber are the familiar two-man type, but more and more the smaller so-called one-man saws are being used. Production with these tools is at least double and sometimes triple that attained with hand tools.

Bunching and Skidding. Most bunching and skidding is done with equipment available on the farm. This is quite often teams of horses, bigger and heavier than those favored in the north woods, but wheeled or crawler tractors are becoming more and more common. Skidding distances to a road that a truck can travel are apt to be short, and very good production figures are frequently obtained. Frequently 5,000 or 6,000 ft of logs are handled by a team and teamster in a day, skidding on bare ground. Scoots and sleds are rarely used, particularly in the southern portion of the area.

Tractors equipped with winches and sulkies are becoming quite common. The wooded country, with its network of old roads, is well adapted to their use, and they increase the capacity and maneuverability of the tractor considerably. Ten to twenty thousand ft of hardwood logs is not an uncommon day's work for such a rig on a skid of ½ to ¾ mile. This ties in well with the output of a four- or five-man chain-saw crew.

Cable skidding, other than bunching with the tractor and sulky, is rarely attempted.

Loading. The skidway, or brow, is also used in this wood-lot country, although suitable slopes for its installation are frequently not available. In such cases logs are sometimes rolled up sloping skids by hand, or more frequently pulled up by a crosshaul. Recently quite a number of trucks have been equipped with a winch under the bed, so that they can provide power for their own crosshaul. Other self-loading rigs for trucks are being developed, such as jib booms behind the truck cab, and flipper arms alongside the bed. These self-loading rigs have made more practicable a "milk-route" system of picking up specialty forest products by wood-using industries along regular routes at scheduled intervals.

Transportation. Practically all of the forest products of this wood-lot territory are transported by truck. Only a few of these trucks are primarily used for transporting logs and bolts. Most of them are light trucks used most of the year on farms and around the towns. Sometimes heavier commercial trucks will attempt a log-hauling job, but this is generally found to be uneconomical because of the loading time involved, and because of the narrow woods roads and light bridges that have to be traversed. Many more of the logging trucks in this area are single-rear-axle affairs than is the case in the north woods, and consequently the average loads are lighter, probably ranging between 1,000 and 2,000 ft of logs and 1 to 2 cords of wood. Tree-length logs in Jersey and piling from the eastern shore are hauled on special truck-tractor and trailer rigs, but these are generally not over the 3-ton class.

Organization of Operations. Logging operations are apt to be somewhat seasonal, largely because of the difficulty of getting year-round laborers. Logging is generally conducted pretty steadily during the winter months, then it shuts down for the mud and planting seasons in the spring, is resumed for a couple of months until haying time, and possibly again between haying and the major harvest season in the fall.

Less and less pulpwood or tanbark is peeled in the spring and early summer because labor is not available. Most pulp mills have put in debarking devices, and some alkaline-process mills have found that they can cook hardwood with the bark still on.

Very little supervision is given the generally independent logging contractors of this area, other than accepting or rejecting their products.

Nature of Cutting Operations. The wood-lot area contains some examples of really high-order forestry. These are apt to be on the lands of descendants from one of the European countries, like the Pennsylvania Dutch, the Swiss, the Swedes, or the Finns, who have adapted practices they learned in the old country to their properties here. More and more, however, credit for these well-managed wood lots is due the state extension foresters or the county farm foresters, who are continually working with the small landowners. Several wood-using industries are also advising landowners on good forest practices.

The general level of forest practice, however, still leaves much to be desired. Too many landowners still sell the products of their wood lots to a timber operator for a lump sum, to cut in any way he wishes. Some landowners even insist that all the merchantable product be removed at the time the operation is going on. As a result, considerable areas of farm wood lot are cut over destructively each year and the productivity of the wood-lot areas has been steadily deteriorating. Minimum standards of merchantability vary with the industry to which the product is being delivered, and the standards of the locality. Pulpwood is generally accepted down to a 4-in. minimum diameter, and saw logs to 8 in. for softwoods and 10 in. for hardwoods. There are numerous specialty products, however, that are acceptable to smaller diameters. Handle stock goes down to 5 or 6 in., so do mine props, and in some areas fuel wood, distillation wood, or bean or fish-trap poles are salable down to 2 or 3 in.

APPALACHIAN MOUNTAIN REGION

The Appalachian Mountain logging region extends from southwestern New York state to north Georgia, extending through western Pennsylvania and Maryland, and along the mountainous boundaries between West Virginia and Virginia, and between Kentucky-Tennessee and the Carolinas. The fertile valleys between these mountains were settled early, and there was considerable timber cutting incident to settlement; but because of the many unmarketable species and the rough topography, much of this territory was not logged commercially until the current century. There are still a number of big logging operations in the Appalachians, working in the remnants of the virgin stands, but these stands will practically all be cut over within 10 years. Much of the land belongs to coal companies, and considerable volumes of timber are cut annually for use in the mines.

The topography is generally rough, with long sustained slopes and numerous high cliffs, especially along the watercourses. Heavily timbered box canyons in the Cumberland Plateau in eastern Kentucky, for example, are frequently surrounded by cliffs 60 to 300 ft high.

The Forest. The northern hardwoods and spruce types are typical of the backbone of the Appalachians at increasing elevations southward. The beech-birch–sugar-maple type is common at 2,000 ft elevation in Pennsylvania, 3,000 ft in West Virginia, and 4,000 ft in the Carolinas. There are some important changes in the composition of this type, particularly at its lower limits. Black cherry is an important component, particularly in Pennsylvania and West Virginia. Hemlock is very common. Paper birch is practically unknown.

Volumes in old-growth stands may run 15,000 to 20,000 ft to the acre, but many cutover stands now being relogged run only 2,000 to 3,000 ft per acre. Below the northern hardwoods, the oaks, black gum, hickories, and formerly chestnut, mixed with white, pitch, and shortleaf pine, are common on the dry slopes. Five to six thousand feet per acre is a reasonably good volume for the oaks, and softwood stands often run twice as heavy. It is in the moist coves, however, that the Appalachian hardwoods reach their best development. The outstanding tree is yellow poplar, but red oak, white oak, buckeye, shag and shellbark hickories, sweet gum, white pine, and hemlock are other important components. Volumes of 30,000 to 40,000 ft per acre are not uncommon, and volumes of 60,000 and 70,000 ft have been recorded. The second-growth stands are similar in composition. As in the north woods, spruce is generally not restocking as well as the hardwoods. Decadent culls and inferior brush of serviceberry, sassafras, silverbell, and sourwood cover considerable areas. An important specialty wood is dogwood, much in demand by textile-machinery manufacturers.

Labor. Logging camps, in the accepted sense, are not common in the Appalachians. The Appalachian mountaineer has always liked to have his family with him, so the loggers usually move into the vicinity of a logging job, erect their own shacks, often from cull lumber, and subsist themselves. Some camps, most of them railroad camps, were formerly run by the operators, but they are now few and far between. Most of the laborers are native mountaineer stock of English ancestry, but there is a considerable admixture of Southern Europeans, many of whom were imported by the coal companies to work in the mines.

The skills required on Appalachian logging jobs cover a wide variety. Skill with loggers' hand tools is required as elsewhere. In addition, there is a demand for riggers, tractor, power-saw, and loader operators, railroad men of all classes, and truckers.

Felling and Bucking. Most felling and bucking of saw logs is done by three-man crews, working with a crosscut, axes, and a peavey. Such a crew will put out anywhere from 2,000 to 5,000 ft of logs a day, depending on their skill and the type of timber being worked. Similar crews are trying to operate chain saws, but they are not very efficient because the amount of swamping and limbing to be done necessitates their leaving the saw lying idle much of the time. Nevertheless, such a crew can put out 5,000 to 10,000 ft of logs a day. Addition of a few additional axmen to the crew frequently makes it possible to almost double this production.

Pulpwood and round mine props are usually cut by two-man crews, using a narrow crosscut and an ax. Two cords of wood, or 150 to 200 props per day, are considered about average production. One pulp company in the Appalachians has been encouraging the use of bow saws, which are so common in the north woods. They have distributed about 600 of them to date, and claim that their use makes it possible to get about half again the production per man-day, because one man can operate the bow saw alone and a great deal of delay due to pinching is eliminated.

A few companies skid their timber in tree length and haul it all the way into the mill in that length, bucking it with a circular saw on the mill deck.

This generally makes possible a greater grade recovery, from hardwoods especially, as well as a greater man-day production. It requires special skidding, loading, and hauling equipment.

Bunching and Skidding. The majority of Appalachian timber is still bunched with animals, and considerable skidding is still done with the same motive power. The common method is to hook 10 to 20 logs together, end to end, and drag them down the long sustained slopes in that fashion. If a level stretch or a slight uphill grade is encountered, the "trail" of logs is split up, and a smaller number of logs is hauled to the next place where a favorable grade is available. In this way 500 to 1,000 ft are frequently brought down the slope of a mile or more at a time by a team, and three to four turns are made per day.

Because of the rough topography and the heavy timber, cable skidding was tried early in the Appalachians, and it has been highly successful in a number of instances. One company in Kentucky is using skyline spans of up to 2,000 ft. In West Virginia and the Carolinas, high-lead skidding is generally more popular. The logs are frequently bunched by horses, and then skidded in loads of 1,000 to 2,000 ft. Individual hardwood logs scaling up to 5,000 ft sometimes have to be handled. These examples are on big railroad jobs. The small operators are also forced to resort to cable skidding to log some areas, particularly the Kentucky box canyons referred to earlier. A number of examples of small skyline rigs, powered by an old automobile motor and depending on the slack in the line and the weight of the block to carry the choker hook back to the setters in the bottom of the canyon, can be found. Production with these setups, of course, varies widely.

Tractor skidding is also becoming more and more common, as it is elsewhere. Some of the bigger operators are successfully using tractors in the 100-hp range, with crawler-track-mounted arches, on steep slopes. Tree-length loads ranging up to 5,000 ft are brought in by these outfits. Smaller tractors are more common, the favorites probably being in the 35- to 50-hp classes.

Loading. Much of the railroad loading is done with old steam-power loaders mounted on railroad trucks. The McGiffert, which is on stilts, with an aperture beneath it through which the empty log cars can be pulled, is something of a favorite. Most of these loaders are equipped these days with tongs rather than a crotch line and end dogs, in the interest of using less man power. Production records vary a great deal, depending usually more on the availability of cars and logs than the capacity of the loader, but one of these machines frequently loads about 50,000 ft or more per day.

For truck loading, the old-fashioned skidway is still widely used. There is also a variety of jammers and cranes, many of them homemade. The truck-mounted crane is fast becoming the favorite tool, because of its maneuverability and speed in loading. Two logs a minute is not an uncommon productivity for one of these machines, when the log dump is not too far away and the ground man and top loader are alert and capable.

Transportation. Railroad logging is still fairly common in the Appalachians. Many of the big logging outfits have their own track and rolling stock, and continue to use them, despite the proven economies of truck logging in this territory. Many of them have trackage rights for their log trains over common-carrier roads, and in some cases their roads are also used for hauling coal.

Some companies, however, have abandoned their railroad equipment and have gone over to the use of trucks entirely. The trucks can bring in logs over a wider territory and are better adapted to hauling from the lighter and more scattered stands in which most of the logging is now being done. Careful cost records also show that they are generally more economical, even when the investment in rail and rolling stock has been pretty well depreciated. Two- and 3-ton trucks with tandem axles, bringing in loads of 3,000 to 4,000 ft of softwood logs, or 2,000 to 3,000 ft of hardwoods, are common.

An exception to the trend to trucks is that several loggers in West Virginia and Kentucky are successfully using light narrow-gauge tramways, and Diesel-powered industrial or mine locomotives. They claim that they can put in their tramroads at a half to a third of the cost of truckroads by using cull 6- by 6-in. ties at 12- to 18-in. intervals, and light 25-lb rail. Two log cars, carrying about 2,000 ft each of short logs, are usually hauled by one of these locomotives, even up 5 to 10 per cent grades.

Organization of Operations. Day labor is much more common in the Appalachians than it is in the north woods, and consequently greater supervision is needed for the logging crews. A walking boss for each loading point is common. A great deal of contract cutting, skidding, and hauling is done, particularly on the smaller jobs.

On one job in West Virginia the felling and bucking crews are operating with electric chain saws, two of which are powered by a generator mounted on an old crawler tractor. A six- to eight-man crew gets from 30,000 to 40,000 feet of tree-length logs a day. This is skidded by a big tractor equipped with winch and arch, and loaded on specially equipped railroad cars to be hauled to the mill. Bucking is done with a circular saw on the mill deck.

Generally logging in the Appalachians is pretty much of a year-round proposition, with slack periods during the mud season in the spring and fall, and when there are deep snows in the winter. The arts of winter logging, as practiced in the north woods, are virtually unknown.

Nature of Cutting Operations. Cutting operations in the Appalachians have been much in the nature of successive high-gradings of the stands.

The high-country spruce was generally logged rather clean, since the smaller stuff was in demand at the pulp mills, but intermixed hardwoods were frequently left uncut. In the northern hardwood cove and oak types further down the slopes, first one species or group of species and then another has been cut, until over sizable areas only cull trees and sprouts remain. Examples of really good forestry practice are few and far between, although now several pulp companies, some of the big landowners (including coal companies), and a few of the sawmilling outfits are actively interested and are beginning to promote good forest management.

The minimum economic size of trees and logs cut has been progressively reduced through the years. Many sawmills are now accepting softwood saw logs 6 in. in diameter at the small end, and hardwoods 8 in. through. Pulpmills generally set a 4-in. diameter inside the bark as their minimum, but accept a few sticks of smaller size. Four inches is usually the minimum also for round mine timbers.

THE LAKE STATES

BY GEORGE BANZHAF
Milwaukee, Wisconsin

It must be borne in mind that, although the three Lake states of Michigan, Wisconsin, and Minnesota are a geographical unit bound together by the waterways formed by the Great Lakes, they are not exactly alike in so far as forest types are concerned. Forest types and woods practices in Michigan and in Wisconsin are very similar, and these two states may be considered together. Forest types and operating conditions differ somewhat throughout the Lake states but are similar in most respects.

THE FOREST

In Michigan and Wisconsin the forests may be classified as consisting of three major types: the pine forests, the hardwood-hemlock forests, and the swamp forests. There are still substantial areas of virgin hardwood and hemlock forests in these two states, although most of them are located in the Upper Peninsula of Michigan. The pineries and swamps, however, have been largely cutover. The three forest types, therefore, must be considered not only in their virgin condition but also with regard to their demonstrated ability to reproduce themselves.

The original pine forests were composed of white and Norway (red) pine and there was considerable variation in the volume of each species present. Because of the fact that the pine forests were cutover during the period before fire prevention was as effective as it is today, the areas were burned over repeatedly and no very satisfactory reproduction has taken place. Vast areas of former pine-forest land are covered with aspen, white birch, and scattered stands of jack pine. These stands of aspen, white birch, and jack pine attain volumes of about 10 cords between 60 and 70 years of age per acre.

The hardwood-hemlock type is present over large areas both as a virgin and as a second-growth forest. In its virgin form the forest is composed of yellow birch, hard and soft maple, beech, basswood, elm, and hemlock, with an occasional mixture of white pine. The principal species from a volume standpoint are hard maple, yellow birch, and hemlock, and the presence of the other species depends largely on the site. The volume of a commercial hardwood-hemlock forest may vary from 4,000 to 18,000 ft per acre. The northern hardwood forest is extremely aggressive in reproducing itself on good sites, provided it is protected from fire. Second-growth forest is likely to consist largely of hard maple, although beech is also extremely reproductive on some sites. Yellow birch, which is the most valuable single species in the forest, reproduces where the soil has been scarified. Hemlock is likely to be eliminated entirely in second-growth forest.

The swamp forest is extremely important both in area and in its production of pulpwood for the large Lake states pulp and paper industry. The chief species is usually white cedar (which is not itself a pulpwood species), with a varying mixture of black spruce, balsam fir, white birch, and an occasional aspen. In some cases there are areas of pure black spruce. These swamps are mostly in second growth, since they were originally cutover 60 to 70 years ago. White cedar, balsam fir, and white birch reproduce themselves as the swamp forests of merchantable size are cutover. Spruce does not reproduce nearly as satisfactorily. A dense cedar forest will continue to increase in volume until it has reached an age of about 200 years. At 100 years of age, and before other intermingling species begin to drop out, it will have a volume of from 1,500 to 2,000 cu ft per acre.

The volumes of timber per acre are shown as a range from a minimum to a maximum rather than an average. The volume of forest areas varies with quality of site, stocking, and age. Hence, the average acre is a hypothetical unit that does not exist except in a table of statistics.

The discussion so far has been limited to Michigan and Wisconsin. The forests of Minnesota differ chiefly from those of Wisconsin and Michigan in that the hardwood-hemlock forest, which is so important in those states, is almost entirely absent as a commercial forest in Minnesota. The forests are largely coniferous and are similar to the coniferous forests of Wisconsin and Michigan.

Topography in all three Lake states is generally level or rolling, although there are certain areas in the Upper Peninsula of Michigan where the topography is extremely rugged.

The most extensive forest areas and those in which harvesting is conducted as specialized operations are the commercial forests in the northern portions of the three states. Hardwood wood lots, however, are common in the southern part of each state, and the economic importance of these wood lots should not be underestimated.

LABOR

Woods labor in the Lake states comes from a variety of racial backgrounds. In all three states the predominant racial strain among woods laborers is Scandinavian. In northern Michigan, the substantial population of Finnish background provides a large number of excellent woods workers. In addition, there is the usual admixture of Central European strains found throughout the northern part of the United States plus occasional workers of French or Indian origin.

On large operations the men live in camps. A logging camp, known among woods workers as a "set of camps," consists of a bunk house (or "men's camp"), a combination kitchen and mess room (or "cook camp"), an office, a stable, and a combination garage and blacksmith shop.

On small operations the men frequently drive to work from farms or nearby towns. In some cases they build small temporary shacks in which one or two men live and do their own cooking.

The skills required of woods workers are the long-established skills with ax, saw, and peavey. Of late years, however, mechanical skills have become increasingly important as trucks, tractors, power loaders and power saws have become standard equipment.

FELLING AND BUCKING

Felling and bucking for the production of sawlogs is done by (1) a crew of two men using an ax, a wedge, and a crosscut saw, or (2) a crew of three men using an ax and a power-driven chain saw.

When power saws are used, two of the men work with the power saw as fellers and buckers and the third man swamps and trims with an ax. In most cases, the saws are powered by a small gasoline engine, which is a part of the power saw itself.

A crew of two men felling and bucking with a hand crosscut saw will produce around 5,000 ft of logs per day, and a crew with a power saw will produce several times that amount. Average daily production, however, is like average volume per acre—it varies greatly with working conditions and with the crew itself.

When producing pulpwood, fence posts, or ties in a swamp forest, a man often works alone with an ax and a one-man saw. A good worker produces a cord or more of pulpwood per day and, here again, his production varies with conditions. Men cutting logs, pulpwood, posts, or ties are usually paid a set rate per unit of production.

BUNCHING AND SKIDDING

Logs are skidded with tractors or with teams. Whether they are skidded singly or with two or three logs chained together depends upon the size of the logs, the skidding power available, and the topography.

Pulpwood, cedar posts, and other short-length swamp products are customarily skidded out on drays hitched to single horses.

It is almost impossible to set a fair amount of skidding to be accomplished in a day. Here again there is a wide variation depending upon the products skidded, the skidding distance, the equipment used, and the topography.

LOADING

Logs are loaded on trucks or cars by means of a derrick-like device known as a jammer. The power is usually a gasoline engine, although occasionally a team is used. The use of a team to provide lifting power is, however, becoming increasingly rare. Where a railroad loading point has been established and logs are being loaded in large volume, a power loader mounted on a flatcar is used. A loader of this type can load upward of 6 carloads of logs per day.

Pulpwood and other short-length products are loaded on trucks or cars in slings, through the use of jammers or by hand.

TRANSPORTATION

Both logs and short-length products (pulpwood, posts, and ties) are normally hauled out of the woods by truck. If the consuming mill is located within a relatively short distance from the woods, the products are often delivered directly to the mill by truck. If the consuming plant is located at a greater distance, the products are hauled to the railroad by truck and are then loaded on cars.

In some swamp operations, products are hauled to the railroad on sleds hauled by either tractors or teams.

Water transport is used relatively little. In the days of the large pine operations, logs were driven down rivers in enormous quantities. On the Menominee River, in Michigan, volumes of logs running as high as 600 million feet were driven in a single spring season during the last decade of the nineteenth century. Hardwood logs, because of their high specific gravity, cannot be successfully driven or rafted. In areas along the Great Lakes, pulpwood is still occasionally driven down rivers and into booms in sheltered harbors. From the booms, the wood is loaded into cargo vessels by means of slings, derricks, or conveyers and is then transported by water to consuming mills.

ORGANIZATION OF OPERATIONS

Most woods operations in the Lake states are seasonal in the sense that there are two major operating seasons—a winter season and a summer season. The winter season extends from the date at which the ground is solidly frozen to the date at which the spring thaws occur. This is, roughly, from December 15 to March 15. The summer season begins when the ground is dry enough to permit truck hauling and ends when the fall rains begin. This varies considerably but is approximately from June 1 to November 1.

Some woods operations are conducted on a year-round basis. Although it is impossible to haul during the spring and fall wet seasons, production may continue during those seasons. This off-season production of logs or pulpwood is then piled or decked pending the resumption of hauling. The additional handling cost is offset by the fact that a greater volume of products is thus trucked in shorter time than would otherwise be the case and by the additional fact that most of the winter hauling is completed before the heavy snows of late winter make heavy snow plowing and expensive skidding necessary.

In the swamps there is, of course, no summer trucking at all. Neither is there much summer skidding. Most summer production is left in piles by the cutters to be skidded and trucked out after the swamps have frozen in the winter.

Operations in the Lake states range in size from a few men working together and driving to work from their homes, to camps employing upwards of fifty men.

Supervision of operations depends upon whether a timber company operates its own camps or whether its timber is cut by logging contractors. If a company operates its own camps, the work is supervised by company logging superintendents, who have subordinates working under them who are responsible for the various phases of the work, such as road construction and maintenance, supervision of log cutters, log scaling, and loading on trucks or cars. The actual organization of supervision varies with different companies.

If, on the other hand, the work is being done by a logging contractor, he normally supervises his own men, although on a large job he may have foremen working under him. In the case of a logging contractor, the timber-owning companies have no control over the conduct of the operation, although their representatives do see to it that the terms of the contracts are carried out and that no logs are accepted and paid for that do not meet contract specifications. A logging contractor is required to remove all of the timber specified in his contract and he is normally paid at a stipulated price per 1,000 ft on the basis of the timber owner's scale at the agreed point of delivery. For the last ten years, the trend has been increasingly toward the use of contractors in woods operations.

NATURE OF CUTTING OPERATIONS

In most cases, logging in the Lake states involves the removal of all merchantable timber in the area. There are, however, a number of large companies operating in hardwood-hemlock forests who conduct their operations on a selective logging basis. The hardwood-hemlock forest, which usually has a satisfactory differentiation of age classes, responds well to this type of cut, from the standpoints of both reproduction and increased growth. The percentage of cut may run from 25 to 65 per cent of merchantable volume, depending upon the planned cutting cycle and the danger of wind throw on the area in question. Of course, since the larger trees are normally removed, the number removed is a much smaller percentage of the total.

The minimum size of products cut is largely determined by commercial specifications. Saw logs are cut to a 10-in. top diameter and in lengths of 10, 12, 14, and 16 ft plus a 4-in. trim allowance. Logging cost, of course, goes down as the number of logs per 1,000 ft decreases. This is sometimes offset by the fact that the lumber-grade recovery on extremely large logs may also decline because of overmaturity and consequent defect.

Although paper mills accept pulpwood down to a top diameter of 4 in. (cut in lengths of 100 in.), the production of small-diameter wood is expensive to the producer because of the large number of pieces handled per cord.

The question often arises as to whether current logging methods meet satisfactory forestry requirements. It is difficult to define satisfactory forestry requirements, as they depend so much upon the individual point of view. Even in Europe, clear cutting is an accepted silvicultural practice provided the area is made to reproduce and remain productive. After all, silviculture on private operations is greatly influenced by markets. A forest owner cannot cut what he cannot sell. In many cases, silviculture is also determined by economic pressures, such as the necessity for liquidating an investment which the owner may feel he cannot afford to carry. Many feel that the requirements of forestry are met if the area reproduces itself and remains stocked and productive. In the Lake states it is fortunate that both the hardwood-hemlock type and the swamp type are aggressive and self-reproductive if protected from fire.

THE WESTERN PINE REGION

BY THOMAS J. ORR, JR.
Weyerhaeuser Timber Company

The Western pine region covers parts of twelve Western states. Obviously, in a region this large there is no single condition of timber stands or of topography that could be considered typical. In parts of the region there are certain characteristic features, however. For example, in the white pine region of northern Idaho, the topography is broken by numerous drainages, the slopes are moderately steep to steep, the timber stand will often reach 40,000 ft per acre. Species which are produced in this region are western white pine, ponderosa pine, Douglas fir, western larch, Engelmann spruce, western red cedar, white fir (*Abies grandis*), and lodgepole pine.

East of the Cascades in Oregon and Washington and of the Sierras in California is an area of 15,000,000 acres of pine lands that is characterized by gentle slopes, light volcanic soils, and more or less extensive plateaus. These stands are virtually pure ponderosa pine, are light and open, with 14,000 ft per acre being a typical volume. According to the locality, sugar pine, white fir (*Abies concolor*), incense cedar, lodgepole pine, Douglas fir, and western larch are associated with ponderosa pine in varying amounts, but these species seldom make up more than 20 per cent of the stand in the main pine belt.

Ponderosa and sugar pine occur in mixed stands in the west side of the Sierra pine belt. The associated species are Douglas fir, white fir (*Abies concolor*), incense cedar, and Jeffrey pine. Stands of 80,000 ft per acre have been recorded on the best growing sites in this region, and it is there that ponderosa and sugar pine attain their maximum size and grow on the roughest ground.

The pure pine stands of the Colorado plateau country in Arizona and New Mexico are open and park-like, with 8,000 ft per acre representing a typical stand density. The slopes in this area are gentle, and the stands very open and free from underbrush. Logging in the central and northern Rocky Mountain region, except for the white pine belt described above, consists of small operations getting out limited quantities of railroad ties, mine timbers, and small saw logs, mainly for local markets.

LABOR

Most of the larger operations maintain camps. Camps are semipermanent. Usually family houses are available for part of the crew. Family houses are built on skids and can be moved to a new location with the logging equipment. Single men are housed in portable cabins, and board at the cookhouse, which is usually company-operated. Crews of smaller operations are usually housed in nearby forest communities and commute to the woods daily.

The skills required are many and varied. For example, locomotive engineers, firemen, trainmen, truck drivers, power-shovel operators, tractor operators, timber fallers, mechanics, blacksmiths, welders, and cooks would be found in the personnel of a crew of 100 men.

There is no longer any predominant nationality among the loggers. In the northern camps, there are perhaps a few more Scandinavians and Finns than elsewhere in the region. In the Southwest there are Spanish-Americans, who work on road and railroad maintenance. Relatively few Spanish-Americans are temperamentally suited to the job of logging. With these slight exceptions, the modern logging camp or logging community is hardly different from any other community in the region. The mixture is typically American.

FELLING AND BUCKING

Crew organization for felling and bucking depends to some extent on the traditions of the region. In California the felling is done as a separate operation by crews which do nothing else. They are followed by a limber, usually one man to each "set" of fallers, and by one or more buckers. All of these men are paid by the scale produced. Elsewhere in the region, each set of fallers does its own limbing and bucking. Output per crew for straight falling varies from 60,000 to 100,000 ft log scale per day, and normally one limber can keep up with his set. Buckers average about half this scale, *i.e.*, from 30,000 to 50,000 ft per day, and normally work alone. Sets making their own logs average about 30,000 ft per day.

Typical falling tools are the 4½-lb double-bit ax, usually with a long narrow blade, and in big timber, a 42-in. handle, a 7-ft crosscut felling saw, several steel wedges and shims, and a 6-lb hammer for wedging. The limber carries a 4½-lb ax, swamping pattern, with a 36-in. handle. The bucker carries a 7-ft crosscut saw with a wide blade, a set of bucking wedges, a hammer, and an ax similar to that of the limbers.

Hand felling is gradually disappearing and is being replaced by power-saw felling. The crew organization may be one of two patterns, that is, felling may be done with the power saw and limbing and bucking by hand, or the power-saw crew may complete the log-making operation. While theoretically it should be possible to exceed the output of hand sets by using power saws, this has not been the case to date. The output is about the same whether hand tools or power tools are used, but the physical effort is less with the power saws. For this reason, the use of power saws has sharply increased in the past few years. Chain saws which are self-contained units are in wider use than the other types.

There are several types of chain saws on the market. Preference in this region is for a self-contained unit, equipped with a fast-cutting chain, which has separate side cutters and router cutters. The development of power felling equipment is proceeding very rapidly, and in time perhaps there will be a single type that is most suitable for use in the pine region. At present the practice is in its early stages of development.

BUNCHING AND SKIDDING

In spite of the varying conditions of topography and timber stands that are found in the pine region, the characteristic skidding tool is the tractor.

The larger sizes, such as the Caterpillar D-7 and D-8, are used, with the largest size preferred because of its greater power. A logging arch is used in conjunction with the tractor, where the ground permits.

The skidding crew consists of a tractor driver and one or two choker setters for each tractor-arch unit. Bunching is done with the tractor unit and is usually combined with the skidding operation. The log that is farthest from the landing is hooked first, and the load is assembled as the tractor moves toward the landing. As a rule chokers are set ahead so that the tractor unit will not be delayed while waiting for the choker setters to dig under the log. The ideal arrangement is for the chokers to be set for the succeeding turn while the tractor is on its way to the landing with its load. The work of setting chokers is made easier if the logs are rolled out—that is, the bucked cuts are broken apart with a tractor assigned to this job. On some large operations it is felt that the labor saved in setting chokers will pay for this extra operation, although the main reason for rolling out is to prevent blue stain when the fallers are working a week or two ahead of the skidding operation in warm weather.

Skidding output varies with many factors but, for example, on ground with slopes less than 20 per cent and a maximum distance to the back line of 2,600 ft, a D-8 tractor and arch will yard about 60,000 ft per day in 32-ft logs.

LOADING

The loading operation in the pine region involves loading trucks or railroad cars, or in many cases both.

Truck loading is most commonly done with a converted shovel or a specialized crane built for that specific purpose. Small operations have numerous other modifications of loaders. The essential part of any of these units is an A-frame, 35 to 40 ft long, and a hoist of some description. The rest of the machine, such as the swing on a crane or shovel, is to speed up the operation. The wheels or tracks are for increased mobility of the unit. The machine is moved out to the landing and the logs are skidded within reach of the boom, where they are picked up and placed on the truck with the loader. The commonly used device for hooking the logs is a crotch line equipped with bell hooks. The hookers on the landing place the bell hooks one at each end of the log.

Loading on railroad cars is done with the McGiffert loader, which is a steam-operated crane so constructed that it straddles the track and the empty flatcars can be run under the loader. Most logs are handled by means of a crotch line and end hooks. Slings are provided for very large logs. The capacity of the machine is limited by the ability of the crew to hook and of the jammer operator to hoist the logs, rather than by the machine itself. The machine will load 300,000 ft per 8-hr day easily, and cases are on record where it has handled as much as 650,000 ft. Most of the loading is done on standard flatcars equipped with bunks and three sets of wrapper chains. In use, the first tier of logs is loaded and the bottom pair of wrapper chains secured. The second and third tiers of logs are handled similarly. The load is topped off with a few logs, which are placed on top the last chain and bind the entire load by pulling the slack out of the chain.

Unloading at the mill pond is done with a power winch and crotch line that spills the logs into the water.

TRANSPORTATION

As more remote timber stands in the region are opened up, there is an increasing tendency to move logs long distances to the mill. This may be done with motor trucks, if the distance does not exceed 30 miles or so, or if the volume of timber to be handled is relatively small. For larger output and longer distances, a combination of truck and railroad hauling is used. This process of trucking to a railhead is characteristic of many of the operations in the pine region. In this manner logs are currently being moved as much as 100 miles from the stump to the millpond.

Water transportation has never been widely used in this region. The one remaining river drive is on the Clearwater River in Idaho. On Flathead Lake in Montana and Cœur d'Alene Lake in Idaho, logs are dumped in the lake and towed to mills along the shore.

ORGANIZATION

Operations that depend entirely on truck transportation are seasonal. In parts of the region there is heavy snowfall, which stops rail logging. Under these conditions, logging is planned so that logs may be decked ahead, either to carry the mill through the winter or to enable the mill to start operations earlier in the spring. Some large operations, which maintain their own logging railroads, plan to log directly to railroad spurs during the winter season and so avoid the winter shutdown.

Operations may be of any size, from a crew of eight or ten men with a single tractor and a homemade loader to a large operation using half a dozen units or more. Whatever the size of the operation, its make-up is influenced by some limiting factor such as the skidding capacity of the tractor, the hauling capacity of available trucks, or the capacity of the loading unit. The organization is built with that daily output in mind, and there is usually an attempt to balance the crew requirements with this output.

For example, to produce about 60,000 ft per day, a crew organization something like this would be required:

1 bull buck, scaler
4 fallers
1 tractor driver
2 choker setters
3 loading-landing men, operating homemade loader
2 or more truck drivers
1 foreman

This adds up to fourteen men, exclusive of the unloading crew at the mill or landing, as the case may be.

A larger organization, operating five tractors and producing about 300,000 ft per day, might be organized in this manner:

Camp Boss

Bull buck		Jammer boss		Truck boss	
Fallers and buckers	20	Hoister	1	Truck drivers	5
Scaler	1	Landing crew	4	Road monkeys	2
Total	21	Tractor mechanic	1	Total	7
		Grease monkey	1		
		Tractor operators	5		
		Choker setters	10		
		Knotters	2		
		Car scaler	1		
		Total	25		

In addition to the operating crew, there would be a crew assigned to construction, one to railroad operation and maintenance, and one to miscellaneous duties of camp operation and crew transportation, under the general direction of the camp boss. The truck boss mentioned above is usually a mechanic in charge of truck operation. Each operation will no doubt have its own version of crew organization.

NATURE OF CUTTING OPERATIONS

Somewhat more than half of the remaining pine timber throughout the West is publicly owned. This timber is under more or less good forest management, and is operated on a sustained-yield basis. The public land is or will be partially cut, which is about as good a practice as is now available for all aged stands. On this land the requirements of forestry are met.

The earlier practice on private lands was to log to an economic minimum-diameter limit—usually ranging from 16 to 22 in. dbh, according to the region. Where plants are on a liquidation basis and are not exchanging their lands under precutting agreements with Federal agencies, this practice persists. In Washington, Oregon, and California the state laws require leaving a certain minimum number of trees per acre as a source of seed and protection of the residual stand. This will assure a second crop of trees, and while it falls far short of meeting the requirements of forestry for the region, this shortcoming is recognized by the state foresters. Amendments to the forest laws appear at each session of the state legislature, and the trend is to strengthen the requirements and the enforcement provisions of the laws.

The policies of those operations that propose to stay in business is in sharp contrast to the foregoing. Most of these operations are practicing partial cutting aimed at maintaining the cut at its present level, at getting the maximum growth out of the reserve stand, salvaging beetle-susceptible timber before it is attacked, and various other silvicultural objectives. The owners of a business committed to sustained yield are much more insistent that an effective forestry program be achieved than a Federal agency would be, and it follows that the standards of forest practice are equally high.

With the strengthening of state laws, and the impetus of the Tree Farm movement, there is little doubt that private forest practice will be consistently improved.

THE DOUGLAS FIR REGION

BY W. D. HAGENSTEIN

Forest Engineer, West Coast Lumbermen's Association and Pacific Northwest Loggers Association

THE FOREST

After a hundred years of logging, the commercial timber stands of the Douglas fir region are still principally old-growth forest. In 1945 the U.S. Forest Survey listed a total of 9,350,000 acres of old-growth forest. Large young timber, principally naturally reforested old burns, occupied 2,200,000 acres. Small young timber, which includes reforested burns and cutover lands, occupied 6,000,000 acres. The average stand per acre on these 17,550,000 acres of merchantable timber is 25,000 ft. The average virgin stand under present utilization standards is 60,000 to 70,000 ft per acre. The lower average for all saw timber is due to the fact that many of the younger stands are incompletely stocked.

The principal species growing in the forests of the Pacific Northwest are Douglas fir (*Pseudotsuga taxifolia*); western hemlock (*Tsuga heterophylla*); western red cedar (*Thuja plicata*); Sitka spruce (*Picea sitchensis*); Port Orford cedar (*Chamaecyparis lawsoniana*); and the balsam or white firs, which occur in order of importance as follows: silver fir (*Abies amabilis*), noble fir (*Abies nobilis*), and lowland white fir (*Abies grandis*). These species comprise more than 95 per cent of the total commercial timber volume. The balance is made up of miscellaneous conifers and several broadleaf species. The most important of the latter are red alder (*Alnus rubra*); big-leaf maple (*Acer macrophyllum*); northern black cottonwood (*Populus trichocarpa hastata*); and Oregon white oak (*Quercus garryana*).

The heavy stands of timber of the region are found on topography ranging from level land through rolling hills to very steep mountain slopes that are heavily indented with countless side drainages. Along the coast there is considerable low-lying land, some of it swampy, which presents particularly difficult timber-harvesting problems. The very infrequent occurrence of gravel in the coastal strip necessitates use of temporary plank roads for hauling logs by truck or the construction of expensive railroad. The more mountainous portions of the region present difficult logging problems because of heavy rock, wet clay soils, and high average rainfall.

LABOR

Since the advent of widespread truck logging operations, there has been a trend toward fewer logging camps. Seventy-five per cent of all logs now make part of their journey from stump to conversion plant by means of trucks. The same roads which are used for hauling logs provide easy access for loggers who commute daily to and from their homes in nearby communities. There still are a large number of camps situated in the woods in the more remote parts of Washington and Oregon.

Logging camps in the region have accommodations for from forty to several hundred men, with modern sleeping, eating, and sanitary conveniences. The men are charged a nominal rate for board and bed and the food served is as good as can be found anywhere.

Many of the camps on large properties are of a semipermanent type with houses for family men and two- and four-man bunkhouses for single men. A few camps remaining from the old days of railroad logging are portable, with their buildings mounted on railroad trucks. Today not more than 25 per cent of the loggers in western Washington and Oregon live in camps.

The loggers in the Douglas fir region are among the highest paid and highest skilled workers of any industry in the nation. Logging has been mechanized to the extreme and because of rapidly moving equipment and lines, the roughness of the terrain, and the volume of timber per acre, work in the industry is classed as extrahazardous; therefore, men of better than average mental and physical agility are required. Mechanical skill is also necessary because of the high level of mechanization.

Loggers in the region are mostly native-born, with a substantial number of second- and third-generation Scandinavians who have proven their adaptability.

FELLING AND BUCKING

Felling is usually done by two-man crews, although sometimes a third man is added in the case of power felling. A two-man felling crew consists of a first, or head, faller and a second faller. The first faller, in the case of day work, is usually paid slightly more and is in charge of the crew. In piecework, they are paid equally, but the head faller still selects the lay for the tree and supervises the work. In the case of power felling, where a third man is sometimes added, the third man chips out the undercut, swamps brush around the tree, and does the wedging, while the others operate the saw.

Hand felling tools consist of a felling saw from 7½ to 11 ft in length, two felling axes, three or four 8-lb felling wedges, an 8-lb maul, a pair of springboards, which are used to equalize the elevation at which each faller stands in case of uneven ground, and a saw-oil bottle.

In power felling, either an electrically driven saw or an internal-combustion saw is used. The cutter bars on these saws vary from 4½ to 10 ft in length and all saws are of the chain type. In addition to the saw, a power felling crew has a felling ax, several wedges, a maul, springboards in some instances, a Pulaski tool, the hoe bit of which is used for removing the scarf, and a saw-oil bottle.

The production per felling crew per day is variable, depending upon tree size, species, density of trees, and the ability of the men. An average for hand felling is 25,000 to 30,000 ft. Power felling averages 75,000 to 80,000 ft.

Bucking is a one-man occupation despite the fact that the timber is among the largest in the world. The reason for one-man bucking is because of the danger of 10- to 20-ton logs rolling after they are cut from the trunk. Practically all bucking is done by hand, as power saws are not yet adapted for this work.

A bucker's tools consist of a bucking saw from 7½ to 11 ft in length, depending on the diameter of the timber, a felling ax, three or four 6- to 8-lb bucking wedges, an 8-lb maul, and a saw-oil bottle.

One man bucking by hand averages from 15,000 to 20,000 ft per day. The work is commonly done on a piece-rate basis, but windfall buckers, who cut logs from the wind-blown trees prior to the felling of the stand, are paid by the day. The most reliable and capable buckers are usually fully matured men in their forties or fifties.

BUNCHING AND SKIDDING

Bunching, which is more properly termed decking in the Douglas fir region, is accomplished by both yarders and tractors. Decking is divided further into hot and cold decking. In cold decking, logs are assembled at a central point for subsequent yarding to a landing at some later time. Hot decking, on the other hand, is the assembly of logs from stump site to a temporary landing whence they are immediately yarded to the point of loading. Decking of both types is usually done by a high-lead yarder or tractor, with or without arch. Yarders used for decking are powered with 75- to 300-hp gas or Diesel engines. Tractors used similarly are 75- to 115-hp Diesels. The output per day in hot or cold decking is so variable, depending on local conditions, that it is impossible to set an average.

Skidding—which, except in the case of the skidder system, is called "yarding" in this region—is done by a number of standard methods. The principal methods used in order of importance are high lead, tight-line skyline (usually North Bend system), slack line, and skidder. Most of the yarders and some of the skidders used today are internal-combustion-powered.

Tight-line skylines are used for swinging logs from hot or cold decks to the point of loading. The other systems may also be used for this purpose, but are generally used for direct yarding from stump to landing. High-lead and tight-line skyline equipment formerly was mainly steam-powered, but now the majority is internal-combustion machinery of from 75 to 300 hp. The machines used for slack lining are special machines with a skyline drum, also formerly steam and now chiefly gas or Diesel. Skidders are machines designed for the Lidgerwood system using either a steel tower or a tree for a spar. Slack-line machines and skidders usually develop 300 to 400 hp and more. All of these machines are used in connection with either truck or railroad primary transportation, except that steel-tower skidders are practically all car-mounted units.

Tractors are also used both with or without arches for direct yarding, and are usually the two largest sized machines.

The average daily output of internal-combustion high-lead and tight-line skyline yarders in direct yarding and swinging from decks is 60,000 to 100,000 ft, depending on the show. Slack lines and skidders generally exceed this daily output by 25 to 50 per cent because of greater yarding distances, necessitating less change of rigging and greater power.

LOADING

The principal loading methods used in truck logging are (1) McLean, or hayrack, boom, (2) heel boom, and (3) crotch line. Two- and three-drum 60- to 125-hp loading engines, mostly gas and Diesel, are used with these loading systems.

In railroad logging the principal loading method is the duplex, or guy-line, system. A special steam duplex loader is used that has reversible engines, thus affording rapidity and flexibility. Some duplexes are attached to a car-mounted unit comprised of a yarder and loader. The car-mounted

Lidgerwood skidders have a loading engine mounted on the frame for heel-boom loading. The trend in truck logging is for lighter and more flexible loading machinery. Many shovel-type loaders, either tread- or wheel-mounted, are being used in truck operations. This type of equipment has a particular advantage in tractor logging, and logs can be landed at any place along a hauling road, thus completely divorcing yarding from loading.

Any of the loading methods with the types of equipment mentioned are capable of loading from 60,000 to 200,000 ft per day. However, because loading output is dependent on yarding output and on truck or car supply, it is difficult to express exactly the output of any loading device. Recently a preloading device for truck logging has been adopted. The preloader is a hydraulically operated false bunk and, by using two or more trailers with each truck unit, a load is ready to be hauled away every time the truck returns to the landing, as the loaders are busy all the time when logs are available.

TRANSPORTATION

Prior to 1930 the principal primary transportation for logs was railroad to dumping points on towable waters, with subsequent flat or crib rafting to the conversion centers. Today 75 per cent of all logs are hauled on part of their journey by truck and trailer over either private or public roads. Many combinations of transportation are used to haul logs from the original loading point to the use location. For example, a typical combination system in use in northern Washington is one where logs are trucked over private roads from 1 to 15 miles to a railroad landing; the loads are transferred intact from truck to common-carrier car with a standing skyline loading system; the cars are then hauled 20 miles on a private railroad and 25 miles on a common carrier to the dump on Puget Sound, thence towed in flat rafts by tugboats to a plant 40 miles away.

Many logs are trucked or railroaded directly to mills that are not situated on towable waters. Some logs are hauled by truck in the Willamette valley, reloaded on the railroad, dumped in the Willamette River, and towed downstream to Columbia River plants.

There are some operations which yard logs directly to portable mills in the woods, in which case there is no primary log transportation. The general practice in this region, however, is to transport logs to the mills by truck, rail, or water.

The relative cost of hauling logs per 1,000 ft per mile by the three primary means is in the ratio 1:5:12, for water, rail, and truck respectively. In as much as many logs are used 200 to 300 miles from their points of origin, the design of the transportation scheme is an important part of operation planning.

Trucks and trailers used in hauling logs on public roads are usually equipped with 8-ft bunks and haul from 4,000 to 5,000 ft. The width of bunk and the number of pounds of gross load distributed on the truck and trailer are established by state law.

On private roads there are trucks with bunks up to 14 ft wide, which are capable of hauling 20,000 to 30,000 ft. The lengths of logs handled by truck are up to 84 ft, common lengths being 24, 32, 40, 48, and 64 ft.

Logging-railway equipment is standard gauge with both geared and rod steam locomotives in use. The principal type of car used is an air-equipped skeleton flat 41 ft long with bunks 9 ft 6 in. spaced 28 ft center to center. A few operations use disconnected trucks for hauling long lengths. Short-length peeler logs and blocks 8 ft and up in multiples of 4 and 8 ft are often hauled in gondolas on common carriers.

ORGANIZATION OF OPERATIONS

Most Douglas fir logging operations are conducted year-round except where adverse winter or fire weather prevents. As operations get farther back in the Cascade and Coast Ranges, seasonality will increase because of heavier snowfall at the higher elevations. Rain does not close down operations unless transportation systems are impaired by swollen streams or clogged drainage facilities. Wind sometimes stops operations, particularly in the cutting department, because of the danger. The state foresters are authorized to close down operations when the relative humidity is expected to drop below 30 per cent, and the average number of days lost in western Washington and Oregon each year for these fire-weather shutdowns is from five to seven.

Operations vary in size from those producing 1 million feet a year to several producing in excess of 200,000,000 ft. In recent years a large number of small operations have sprung up as a result of the truck and tractor, which allow taking the transportation to the logs instead of bringing the logs to the transportation, as is the case in railroad operations. The adaptability of small operations to changing demands for species and grades, to changing weather conditions, such as dropping down below the snow instead of closing down, and the ease with which their light equipment can be moved all make for economic advantage.

While many large properties are now being managed on a permanent basis, it is thought that the future will see their log harvests conducted by small operations, rather than by the large units which removed the original crop.

The number of men per operating "side" and per camp depends on the number of sides and the kind of equipment being used. To show comparatively the number of men used on different kinds of operations, the average number per side for each of the following is given:

Equipment	Number of Men
Railroad:	
Steam high lead	13–15
Steam skidder	23–25
Steam slack line	18–21
Gas or Diesel high lead	11–12
Truck:	
Gas or Diesel high lead	9–10
Tractor	7–8

This includes the yarding and loading crews for one landing or side only. The number of men used in the cutting crew depends upon the production capacity of the side, to ensure that sufficient timber is felled and bucked in advance to keep the rigging crews busy. The average number of fallers and buckers per side is about four sets, a set being two fallers and two or three buckers. In addition, there are one or two windfall buckers who precede the fallers for each side. One woods scaler looks after from two to four sets.

The supervision of operations in the region is generally set down about as follows:

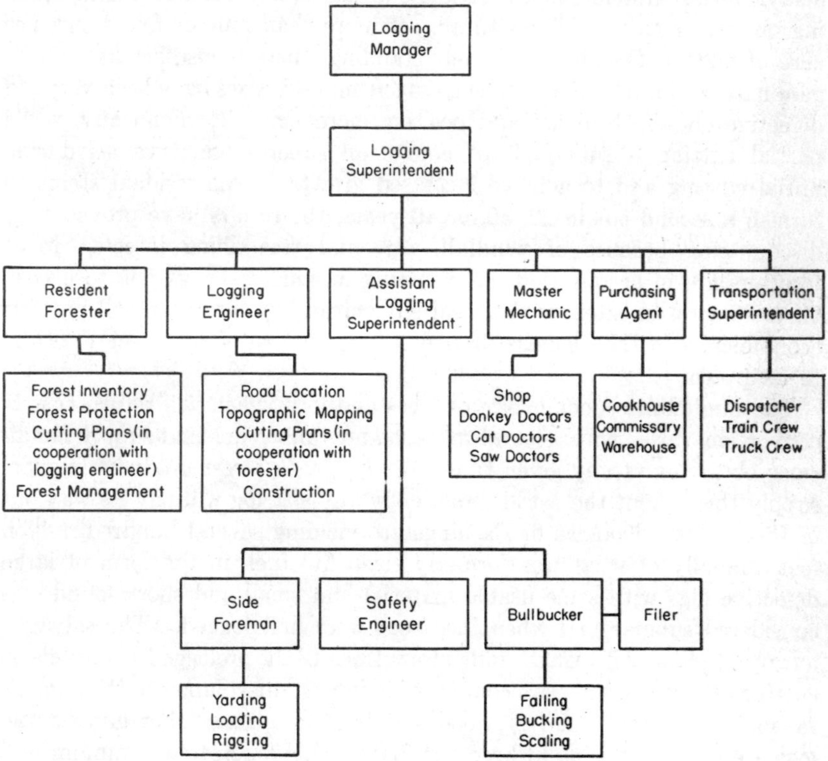

The foregoing chart is taken from a typical large operation producing in excess of 100 million feet annually. Many variations in operation organization are effected to fit the needs of individual companies.

NATURE OF CUTTING OPERATIONS

Clear-cutting is the main system used in harvesting old-growth timber in the Douglas fir region. Fortunately, from both the silvicultural and engineering standpoints this is the most desirable method for handling Douglas fir, if the object of management is to reproduce this species. Clear-cutting has been modified by the wide adoption of the "staggered-setting" system, which cuts one setting and leaves the adjacent one standing and so on along a railroad or truck road throughout a logging unit. The principal advantages accruing from this system are the better fire protection afforded because of smaller cutover areas in one contiguous piece and prompt restocking from the seed-bearing trees in the nearby uncut or residual setting. This system, which began on private lands, is now common timber-sales practice of public-forest agencies.

Partial cutting has been tried in old-growth Douglas fir stands as both a silvicultural system and an economic measure, but has been found wanting on both counts. Silviculturally, if more than four or five stems per acre of mature Douglas fir are left standing, there is insufficient light to gain natural reproduction. The trees left in most cases have been very old defective ones and do not add positive increment. Economically, while partial cutting is supposed to remove all supermarginal values during initial cutting and to achieve increased growth in the residual stems to furnish a second cut in 20, 30, or 40 years, the results have proved very discouraging because of windfall, rot, and tree-killing insects. Most foresters are in accord that clear-cutting in small settings not to exceed 60 to 80 acres in extent is the ideal silvicultural system that will pay out economically in the Douglas fir region, irrespective of the species desired to be grown.

The minimum economic size of trees and logs which will return cost in conversion varies with general market and timber demand-supply conditions obtaining at any given time. Today, with forest products in short supply throughout the world, practically any size log will pay its way out of the woods. Douglas fir slashings are yielding several hundred million feet annually of what was formerly slash fire fuel, in the form of large defective logs with some usable material and small and short sound logs considered submarginal when the areas were first logged. The salvaged leftovers, along with small understory trees being prelogged before clear-cutting of virgin stands, are contributing important volumes to the region's sawmills, pulp mills, and plywood and treating plants. For lumber use, logs are currently being utilized the first time through to a minimum of 8 to 10 in. in diameter and in lengths as short as 8 to 12 ft. Pulpwood relogging is taking material down to 4 in. in diameter in 4- and 8-ft lengths.

Oregon in 1941 and Washington in 1945 enacted similar forest-practice acts, which require all private forests to be cut with a reservation of a specified area of seed-bearing timber to reforest each harvest area. In Oregon the unit of compliance is 160 acres and at least 8 acres in each 160 must be left with standing seed-bearing commercial species after slash disposal. In Washington the operator must reserve seed-bearing timber on 5 per cent of each area clear-cut irrespective of size and bring such seed source successfully through slash disposal. In case of violation the states may plant the area not restocked and assess $8 per acre against the landowner.

The Washington and Oregon state foresters have stated that compliance with the minimum cutting-practice requirements of these laws is leaving 90 per cent of the lands harvested in productive condition. They both have a force of competent foresters working with the operators to assist them in the preparation of and the carrying out of cutting plans. Beyond these minimum requirements for all operating private lands are the 2½ million acres which are certified as Tree Farms. The forest practices on these timber-growing properties go far beyond the basic requirements of the forest-practice acts. More than two hundred foresters are working toward the ultimate goal of permanent forest management on these Tree Farms.

Forest protection in logging operations in the Douglas fir region is outstanding in the United States and the operators are spending 3 to 4 million dollars annually at present to combat the number-one enemy of permanent forest enterprise.

CALIFORNIA REDWOOD

BY EMANUEL FRITZ
University of California

The redwood region, from a commercial lumbering standpoint, covers a narrow discontinuous strip along the California coast from Santa Cruz to the Oregon line, a distance of about 350 miles. However, present-day large-scale lumbering centers in the west central parts of Mendocino and Humboldt counties, and on the lower Klamath River in Del Norte County.

THE FOREST

Coast redwood (*Sequoia sempervirens*) is the predominant species in the forest. The great size of the tree, its great weight, the dense stands, the rough terrain, the tendency of the tree to break as it strikes the ground, the mixture of all sizes on a single forty, the long dry summer season, and the heavy winter rains give redwood logging a distinctive character and make it particularly difficult.

On the river benches, the forest is generally pure redwood. On the better slopes redwood commonly makes up more than 95 per cent of the stand in volume and in numbers of trees. On the fringes of the redwood belt the percentage of redwood may drop to considerably under 75 per cent. Most common associated species are Douglas fir and lowland white fir (*Abies grandis*). In the northern part of the region there also may be coast hemlock, Sitka spruce, a sprinkling of western red cedar, and Port Orford cedar. When hardwoods are present, they include tanbark oak, chinquapin, big-leaf maple, madrone, and, along the streams, cottonwood and alder. The tanbark oak is occasionally harvested for its bark prior to the logging of the redwood. The hardwoods play no important part in the logging.

Until very recent years only the redwood was utilized, except that the Douglas fir was cut and logged for company use. Since just before the Second World War the Douglas fir has come to be included, and during the war the white fir also was harvested.

Redwood trees are not all big and old, as is commonly believed. The forest is made up of all ages and sizes. On a single forty in central Humboldt County, 60 per cent of the trees may be 12 to 36 in. in diameter, 25 per cent in the 36- to 60-in. class, and 15 per cent may be 60 in. and over. The latter 15 per cent may account for 50 per cent, often more, of the total volume. But the forest varies greatly. In Mendocino County, a single operating unit may not average more than 40,000 ft per acre over a 50,000-acre area. In Humboldt County a single operator averaged 110,000 bd ft per acre on a mill-tally basis for 20 years of operation. Several large stands exceed even this figure. Stands exceeding 250,000 ft per acre over small areas occur in only a few localities.

Trees exceeding 10 ft in diameter are not limited to rich river benches. They may be found on the tops of ridges as well as on the slopes and in Santa Cruz County as well as in Humboldt and Del Norte counties. In general, heights vary from 150 to 350 ft. Tops are very limby and frequently, in a 250-ft tree, may be 100 to 150 ft long and so rough as to be unsuited for the manufacture of lumber.

TOPOGRAPHY

The topography varies from moderate to very steep slopes. The southern half of the redwood belt has the steepest terrain but the timber averages smaller trees and lighter stands. Throughout the region some slopes are too steep for tractors. Prior to the advent of tractors, when wire-rope steam-donkey systems of logging prevailed, the slopes often aided these systems rather than hindered them, and sometimes a canyon was spanned with a skyline. A redwood-logging operation may extend over an altitudinal range of more than 2,000 ft, or from a stream bottom 150 ft above sea level to 2,500 ft on a ridge.

LOGGING SYSTEMS

Almost every method ever devised by loggers has been used in the redwood region. Earliest logging, along stream banks, was largely by man power aided by heavy jackscrews. Jackscrew logging became a skilled art and even after animals and power were introduced, jackscrews were used for positioning logs. Animals—oxen or horses—soon became quite common and were used into the present century, although the first steam power was introduced experimentally in about 1882. Manila rope used at first was soon found unsatisfactory and was displaced by wire rope. The Dolbeer logging donkey was developed by John Dolbeer, a copartner of Dolbeer and Carson Lumber Co. of Eureka. This donkey, the forerunner of all logging donkeys, had a vertical engine and a vertical spool, but no drum.

Donkey-engine wire-rope systems gradually became more powerful and more complicated, developing from the direct ground-skidding system, or ground-lead system, through the high-lead and to the skyline systems, the ultimate being the slack-line system.

As early as 1924 or 1925 a tractor of the Holt farm type was tried out in Mendocino County but was soon replaced with a Caterpillar 60, and other companies in Humboldt county purchased a few. These machines proved to be too light for yarding the heavy redwood logs. In 1933, a Caterpillar Diesel 75 was used on Noyo River on a trial basis but it was not until January, 1935, on Van Dusen River, that the crawler type of tractor came into permanent use. Since then every operator, small and large, has acquired fleets of the heaviest types for direct yarding and for road building and other purposes. Tractors have pretty well displaced steam systems, except for winter logging and on very steep slopes; however, Diesel-engine-powered high-lead systems are becoming popular.

FELLING AND BUCKING

Until long two-man handsaws were manufactured, redwood trees had to be felled with axes. Hand felling with crosscut saws is still common but power saws are used very extensively and may displace handsaws completely. The large amount of chopping for the undercut gave redwood fallers[1] a distinctive name—"choppers."

Choppers work in pairs, or sets. Each set is equipped with a larger array of tools than is used elsewhere. A complete outfit would include four

[1] Felling is called "falling" in the redwood region, and workmen are called "fallers" as well as "choppers." To fell a tree is to "fall" it in the local vernacular.

to six falling saws; a set of eight to twelve wedges and shims; sledges; axes; a "gun stick"; a screw jack; a mattock; a shovel; a few sticks of dynamite in a sack; caps; a modified crowbar; and a plumb bob. The falling saws vary from 4 to 10 ft. When a particularly large tree is encountered, a longer saw is taken out from camp on the days it is needed. Occasionally, two saws have had to be welded together. The shims, necessary for placing above and below the wedges because of the soft wood, are of several thicknesses; the thinnest is made of old band-saw steel, and the thickest are ⅜- or ½-inch boiler plate. Thick shims are used on leaners and the wedges are occasionally doubled to "lift" the tree away from its lean into a better layout.

Typical falling axes are double-bitted and have narrow thin blades and 40-in. handles. The handles are commonly bent to protect the hands and to permit a deeper reach into an undercut. Ax makers provide a special blade known as a redwood falling ax. The gun stick, built like a pantograph, is used for gauging the back edge of the undercut to make it at right angles with the proposed line of fall. The plumb bob is for checking the lean of the tree when it is not easily determined by eye. The dynamite and caps are used for shooting "suckers" from large trees. Suckers are small trees sprouted from the stump and closely appressed to the larger trees. The screw jack is used for rolling windfalls and chunks out of the layout or for rolling chunks into hollows. This device, though desirable, is now seldom used.

Fallers usually own their equipment, except for saws of 12 ft or longer. Each pair of fallers is assigned a strip by the chopping boss. Starting at the foot of the slope the men work up hill, felling the trees progressively. The head faller examines the timber ahead of him and occasionally finds it necessary to fell a leaning tree some distance ahead of his general operation. Redwood is always felled uphill to reduce its tendency to break. When encountered, Douglas fir, being stronger, may be felled down hill, particularly if the yarding landing is below. Crossing of trunks is avoided on a well-supervised operation because of the excessive breakage that may result.

A redwood 200 to 300 ft long requires considerable clear space for a layout, and because of its length and great mass and its tendency to break, the layout must be prepared if there are obstacles, humps, or hollows over which the tree will fall. Old-time fallers frequently spent as much time preparing the layout, or "bed," as it is often called, as they did felling the tree. Humps would be leveled off with a mattock and shovel, an obstructing windfall would be sawed through and a piece rolled out with the jackscrew, and low places would be filled. The aim is to get a falling surface that is a straight line for more than 100 ft. On flat ground, the fallers would sometimes roll in small logs to form three, four, or five points on which the tree is to land. Present-day fallers will not expend so much effort to prepare layouts. To keep their breakage within bounds, they are paid, when on a contract basis, only for the log scale in the trunk up to the first break.

For hand falling the usual sloping undercut is made with axes, and the backcut, several inches higher than the floor of the undercut, is made with the saw. There are variations, however, and redwood fallers have attained uncanny skill in applying them to cause the tree to fall exactly where intended and without shattering. Sometimes the sloping cut is in the stump and is cut upward, instead of in the butt log. This is called sniping and lets the trunk slide off the stump. A "Humboldt" undercut is also used. This requires very little or no sloping cut, but two saw cuts, one 4 to 6 in. above the other, are made and a block is pried out with a bar.

In 1928, when one company put its fallers on a contract basis, it was not long before a contract shopper experimented with the use of a dragsaw powered with a reciprocating gasoline engine for felling. This machine already had been in use several years for bucking. The scheme worked well and very soon many fallers were using dragsaws. The standard bucking dragsaw needed only to be turned on edge, the gas and water tanks had to be reset, and supporting spikes had to be driven into the tree. The dragsaw is still used but is being displaced by chain saws, either electric or gasoline-driven. Chain saws came into the redwoods in about 1941 and they are now quite common. Many fallers own their power saws.

When power saws are used the tree is felled in the usual manner, except that stumps are lower and the undercut is generally a Humboldt, as described above. A lower stump is often a questionable gain since the lowest portion of the trunk of a large tree is generally tough and springy and does not make good lumber.

Hand fallers prefer to work from a staging, and even when a low stump is called for, the "driver," or springboard, is set in a root buttress. Humboldt County fallers place two driver boards for each man and place two planks across them, making a crude stage. Mendocino County fallers generally work by standing on the narrow driver board and dispense with the platform. Power-saw fallers work from the ground, although occasionally springboards are used.

Bucking, when done by hand, is a one-man job, although the saw used is designed for use by two men. This saw has different teeth from the falling saw and the blade is wider. One handle is removed. The bucker first prepares a secure foothold on the ground and starts his saw at the cutting mark and then patiently draws the saw back and forth—sometimes for the best part of a day on a large log. Buckers must study the lay of the trunk carefully because when a log is cut through, it might roll and crush him. A good bucker will carefully secure the log with chunks and blocks for safety. The bucker's equipment includes a set of saws of several lengths, a set of bucking and "hanging" wedges, a sledge, an ax, a bar, and sometimes a mattock and shovel. Bucking wedges are shorter, wider, and blunter than falling wedges. Bucking wedges are usually provided with a hole for a short length of ¼-in. rope and two wedges are tied together. To prevent a log from turning even slightly and binding the saw when the cut is near completion, a hanging wedge is driven parallel to the grain across the cut.

On some operations fallers specialize only in felling, while crosscutting, or bucking, is performed by others. On other operations the fallers also do the bucking. This system is believed to cause the fallers to fell the trees more carefully, so that bucking is more easily accomplished for the best log lengths. A set of tools for a combination faller-bucker is augmented by the bucking sets and special wedges.

When fallers and buckers work separately the trunk may be scaled twice, first for the fallers and then for the buckers, but this depends on the basis of payment. Fallers are paid for the net log scale content of the trunk to the first break; buckers may be paid on a basis of square feet of surface sawed or on a per 1,000-ft basis.

PEELING

Redwood logs must be peeled before they are ready for sawing in the mill. The bark is so tough and stringy as to interfere with the manufacture of uniform lumber. Traditionally the peeling is accomplished in the woods, sometimes before bucking, sometimes after. Peelers work singly, in pairs, and, in large timber, sometimes in groups of three or four. Their equipment is simple—an ax and a peeling bar for each man. The bar is similar to a crowbar except that its spud is chisel-like. It is 5 ft long and weighs 13 lb. The log is first ringed with an ax at about 8-ft intervals. The bark is then pried loose. On thick-barked logs (the bark may be 6 to 12 in. thick on butt logs) the men work together, coordinating their prying to loosen the bark quickly in long strips. Bark peels readily in the spring, but sticks tightly at other times.

Since 1935 the annual Redwood Logging Conference has attempted to encourage yarding of logs with the bark in place, and doing the peeling at a central place at a railhead or at the mill site. In 1929, one large company, having developed a market for the bark, built a specially designed peeling plant on the edge of the mill log pond. Without a sufficient market for all the bark of the region, other operators have felt that centralized peeling would be too costly. However, during the wartime man-power shortages it was difficult to obtain peelers and many logs were sawed only partly peeled. At the same time peeling was undertaken at woods landings and at the mill. There are many advantages to centralized peeling, among which are the following:

1. Woods peeling is hard and dangerous. When centralized, better and safer working conditions can be developed; fewer men are required and better employees can be obtained; the logs can be turned with special equipment and thus peeled all around. (In the woods the under side cannot be reached.)

2. The logs arrive at the mill in better condition. Protection of the bark, while the logs await yarding, prevents surface checking and consequent loss of lumber yield. The bark also absorbs the shocks of yarding and reduces bruising, and prevents grit and gravel becoming imbedded in the wood. Consequently, the logs on being placed on the mill carriage for sawing are cleaner; defects are more readily seen and judged by the sawyer; saw blades receive less wear; and more lumber is obtained. Once the sapwood of redwood was regarded as a defect. At present, boards that are free from knots (select grades) but contain sapwood command only $8 to $10 per 1,000 bd ft less than the top grade—clear, all heart. The sap grades can be saved better when the logs are peeled centrally.

3. Having all the bark centralized, its great volume—varying from 10 to more than 20 cords per acre—becomes conspicuous and is more likely to provoke the ingenuity of inventors toward discovering wider uses for it.

4. Since the bark waste adds materially to the fire hazard in the woods, an added benefit of removing it with the logs is obvious.

Centralized peeling requires a well-planned handling station, probably heavy cranes for spotting and turning the logs and for removing them after peeling, or a permanently assigned bulldozer for the same duties. In addition, a bark-disposal system must be developed. Such a peeling plant may be expensive, but its cost should be justified by the direct and hidden costs of the conventional system of peeling in the woods. Peeling may add $1 per 1,000 ft to the cost of the logs.

YARDING

Present-day yarding methods include cable systems, operated by steam or Diesel donkey engines, and crawler-type tractors. The cable systems include the high lead, the North Bend skyline, and the slack line. Tractors are rapidly displacing the cable systems, except in the winter rainy season and on steep ground, in both of which cases tractors cannot operate to advantage.

Wire-rope systems require that redwood be yarded uphill to reduce breakage and the railroad or truck roads must be built to the spar trees. Railroads reach the landing levels by inclines and laterals. The longest incline was 7,200 ft long. Main truck roads follow up the slopes on grades up to 10 per cent, with branches attaining 20 per cent. When tractors are used, yarding is downhill and the main roads are at lower levels.

Spar trees for cable systems are topped redwoods or Douglas firs. The trees may be topped at between 150 and 200 ft from the ground. Being so high, the spar tree is often guyed at the middle (with buckle guys) as well as at the top. The head, or lead, block is hung from near the top of the spar tree. The tail block is secured to a stump or is hung from a tree on the edge of the distant uncut area. The spar, or head, tree, also called the high pole, is selected before the timber is felled and when the location of the yarding setting has been determined.

The setting[1] is established with reference to the topography and to the density of the stand, the idea being to cover the entire property with as few settings as possible and yet have each one economically placed, and to balance the timber taken to each pole. On long slopes, and for deep corners, it is sometimes not economical to extend a road to the furthermost setting. Consequently the logs are yarded to two separate poles, in effect two separate logging units. The distant logs are yarded by a high-lead or skyline system or by tractors to a cold or a hot deck at the first pole, and from there they are roaded or swung by high lead or slack line to the main landing at the railroad track or truck road. In rare instances there may be two swings, *i.e.*, three donkey-engine units. Now that steam power is being displaced with Diesel power or tractors or both, a promising system is a combination of a Diesel-power swing unit using a high-lead or slack-line rigging, while tractors yard the logs to the tail tree of the swing. This system can be used in wet weather because the tractors do not have to follow the same trail repeatedly and suffer the possibility of becoming mired. Mileage of track or road is thus reduced, offering an additional advantage.

The high-lead and skyline systems usually result in a clear-cut area. Trees too small to fell and utilize are pulled down as the main line is "tight-lined" from one "tail holt" to the next, *i.e.*, in changing from one yarding lane, or road, to the next. These systems cause considerable breakage in the soft but heavy redwood logs. The slack-line system is by far the most destructive, but it is considered a good "log getter." It is debatable, however, that its cost for rigging, cable, and power, and the loss of the merchantable footage involved is compensated for by the more rapid yarding. A slack-line donkey engine, commonly called "slacker," is the heaviest type available, its cylinders being 12 by 17 in. or 13 by 14 in., and it must be equipped with an extra drum for lowering and lifting the skyline. Some operators, not in position to purchase a slacker but desirous of using the slack-line system, use two ordinary high-lead yarding donkeys, one employed to handle the skyline only and the other for handling the main and haulback lines. This arrangement is called a "ukele."

[1] The term "setting" is applied to an entire setup of landing, spar tree, and the area to be yarded to the spar tree. The landing includes the spar tree, the yarding and loading donkeys, and the immediate parts of the track or road.

High-lead logging may reach out 1,000 ft and occasionally 1,500 ft, whereas skylines, particularly the slack line, can reach out 2,000 ft. Occasionally the reach is to 2,500 ft and instances of 3,000 ft for long corners have been reported. The possible reach varies with each setting and depends on the height of the main block on the spar tree and the deflection. Including guys, straw, sky, main, and haulback lines, a slack-line system may require 13,000 ft of wire rope varying in diameter from ½ in. for straw lines to 1⅛ in. for the main line and 2 or 2⅛ in. for the skyline. To look up through the rigging of a slacker pole is reminiscent of the mast of a full-rigged ship. On an ideal setting, a rarity, the slack-line system can "fly" the logs clear of the ground for much of the full length of the road, but generally the ground and stumps are struck regularly, with resultant damage to the logs. Tractors can reach out 1,500 ft economically and occasionally go more than 2,500 ft in "long corners."

The high-lead system requires for its operation eleven men, the North Bend sixteen men, and the slack line seventeen men. Each, of course, requires additional personnel for rig-up crews. A tractor setting including one bulldozer and two tractors requires eleven men and one additional man for each added tractor on long pulls. The high-lead system can yard an area of 40 to 80 acres, the North Bend 300 acres, while the slack line can handle 450 acres, depending on the length of the arc covered. Occasionally more than 45 million feet Humboldt log scale (the Scribner less 30 per cent) are yarded from a single slacker setting. Obviously, at the rate of 125,000 ft per day, a single slacker setting is in use many days and changes are not so frequent as in regions of lighter stands. The high-lead system will yard about 110,000 ft per day.

When a Diesel engine is used as the power unit, the engine is of 400 hp and is equipped with a torque converter. It is generally used in connection with the high-lead system of rigging and the yarding distance may be 1,000 ft—more often less when less power is used. A fair job of selective logging can be done with such an arrangement.

LOADING

Loading in the redwood region is commonly accomplished, at long-life operations, with steam donkey engines operating cables rigged for crotch-line, McLean-boom, or heel-boom systems. Each requires a high pole for carrying the rigging and blocks. One large company operates a self-contained track-type mobile heel-boom loader, or jammer. Small operators favor the crotch-line loading system operated by a tractor equipped with drums or by a light gas engine and drum. In sections of smaller timber, various types of A-frame loaders and guy-line loading systems have been used. Loaders should keep up with the yarders. More likely the loaders have to wait on the yarders when the yarding distance is great. Loading crews usually have three men, a top loader, a groundman, and an engineer. Sometimes an additional man is used.

Light logs are loaded with one or two heavy tongs; heavy logs require wire-rope straps. Some operators use straps for all logs since tongs puncture the logs rather deeply and affect the lumber yield.

TRANSPORTATION

Every manner of log transportation has been used in the redwood region—from oxen and river driving to railroads and trucks. The last river drive occurred on Big River in Mendocino County on June 23, 1936, from an up-river camp above the railhead. Even small creeks were driven in the early days, the head of water being obtained from numerous splash dams. Sinker logs sometimes were several years reaching the mill.

Railroads were used as early as the 1860's and became the standard means of log transportation until very recent years. Trucks were not used in a serious way until 1935, and today they have caused the abandonment of most railroads particularly branch lines. Incidentally, all the original steel and rolling stock for all the logging railroads in the redwood region were delivered by coastwise schooners, there being no common-carrier rail connection at that time. Only five operators still use railroads and then only to a railhead to which logs are delivered by motor trucks. All others depend on motor trucks entirely, from landing to mill dump. One railroad is narrow gauge, the other four are standard gauge. Locomotives vary from 50-ton shays to 90-ton rod engines. One company, with 23 miles of narrow-gauge track, used a 20-ton locomotive until the railroad was abandoned in 1946.

Logging cars once were only of the stub type (two short wooden four-wheel trucks coupled together with a wooden reach 10 ft or more long). At present they are of the standard logging-car type, with two wood or steel bunks, safety chocks, and airbrakes. Loaded with 40-ft logs, a car will carry an average of 6,500 ft Humboldt log scale.

Motor logging trucks in the redwood region are of every type, from small Fords to 18-wheeled jobs designed especially for logs. Opinion varies as to whether a truck fleet should be entirely of the heavy type, of the moderately heavy type, or if they should be mixed light and heavy. Some companies own and operate their own fleets of trucks, others contract the hauling to trucking contractors who may use one or a few trucks or a fleet. Smaller trucks are driven by gasoline engines and the larger by Diesel engines. Trucks operating on public highways are limited as to load and to an over-all 8-ft load width. Loads are about 4,200 ft Humboldt log scale, on standard 8-ft-bunk trucks.

Loading crews must exercise ingenuity, care, and planning to get all trucks loaded safely and to capacity. Nevertheless it frequently happens that a single log is not large enough to make a pay load and too large to permit smaller logs to be loaded with it. Highway regulations penalize unsafe loads and loads above the legal weight limit, though within the truck's capacity; consequently the loading crews and particularly the driver must watch the loading. It is the driver who gets the traffic-violation ticket, but the owner pays.

Truck roads are built by the operators' own crews using all the usual equipment involved in road building—bulldozers, power shovels, carryalls, blasting equipment, gravel trucks, and the like. The roads are engineered for shortest length and to keep the grades low. Typical truck-road grades are under 10 per cent with the load and 4 per cent with an unfavorable load. Spurs may go up to 20 per cent favorable. Road width is generally 14 ft on a 20-ft base before the ballast is added. For all-year use the bed is rocked with river gravel or native crushed rock to a depth of 20 to 24 in. Roads used only in the dry season are unpaved.

Maintenance of roads, particularly drainage, is given close attention. For keeping the surface intact, power graders and sprinklers are kept in constant service.

LABOR

Originally redwood-logging labor was recruited from New Hampshire, Maine, New Brunswick, Newfoundland, and eastern Canada. Since the turn of the century, Scandinavians, Finns, and Italians were added in increasing numbers. During the Second World War a considerable number of Oklahomans and Texans were employed. In general, labor is quite steady and the bulk of it is permanently established. Until recent decades even common labor was not migratory. Labor troubles were rare until about 1935, when, upon outside agitation, the difficulties culminated in an unsuccessful strike in 1946. Considering the year-round employment, the wages are in line with other logging regions and are above the wages paid in other lines of work in the immediate region.

Logging camps built directly on the logging area, once common and numerous, are now the exception and provide for but a small part of the entire crew, usually only the single men and floaters. The camps that remain are permanent and located near a highway. Since automobiles became common and logging truck roads made it possible to reach the job in a private car, many employees live in nearby towns and drive to work. On the larger operations, where main-line railroads are still used to a railhead, the company vehicles provide transport from the main camp. Cookhouses are operated at the remaining camps but men living at home carry their lunches. Existing camps are quite modern, comfortable, and neat, and have all modern conveniences. The food is excellent in quality and extravagant in quantity.

Where no general camp exists, there are usually a few houses for watchmen, timekeepers, mechanics, and others who prefer to live closer to the job or for those who are needed for property protection. These provide their own eating facilities. The designation "camp" is used even if there are no cookhouses or general living quarters. Actually it is only a headquarters for the logging organization and includes an office, a filing shed, a blacksmith shop, repair shops for trucks, tractors, and other equipment, a storehouse, and watchman's cabins.

ORGANIZATION

Redwood-logging operations are on a year-round basis and there is little time lost except during excessively wet periods. There are the traditional Fourth of July and Christmas layoffs for from 1 to 2 weeks for repairs.

On large operations, all logging and woods-planning work is under a logging manager or logging superintendent. Under him are the following departments, each under its own chief or foreman: *engineering*—charged with surveying and mapping; *construction*—charged with building camps, roads, bridges, etc.; *logging*—charged with the felling, bucking, peeling, yarding, and loading operations and rigging up spar trees and preparing landings; *transportation*—charged with transporting the logs to the mill; *safety*—charged with providing safety measures, seeing to it that they are used, and investigating accidents; *maintenance and repair*—charged with maintaining roads and equipment; and *scaling*—charged with scaling behind the fallers and buckers and on the cars or trucks. Detailed organizations differ, but generally the logging superintendent has a camp boss, unless he performs this function himself on the smaller operations. On large operations the superintendent is at the main headquarters at the mill and the camp boss directs all activities in the woods. Under the camp boss, on large operations, will be a side boss—one for each setting; a chopping boss, who looks after the felling, bucking, peeling, and the pieceworkers' scaling; a rig-up-crew foreman for rigging spar and tail trees and preparing landings; hooktenders are under the side boss.

On operations producing 30 to 35 million feet of logs per year the logging superintendent is directly on the ground and supervises all the departments through a side boss, chopping boss, hooktender, construction boss, repair-shop boss, and a maintenance boss.

Logs are transported directly to the mill dump but on small operations, likely to be tied up in wet weather in the woods, logs are cold-decked at the mill, with an occasional reserve cold deck close to a paved road in the woods. Cold decking at the mill is a development of the past decade and became a necessity under higher costs to assure steady mill operations. Storing too large a supply of logs in salt water is unsafe because of marine-borer depredations. One large salt-water operator in 1947 developed a 70-acre fresh-water reserve storage pond at the woods headquarters.

NATURE OF CUTTING OPERATIONS

In the days of oxen and very light power logging, many trees under merchantable size were left standing. On such cutover lands there is now an excellent stand of young timber, predominantly redwood, originating from stump sprouts and natural seedlings.

When power and speed were increased, the land was logged quite clean. Undersized trees were pulled down and left with the logging debris. The resulting cutover land quickly became covered with shrubby species and was dotted with redwood sprout clumps. This kind of logging obtained for about 30 years—longer in some localities. In the 1920's a comprehensive artificial planting program was undertaken and nursery-grown seedlings were planted on the cutover land. This program continued for about 8 years, but the per cent of survival was unexpectedly low and it came to an abrupt close in early 1930 when the business depression also severely curtailed regular logging.

In 1934, efforts toward leaving seed trees were begun and in the following year a start was made on selective logging.

As tractor yarding increased after 1935, the amount of selective logging increased correspondingly. At the present time the area logged clean is probably under 15 per cent of the total and will continue to decrease. It may not reach zero because of the necessity of using cable systems of logging in the wet season and at all times on very steep slopes.

Redwood reproduces readily from stump sprouts. However, since the parent stumps generally number under fifty per acre and since all stumps do not sprout, the resultant sprout reproduction is too open to provide for a fully stocked stand.

To obtain natural reproduction, seed trees must be left in such abundance that, in addition to seed, they will also provide enough shade and protection against strong, dry winds to prevent desiccation of the top few inches of soil. As will be pointed out later, the soil is quite bare after logging and burning; it is generally a heavy soil susceptible to drying to brick hardness during the 6-month dry season. Redwood seedlings are quite sensitive to soil desiccation, particularly since they are unable to sink their roots quickly enough below the top 4 in. of dense soil before it dries out. Douglas fir does much better than redwood in becoming established. If a redwood seedling survives its first two dry seasons, it can be considered to be established.

A California law passed in 1943 requires that no trees under 18 in. in diameter at a point 6 in. above the ground be cut. The State Forest Practice Act passed in 1945 provides that seed trees of any conifer species, and undamaged, be left standing at the rate of four trees per acre, based on a 5-acre count. These seed trees must be 24 in. dbh or more outside the bark. In actual practice, the operators, with a few small-unit exceptions, are exceeding the state requirements, leaving often six to ten or more trees per acre and all well above 24 in. dbh. One company, operating in very large timber, has a minimum-diameter cutting limit of 36 in. Other operators in the northern belt follow a similar policy but their diameter limit is flexible above 24 in. Pole-size trees are generally difficult to save when they occur in dense stands. Operators in the southern half of the belt, having smaller timber, leave more than the required four 24-in. trees, but not so large an average diameter. In addition they are able to preserve more of the trees under 24 in. The volume left standing varies from 10,000 to 40,000 ft, depending upon the original density of the stand and the number of submarginal trees.

At the present time, selective logging is purely on an economic basis and silvicultural requirements are not considered as to the specific trees left uncut. There being no tree-killing insect or disease pests, no sanitation cuttings are necessary. Malformed trees above the merchantable diameter limit are not always felled. Nor is it necessary or desirable that they be cut. Such trees generally make good seed trees and, by virtue of improved growth following logging, may develop into merchantability for the second cut. In addition, they add to the windbreak effect of the residual stand.

There is a general realization among long-life companies that the value of cutover land is enhanced by the presence of a residual stand. Accordingly they are interested in having their crews avoid the felling of or damage to trees under the merchantable diameters. Residual trees, in spite of several hundred years of suppression, have the capacity of recovering promptly and making accelerated growth. Trees, 20 to 24 in. in diameter, if they are not injured during logging or slash burning, are capable of doubling their diameter in 20 years. Several companies have under consideration setting up their operations on a continuous production basis and are assembling the necessary basic information.

SLASH DISPOSAL

Slash disposal is one of the most difficult problems of the redwood logger who is interested in not only removing a serious hazard but also in preventing damage to the residual stand and excessive deterioration of the soil. The volume of slash is extraordinarily large because of the large crowns of the merchantable trees, the underbrush, broken and shattered logs, rotten logs, and small trees crushed by the larger ones. During a dry period, winter or summer, a fire once started and not promptly extinguished may get out of hand. September and October are months of strong, dry, hot winds and make the fire hazard exceptionally great. Slash must therefore be disposed of, and, until a better way is found, it must be burned. At one time, in the days of general clear-cutting, some operators set slash fires after felling and peeling but before bucking; others set them shortly after felling and before peeling; still others logged "in the slash" and burned after logging was completed.

Now that seed trees are left and selective logging is well under way, slash burning becomes a real problem. Fortunately tractor roads break up logging areas into so many small units that broadcast burning need no longer be practiced and smaller fires can be set unit by unit, beginning with the uppermost ones. Fire protection in the redwoods, once receiving only casual attention, has become an important department of the logging operation.

SOUTHERN HARDWOODS

BY ED R. LINN

Oklahoma Agricultural and Mechanical College

The southern hardwood area is contained mainly in the states of Maryland, West Virginia, Virginia, Kentucky, Tennessee, North Carolina, South Carolina, Georgia, Florida, Alabama, Mississippi, Louisiana, Arkansas, Missouri, east Texas, and east Oklahoma.

Within these states there are 226 million acres of commercial forest land, or 54 per cent of the total land area. Hardwoods predominate in the northerly part of this area and along the streams of the Deep South. Softwoods occupy the higher lands of the Deep South either in pure stands or in mixture with hardwoods. Along the coast where drainage is poor and in inland areas with very poor drainage (swamps), cypress and associated hardwood species predominate.

Logging or harvesting practices in southern hardwoods vary widely over this vast area. The methods used are influenced by several factors, some of the more important of which are topography, elevation, latitude, tree species, the nature of industry in the locality, markets for the products, kind of ownership, and the cutover condition of the forest as a whole.

THE FORESTS OF THE HARDWOOD REGIONS

Harvesting practices in Southern hardwoods can best be treated by considering each of the several logical divisions of this large area separately. These hardwood regions are the southern Appalachian; the Southern bottomland; the south-central, or Cumberland, plateau, and the cypress-tupelo swamp.

These several regions merge into each other so that logs and other harvested products from two or more regions are often processed at some of the larger mills in the South.

Southern Appalachian. The southern Appalachian hardwood region comprises the forest lands of western Maryland, West Virginia, western Virginia, the western parts of North and South Carolina, a small part of northern Georgia, eastern Tennessee, and eastern Kentucky. The region includes the high mountains in the Allegheny and Cumberland ranges on the west and the Blue Ridge and Smoky Mountains on the east. Elevations range from 1,000 ft to over 5,000 ft. Steep slopes and narrow valleys are characteristic of the topography of the region.

The principal tree species considered commercially important are

Ash	Hickory (spp)	Persimmon
Basswood	Locust	Poplar (yellow)
Birch (black-yellow)	Maple (hard and soft)	Sycamore
Buckeye	Oaks	Walnut
Butternut	black	Fir (balsam)
Cedar	chestnut	Hemlock
Cherry	pin	Pines
Cucumber	post	pitch
Dogwood	southern red	shortleaf
Elm	northern red	Virginia
Gum	scarlet	white
Hackberry	shingle	Spruce (red)
	water	
	white	

The timber stand of the southern Appalachian region is predominantly second-growth cutover northern hardwoods with some northern spruce, fir, hemlock, and white pine on the higher elevations. There is very little virgin, or old-growth, timber left in this region. Kentucky at the beginning of the year 1945[1] had 145,000 acres of old-growth, or virgin, timber and West Virginia had 84,000 acres of old growth. North and South Carolina have no virgin timber left.

Most of the cutting, or harvesting, in Kentucky and West Virginia occurs on the 8.4 million acres of second growth and old growth. The remaining commercial forest land in these two states, amounting to 13 million acres, is classed as pole-sized timber, which is harvested in some localities for a variety of products or is being allowed to grow.

The timber stands in the remainder of the southern Appalachian area are in comparable conditions to the aforementioned states. In so far as hardwoods are concerned, about half the area is in second-growth trees of saw-timber size and half the forest land is occupied by pole-sized trees, seedlings, and saplings, with some poorly stocked areas.

[1] "Basic Forest Statistics for the United States," U.S. Forest Service, 1945.

Stands per acre in the saw-timber areas run from 2,500 to 5,000 ft per acre, and in some locations there are acres containing 10,000 to 12,000, and up to 25,000 and more feet.

Southern Bottomland. The Southern bottomland hardwood region occupies the stream banks and flood plains of all the river systems from Virginia to Texas. Along the Atlantic and Gulf coasts it merges into the cypress-tupelo swamplands. Further inland the abundant presence of red gum and the several species of white and red oak broadly characterize the region. Elevations range from almost sea level to several hundred feet, depending on the stream and its flood-plain conditions. The Mississippi River from southern Illinois south and its tributaries, for many miles of their length, flow through the largest bottomland hardwood area. However, all the important river systems in the South have bottomland hardwoods on their banks and flood plains. The over-all topography is flat, and consequently it is poorly drained. The millions of acres of the delta lands in Mississippi and similar overflow lands in Arkansas and Louisiana, now some of the finest agricultural land in the world, once were occupied by a magnificent hardwood forest. Today the forests are largely confined to the most poorly drained areas of this region.

The principal species considered commercially important are

Ash
Basswood
Beech
Cherry
Cottonwood
Elm (soft and rock)
Gum (red and sap)
  black gum
Hackberry
Hickory (spp)
Holly
Locust
Magnolia (evergreen)
Maple (soft and hard)

White Oak Group
  white
  post
  overcup
  cow
  bur
  chestnut
  swamp white
Red Oak Group
  southern red
  cherrybark
  shumard
  black
  water
  willow
  laurel
  red
  pin
  scarlet
  blackjack
  nuttall

Pecan (sweet and bitter)
Persimmon
Poplar (yellow)
Sassafras
Sycamore
Tupelo
Walnut
Willow
Cypress (red and yellow)
Cedar (white and red)

The timber stand of the Southern bottomland hardwood region is mostly uneven-aged cutover second growth. The predominant species are red gum and the red and white oaks. Most of the virgin timber has been cut. Until a few years ago the Singer tract in northeast Louisiana was one of the few remaining large bodies of virgin hardwoods. Another large tract of uncut old-growth bottomland hardwood is located along the Congaree River in South Carolina. In the late 1930's the Forest Service acquired some 13,000 acres of southern hardwoods in the southern part of the delta lands of Mississippi that carried an average stand of 9,000 ft per acre. Both areas have been cut into in recent years. Present-day stands per acre range from a 1,000 ft in the cutover areas to average stands over some areas of 4,000 to 9,000 ft. Individual acres and small tracts carry stands up to 30,000 ft and more per acre.

South Central. The south-central hardwood region has mixed characteristics of both the bottomland and Appalachian regions. The plateau lands and the steep valleys have stands similar to those of the Appalachian region except that red cedar and poor-quality oak grow on the thin soils of some of the upland and on southern slopes. Where the streams have flood plains, the forests are quite similar to those of the bottomland region. Most of the commercial hardwood species found in the two adjoining regions, with a few exceptions, are found in the south-central region. The elevation of the region varies from a few hundred feet up to several thousand feet. The forests have been cut over several times in the last 150 years and now consist largely of uneven-aged second-growth stands, with the bulk of the saw-timber volume of small diameters. Stands per acre of several hundred feet to 4,000 and 5,000 ft are logged.

Ozark. The Ozark hardwood region lying in northwest Arkansas, eastern Oklahoma, and southern Missouri at elevations of 500 to 2,000 ft has a hardwood forest with northern characteristics. The north slopes and deep valleys contain hard maple and oaks. The fine white oak of this area has for years been an important source for staves used in tight cooperage. Extensive railroad-tie operations in the second growth have left an uneven-aged second-growth forest of rather small-sized trees. Little or no virgin timber is left in the region.

Cypress-Tupelo Swamp. Cypress and tupelo forests occupy the low swampy areas from the Dismal Swamp in Virginia to the marshlands of southwest Louisiana and southeast Texas. These tidewater cypress areas, practically at sea level, are found scattered along the coast and may be found as far as 100 miles inland.

Early logging operations took most of the tidewater cypress and only a few large companies are now operating in the deep swamps. Tupelo is logged by both the cypress companies and by the hardwood concerns in or adjacent to this region.

Cypress occurs inland far beyond the 100-mile distance arbitrarily set off to mark the tidewater cypress area, and is found as far north as southern Illinois. Stands of cypress run from 6,000 to 10,000 ft per acre and as high as 30,000 to 40,000 ft per acre in restricted areas. Tidewater cypress logging is in virgin timber and the trees are often over 1,000 years of age. Tupelo logging frequently is carried on over areas where the cypress has been removed some years ago. The first cutting of these areas passed over the tupelo, which at that time was unmerchantable.

LABOR

Woods labor in the Southern hardwood areas varies, in race and color, according to geographical location. In the southern Appalachian region, the south-central region and the Ozark region, white native-born local labor is employed. In the bottomland hardwood region, common labor on the sizable logging jobs is largely negro. Smaller operations may have white crews made up of local farm people. The more skilled jobs such as truck drivers usually are filled by white men, and the foremen or others in charge are generally white. The cypress-tupelo logging crews include both white and negro labor. The Cajun, or native of French descent, works in the swamp getting out cypress, tupelo, and other hardwoods growing in southwest Louisiana.

The large and semipermanent type of logging camps are becoming less common in Southern hardwood areas as the larger lumber companies cut out their own timber and turn to the buying of logs or small tracts of timber to supply their mills. Many logging crews consist of workers who live on nearby farms or at the mill site and either provide their own transportation to the harvesting operations or are transported to the job in company-owned trucks. The camps in use are commonly of the temporary type made of rough lumber and tar paper.

FELLING AND BUCKING

The two-man crosscut-saw crew is commonly used in felling and bucking hardwood logs throughout the Southern hardwood region. Whenever possible the logs are cut 16 ft in length, with an allowance of 4 in. for trim. Shorter logs of 14, 12, and 10 ft are cut when these log lengths yield a better outturn of boards in sawing or when the given tree will not make all 16-ft logs or when defects occur that prevent making 16 ft logs. On the more conservative logging operations, efforts are made to leave trees smaller than 14 in. dbh, although ash is cut to smaller sizes.

The chain saw is gaining in popularity in the region. As an example of what is here at the present time and also what the future holds, the TVA found that in hardwood timber in east Tennessee, a chain saw, with a four-man crew, could cut from 5,000 to 21,000 ft of logs per day, depending upon the size of timber. The minimum was for trees 9 in. dbh, while the maximum was for trees 32 in. dbh. Within this diameter range, the volume cut per 8-hr day varied directly with diameter. Where the average tree was 15 in., the daily output was approximately 12,000 ft of logs. It is probable that very little hardwood timber will be operated when the average dbh is less than 14 in. because of the little likelihood of profit from trees smaller than this.

This same investigation showed that a two-man crosscut-saw crew would cut from 1,250 to 5,500 ft in the same timber. This means that a two-man crosscut-saw crew has approximately 25 per cent the output of a four-man chain-saw crew. On this basis, four men with a chain saw would cut twice as much timber as the same number of men with crosscut saws.[1]

BUNCHING AND SKIDDING

Skidding is done by horses and mules, the former being favored in the mountain regions and mules in the bottomland areas. On very steep slopes logs are sometimes peeled on the running side and skidded by hand to less steep gradients and then skidded by animals. In the bottomlands, when the ground is dry enough to permit, trucks go directly to the logs and no skidding is necessary. Some of the more highly organized operations make use of crawler-type tractors equipped with a winch and a fair-lead arch to skid and bunch the logs. There are several types of equipment mounted on trucks with which a three-man crew can skid, bunch, and load from 10,000 to 25,000 ft of logs per day, depending on the type of timber, at a cost of $1 to $2 per 1,000 ft. In the deep cypress swamps an overhead, or high-lead, cable system with rehaul is used to skid the logs to railroad tracks or to barges.

LOADING

On many small and some large logging jobs in areas where there is sufficient slope, gravity loading is done from skidways. When the ground is comparatively flat, log trucks are loaded by the crosshaul method using a team or tractor. A few hardwood companies operate logging railroads and use specialized power loaders to put the logs on the cars.

The several skidder-loaders mentioned before are used on the larger operations. They can load three or four trucks per hour with hardwood logs, averaging 800 to 1,500 ft per load. Another method of loading is the A-frame loader, often built at a local machine shop. The equipment consists of a truck with a power take-off, a loading boom, a small winch, block, and wire rope. The A-frame loader is not as readily portable as the ready-made units, nor can it load as much per hour.

TRANSPORTATION

The 1½-ton truck, usually dual-wheeled, is used extensively in hauling hardwood logs to sawmills in the Southern hardwood area. Larger trucks and the trailer-type truck with tandem-dual wheels on both truck and trailer are also used. Some of the trucks are equipped with self loaders.

[1] Carpenter, R. D., "Organizing for Efficient Logging," Southern Forest Experiment Station.

Log- and stumpage-buying mills have hauled logs as far as 75 miles, where the main part of the haul was on a good paved road. Costs per thousand for hauling vary greatly. The sawmill operator tries to average his hauling costs so that his cost of logs at the mill will enable him to make a profit. The species and quality of the logs influence the distance and the transportation costs that can be borne.

The transportation of hardwood logs by logging railroads is giving way to trucks. Some companies maintain a main-line railroad for hauling the logs from the woods to the mills, but most concerns have discarded railroads in favor of trucks for hauling logs. The common-carrier railroad systems have special log freight-rate classifications and millions of feet of hardwood logs are shipped to mills in gondola or coal cars when such rates are not too high.

Hardwood logs cut along the Mississippi and other large rivers are often transported to the mills by barges. Some of the light species are rafted. One company floats hardwood logs to the mill by placing the heavy logs between metal pontoons. The logs are stored at the mill by releasing the pontoons and allowing the logs to sink to the bottom of a shallow bay.

ORGANIZATION OF OPERATIONS

Logging operations in the mountainous and plateau section of the Southern hardwood region continue the year round. Extremely bad weather, snow or excessive rain, and local high water slows down or halts operations at times.

In the bottomland region logging starts in the spring as soon as the backwater is out of the lowlands and the ground is sufficiently dry to permit wheeled vehicles, animals, and tractors to operate. Logging continues until the fall rains again make logging impossible. Hardwood mills endeavor to bank sufficient logs at the mill to enable them to saw through the winter months. Logging crews are usually small, since the typical large mill consists of a single- or double-band mill, a unit having a rated capacity of about 3,000 bd ft per hr. One large operation in West Virginia has a triple-band mill, and it is classed as the largest hardwood sawmill in the world. The hardwood logging boss on the larger operations must know hardwood sawmilling, as the correct selection of log lengths increases the profitability of the entire operation. He watches the felling and bucking of logs closely, especially when the labor is not skilled. On small operations one or several two-man crews carry on with little supervision.

NATURE OF CUTTING OPERATIONS

Southern hardwood logging in the past consisted of cutting the high-quality and accessible trees for saw logs, veneer logs, and stave bolts, leaving many of the inferior-quality and low-value species standing. In many places this cutting was followed by tie-cutting, slack-cooperage, chemical-wood, and, in some instances, pulpwood and mine-prop operations.

Millions of acres of fine hardwood forest land in the bottomland area have been logged and then cleared to make farms.

Operations today on land that is to be kept in forest usually harvest the better formed trees or practically all species down to a 14-in. stump diameter and leave the poorly formed and defective trees. Log sizes have been decreasing for years. Before the Second World War a 14-in. diameter for most hardwood trees, except ash, which was cut to 9 or 10 in., was considered the minimum profitable size. The demand for hardwoods during and after the Second World War has lowered the economic size and the quality of trees and logs. At this time any hardwood species that will saw out boards is merchantable if the grade outturn justifies putting the log through the saw.

A number of companies operating in hardwoods are keenly aware of the need for better forestry in Southern hardwoods. Many such companies employ foresters to work on their own lands and some are working with farmers and other small wood-lot owners for better cutting practices when sales of logs or stumpage are made. The Appalachian Hardwood Manufacturers Inc. have a Division of Forestry and have issued suggested cutting-practice rules as a guide for their members and others. The late war and the continuing demand for hardwood products, lumber, ties, mine props, veneers, flooring, acid wood, staves, boxes, and furniture made a heavy drain on the Southern hardwood forests. This drain, however, enabled many operators to remove poorly formed trees and less desirable species, so that the resultant condition of many acres indicates a better future forest it good forest practices continue to expand in Southern hardwood forests.

THE SOUTHERN PINE REGION

BY W. C. HAMMERLE

Forester, Southern Pine Association

THE FOREST

The Southern pine region comprises 30 per cent of the total forest area of the continental United States. It includes the states of Virginia, North Carolina, South Carolina, Georgia, Florida, Alabama, Tennessee, Mississippi, Louisiana, Arkansas, and east Texas and southeast Oklahoma. With only 2 per cent of the forest area classed as noncommercial, the region contains 183,266,000 acres of commercially valuable forest land. This is the largest commercial forest area of any of our forest regions.

The Reappraisal Survey (Jan. 1, 1945) of the U.S. Forest Service classifies the commercial forest area according to character of growth as follows:

Forest Condition	Acres	Per cent
Sawtimber	98,377,000	53.7
Pole timber	29,322,000	16.0
Seedling and sapling	22,873,000	12.5
Poorly stocked and denuded	32,694,000	17.8

Of the saw-timber area, less than one million acres are classed as virgin growth. The remaining area is composed of second-, third-, fourth-, and even fifth-growth timber. This condition, coupled with fast-growing species, favorable climate, and long growing seasons, is responsible for the South having the highest saw-timber growth rate of all forest regions. The annual growth rate averages 6.7 per cent for all softwood saw timber; 4.9 per cent for all hardwoods; and a combined total average of 5.9 per cent.

Pine- and pine-hardwood-forest types comprise approximately 60 per cent (110 million acres) of the South's total commercial forest area. Estimates indicate that for these two major types approximately 50 per cent is classified as saw-timber area, 32 per cent pole-timber area, and 18 per cent is in the two remaining forest survey classifications. The southern pine saw-timber stand is estimated as 182,372 million feet. It accounts for over 94 per cent of the total softwood saw-timber volume and 54 per cent of the combined total of softwood and hardwood saw timber. Only about 5 per cent of this pine saw-timber volume is found on other than "saw-timber areas." According to the Forest Survey, the southern pine saw-timber stand averages about 3,300 ft per acre for the pine and pine-hardwood types. Composition of the total pine saw-timber stand by species is approximately: loblolly pine—48 per cent; shortleaf pine—28 per cent; longleaf pine—11 per cent; slash pine—7 per cent; all other pine—6 per cent.

LABOR

Southern lumber—southern pine, hardwood, and cypress, which, as to operations, cannot well be separated one from the other—provides jobs for approximately a quarter million people in average years. In the prosperous years of 1946 and 1947, employment was placed at 300,000.

The labor employed in harvesting southern pine timber never lived in logging camps to the same extent as workers in other lumbering sections of the country. With the transition from large to small mills and the almost universal adoption of truck logging, the logging camp has almost entirely passed out of the picture. Woods workers in the South live mainly, as do those engaged in agriculture, in small towns or sparsely settled rural regions.

An ample supply of native labor has always been available. The workers have been stable and independent, many preferring to retain some association with agriculture, from which most of them came. Sawmill and woods labor revert frequently to farming; part-time employment by the lumber and logging industry is common.

The lumber industry of the South employs more negroes than any other industry, with the exception, of course, of agriculture. While no recent statistics are available, it is estimated that more than three-quarters of the work force in the Southern lumber industry is negro. Negroes outnumber the whites in both the mill and the woods, but there are proportionately fewer colored workers in the forests. A substantially greater percentage of whites are in the skilled occupations in the plants and the woods.

Mechanization in the Southern lumber industry has not developed to the full extent possible because of the existence of an ample supply of common labor. As a consequence, in spite of unit labor costs equal to other lumbering regions, hourly rates of pay in Southern lumber are not as high as in the West or in the Lake states; unskilled labor is predominant, and man-hour production therefore is less.

Men harvesting timber in the South face a threat to steady employment not present in timber-producing sections where logging is seasonal. The Southern states have the heaviest rainfall in the country and logging ceases at frequent intervals throughout the year when the ground becomes too soft or the bottoms flooded. The break in the log supply in turn causes many of the small sawmills to shut down. To some degree, the wage-hour law has limited the earnings of woods workers through its penalty of time and a half for work in excess of a 40 hr week.

Wage rates disclose a rising trend through the years, marked by an abrupt ascent from the beginning of the Second World War. The minimum wage established for the Southern lumber industry by its code of fair competition under the National Industrial Recovery Act (August, 1933, to May, 1935) was 24 cents an hour, which was an increase over the great-depression scale and comparable to the amount generally paid in the middle and late 1920's. A slow, steady rise was noted in the late 1930's, gaining momentum in the early 1940's. At the close of 1947, the rate for common labor was 60 cents. The work week at most sawmills is 40 to 50 hours, and the woods workers usually get the opportunity to earn an equal number of hours' pay. Many of the large companies provide low-rent housing, garden plots, and free fuel, water, and light for employees.

Although most woods labor is paid an hourly wage, a common practice is the remuneration of cutters or sawyers by a specified piece rate per 1,000 ft of logs cut.

It is not possible to estimate reliably the number of people hired by logging contractors, as distinguished from sawmill companies, but the figure is substantial.

Labor-union locals were organized at more than a hundred Southern operations at the end of 1947, but few woods workers were numbered in their membership. These employees work in such small groups and live over such a wide area that the difficulties of unionizing them appear insurmountable.

FELLING AND BUCKING

Crew Organization. Men hired for woods work should be strong, rugged, healthy, and capable of handling tools and equipment used in logging. Due to improved roads and present-day transportation facilities, it is seldom necessary to maintain logging camps at southern pine operations; thus logging crews are able to report to and from work daily.

Felling and bucking is usually done by two-man crews using conventional crosscut saws. Crews do their own saw filing. Operations employing six or more cutting crews provide additional men to file the saws. A log scaler, who also blazes the strip lines for each saw crew, completes the felling and bucking detail.

Tools Used. Manually operated crosscut saws are in general use. This method is gradually being supplanted, however, by power-driven chain- and circular-saw equipment. Approximately thirty different types of power-driven apparatus for felling and bucking are available to lumbermen who desire to mechanize sawing operations.

Crews using power-driven chain saws usually comprise four men, but crews of three, five, and six men are also used. The general procedure is to have two men operate the chain saw, with one man close at hand to wedge or pry up on a log where required to relieve a pinch. Additional crew members are kept busy notching, limbing, and occasionally assisting near the saw.

Besides saws, the cutting crews require the following tools and equipment: axes, wedges, mauls, files, hones, log scale sticks, oil, gasoline, and kerosene.

Production per Day. Experienced two-man cutting crews using conventional crosscut saws can produce 5,000 ft in an 8-hr day. Crews of four men using power saws will produce four times the output of a two-man crew using manually operated crosscut saws.

Mechanization of logging operations has been handicapped by the type of labor available—predominantly unskilled common labor. The industry is faced with the job of training this labor in the operation, care, and use of mechanized equipment. Such programs have already been inaugurated.

BUNCHING AND SKIDDING

Cable. Power-line rehaul yarding is an inexpensive method of handling logs in the South. Cable tongs bring in logs from distances up to 600 ft. Diesel and gasoline engines are supplanting steam yarders generally. This method can be readily varied to meet special conditions such as taking logs from swamps, across river bottoms, and up or down hillsides.

Tractors. There remains no question as to the efficiency of track-laying tractors equipped with winches for bunching and skidding logs where the going is rough. This equipment will snake logs out to the bunching ground and then skid them to the skidways with a minimum of effort and expense. Where bad swamps or lakes have to be logged, the winch is used to line-log the timber out of inaccessible places. Long wide tracks keep the tractor on top of soft boggy ground and gripping grousers give the needed pull to make large skid loads possible.

Teams. Skidding by horses and mules is still popular in mountainous country or when timber is small and distances are short, particularly on small operations. Horses and mules can be worked for about five years and the initial investment is comparatively low.

Oxen perform better than other animals in boggy terrain. Their broad hoofs and sturdy legs enable them to negotiate swamps that would be considered impossible with any other kind of animal team.

Equipment. There has been a marked tendency in recent years to mechanize woods operations. Besides cable logging systems, tractors, and animal teams, various types of allied equipment are in general use, such as loading cranes, logging sulkies, wagons, arches, and skid-loaders. Each serves its particular purpose in the varied terrain found among southern pine operations.

Output per Day. Output will vary with different log sizes, distances involved, and varied types of terrain. These three factors considered equal, it is generally conceded a mechanized operation will handle approximately seven times that of a crew using teams. This ratio will vary with any appreciable change in distance. Output increases when larger sized logs are handled.

Daily output for a tractor equipped with a winch, handling average loads of 1,000 ft on hauls up to 600 ft, will run approximately 20,000 to 30,000 ft.

LOADING

Methods. In the woods numerous methods are used to load logs on wagons, trucks, and railroad cars for transportation to the mill. In some instances logs are rolled onto trucks and wagons at skidways where special loading ramps have been erected. The ramps or docks are usually lower at the loading end, and thus gravity is employed to facilitate loading.

Crosshaul loading is usually employed by small operators using either animal teams or small tractors. Logs are usually bunched along roadsides or other accessible places that will permit trucks or other conveyances to be driven up beside them. In crosshaul loading, two poles or stakes on which to roll the logs are placed against the side of the truck at the front and rear, from which the logs will be loaded. A cable or chain drawn by either team or tractor is looped midway around the log, causing it to roll up the poles onto the truck. This operation is repeated until the truck has been fully loaded.

Sky-hook loading is another method commonly used by both small and large operators. A cable equipped with a pair of tongs is passed through a sheave suspended from a nearby tree, and a tractor winch or tractor donkey provides the power for lifting the logs onto trucks or other conveyances.

Various types of mechanical apparatus have been developed to load logs. Gasoline- and Diesel-powered machines have replaced steam loaders in most instances.

Equipment. Track-laying tractors equipped with winches are indispensable to present-day logging. Truck- and tractor-mounted cranes are in general use. Other mechanical loading equipment includes Hystaway loaders, Drott Skid loaders, Timber Tosser, Loggers Dream, and Champion Skidder loaders.

Capacity per Day. Increased production and lowered costs come not from the use of power alone but rather from the planned use of the right type and size of power. Ordinarily it takes 30 to 45 min. to load a truck by crosshauling with mules, whereas tractors equipped with winches using a snatch block in a tree can load upwards of 20,000 ft of logs in an 8-hr day. Tractor cranes or speeder-loaders will load 40,000 to 60,000 ft per day.

TRANSPORTATION

Rail. Railroad logging is a feature that characterized the marked changes in logging methods that took place in the '80's as the principal source of lumber production shifted to the South from the Lake states. Railroad logging has passed its peak of importance in the South, and has rapidly given way to truck logging. As present rail equipment wears out and as new operations are started, it seems unlikely that privately operated logging railroads will be used regardless of the size of mills erected.

Trucks. Practically all mills use trucks for hauling logs, even those operating their own logging railroads. Although truck logging is a relatively recent innovation among some of the larger mills, many operators, large and small, have been using trucks for several years. A wide range of heavy-duty gasoline- and Diesel-powered logging trucks have been developed to take the toughest logging job in stride. There have also been designed and developed improved logging trucks having self-contained loading equipment and other special features.

Specially designed heavy-duty tires have been developed to increase efficiency and reduce hauling costs. Various kinds of specially constructed trailers for log hauling have also been developed. Both truck-tractors and trailers are equipped with air or hydraulic brakes, as well as other safety features.

It is probably safe to assume that logging trucks of the future will be lighter but stronger, with an increased ratio of pay load to gross weight and with a greater ratio of power to gross weight, resulting in not only greater pay loads but also in lessened power-plant labor and wear and consequently less time in the shop, with lower maintenance expense.

Water. Water-borne transportation of saw logs in the South is practically nonexistent. Only in a few widely scattered sections, principally in low bottom land, can facilities be found for transporting logs by water. Barges are usually employed to do the hauling instead of rafting, which was the common practice during the early years of sawmilling in the South.

ORGANIZATION OF OPERATIONS

The organizational structure of logging operations in the southern pine region is as amorphous as can be expected in an industry operated in more than a dozen states by perhaps 10,000 individualistic entrepreneurs, under extremes of climate and in terrain ranging from swamps and river bottoms to hilly country.

Southern pine sawmill and logging operations are so integrated that logging cannot be described without simultaneously portraying the milling part of the business. Many farmers operate tractor mills in their off season to get additional cash income, to provide employment for their hands, and to furnish rough lumber for use on their own farms. Their logging is done with the sawmill crew and consists of two to six men who perform all woods work. Even in the aggregate, however, the production from this type of operation is negligible.

A larger kind of sawmill, either portable or stationary, which, while it operates intermittently, nevertheless cuts lumber approximately 200 days a year. Mills of this character get their logs through purchase or through buying standing timber and having it felled and brought to the mill either by their own crews or by logging contractors. Among the portable sawmills of this variety, employing from six to twenty men at the plant, practically all buy logs or stumpage and very few have standing timber sufficient to supply their operation for more than a few months.

The portable mill generally has no log yard and only limited facilities for log storage, usually enough for 3 or 4 days. It is dependent upon a steady flow of logs from the woods direct to the mill for continued operation.

Stationary operations have a greater tendency toward the acquisition of land and stumpage, although most of them also buy logs and standing timber for immediate cutting. A survey by the Bureau of Labor Statistics in the fall of 1946 revealed that approximately half the Southern pine mills have no integrated logging operations, which implies that they buy logs or timber for prompt felling by contractors. The type of mills just described contribute the bulk of the southern pine lumber production.

During the transition of the Southern pine lumber industry from large to small mills, "concentration yards" began to appear in substantial numbers in the 1920's. The concentration yard is an operation with kiln or air-drying facilities and planing and finishing machinery, which takes rough green lumber from feeder mills and prepares it for market. The concentration plant usually does not have a timber supply of any size of its own, but relies on feeder mills within a radius of perhaps 50 miles to keep it supplied. While the concentration plant may own a smattering of timber acreage, it generally is not sufficient to keep the plant going for more than a few weeks or months.

The integrated type of operation, which includes everything from growing the trees to placing the finished lumber in the car, is usually a medium-sized or large firm, which owns thousands of acres of forest, and when logging on its own lands, does so with its own employees or through independent contractors.

The logging crews of independent contractors resemble those composed of the employees of integrated operations. There are specialists for each task: swampers, teams of sawyers or cutters, "cat" drivers, loggers, loader operators, tong hookers, teamsters, truck drivers and, in the case of the few mills still logging by rail, section hands and rail crews.

The degree of supervision required in the woods is not so great as in the early days of the century, when stands were dense, woods work was concentrated, and power skidders and loaders were prominent. Since woods crews nowadays are small in most cases, the principal supervision is by a working foreman. Even in large crews, supervision today is also remote, particularly where "flatheads" (sawyers or cutters or fallers) are paid by piece rate, *i.e.*, a specified amount per 1,000 ft cut. A woods scaler measures their day's work.

According to a study by the Southern Pine Association in 1946, transportation to the working site in the woods is provided by approximately half the employers, with the exception of the state of Texas, where exceptionally few woods workers go to the job site in their own conveyances. Approximately half of the companies which log through independent contractors also do some logging themselves with their own equipment and crews.

The woods population in logging is predominantly sawyers (fallers), who outnumber the truck drivers by three to two and swampers, teamsters, and tong hookers by nearly three to one. At the end of 1947, the ratio was more than ten handsaw sawyers or fallers per man in similar jobs with power saws. Tractor operators, loader operators, section hands, and handsaw filers were also numerous.

Logging in the South is beset by a factor absent in other sections of the country where the work is seasonal. Rainfall is the nemesis of logging in the South. A two- or three-day sustained rainfall, which is likely to occur any time during the year, will force a shutdown of logging perhaps for a day, maybe for a week or 10 days. It is not a totally seasonal incident in the South, as it is in the Northwest and Northeast. The U.S. Department of Agriculture statistics show that the states in the southern pine region practically all have more than 50 in. of rain a year. Rain makes the woods floor soft and therefore difficult of ingress and egress; the slick clay and mud roads frequently do not permit traction for logging trucks. Low lands flood quickly and most of the integrated operations and the larger logging contractors have cats and specialized trucks to do bad-weather logging.

NATURE OF CUTTING OPERATIONS

Forest management is still young in the South, although remarkable progress has been made, particularly in the last decade. The development of precise cutting practices under planned management is in process. Partial cutting, seed tree, and so-called clean cutting have all yielded both satisfactory and unsatisfactory results in various sections of the region. These variations, occurring within the same species and forest type, appear to be primarily influenced by stand composition and condition, and soil type. Major management efforts at present are centered on practices—primarily improvement cutting—designed to develop more fully stocked stands of the principal pine and hardwood species.

Present demand for lumber and existing market conditions have caused highly competitive bidding for stumpage, both within the lumber industry and with the pulp and paper industry. As a result, timber is purchased down to 10 in. and even 8 in. in diameter. Timber of this small size is well below the minimum economic size for the lumber industry under normal operating conditions. The predominance of small transitory mills tends to maintain stumpage purchases to low diameter limits under abnormal conditions. Operators of permanent-type mills continually strive to discourage the sale and purchase of small timber.

Low-diameter-limit cutting is by far the most prevalent practice in effect. It occurs mostly on lands of small private owners, from which the bulk of the lumber produced is cut. Although termed clean-cutting, or clear-cutting, in actual practice the results are not as drastic on the majority of the area cut over as the term implies. Fortunately, the character of the stands is such that sufficient reproduction and young growth, below the cutting diameter, remain to develop another crop of timber—provided, of course, that the ever-present problem of fire is controlled. There is, moreover, a definite trend to leave seed trees of larger sizes to assure reproduction.

Each year more and more saw timber is being harvested on a partial-cutting basis. Lumber companies, public forestry agencies and consulting foresters are rapidly expanding their services in marking timber prior to harvesting for private landowners. The greatest progress in this respect has been made on forest industry lands and on those of larger nonoperating owners with holdings of 5,000 acres and over. This development was shown to some extent in the cutting-practice data secured by the U.S. Forest Service in the 1944 reappraisal. This study indicated that cutting practices on 77 per cent of the lands in ownerships of over 50,000 acres and on 37 per cent of those in ownerships of 5,000 to 50,000 acres were rated as meeting or exceeding the standards of the survey. These standards were the "Suggested Rules of Forest Practice, The Southern Region," prepared in 1944 by the U.S. Forest Service. According to the Service, they were "designed to attain standards of practice higher than the minimum measures necessary to keep forest lands productive." Obviously, the percentage of private lands cut under practices meeting minimum forestry requirements would be much larger.

No discussion of cutting practices in the South would be complete without reference to the constructive efforts in effect to secure the adoption of sound harvesting practices on private lands. The lumber and other forest industries have undertaken concerted action with public and private forestry agencies to educate and sell the vast multitude of private landowners on the advisability of growing continuous crops of timber. Tree Farm and Keep Green programs, primarily financed by the industry, provide cooperative, educational means of reaching both large and small landowners. The most intensive forest-management selling campaign ever attempted was inaugurated in Alabama in the fall of 1947. Through it practically every private forest landowner in Alabama will be reached and opportunity afforded him to secure technical service and advice in managing his woodlands. State and Federal agencies, the forest industries, and private foresters are all cooperating in this intensive program, which will gradually be expanded to include the other states.

Technical services in selecting timber for cutting and in managing woodlands are available, in varying degree, in all the Southern states. Many lumber companies provide marking service for landowners to encourage proper cutting practices. The development of such technical services is rapidly being expanded and will materially affect harvesting practices on all classes of lands in the immediate future.

SELECTED REFERENCES

Anonymous, "Feeding Men in Camps," Natl. Business Pubs., Ltd., Gardenvale, Quebec, Canada, 1936.

Anonymous, "Logger's Handbook," Pacific Logging Congress, Portland, Oreg., 1937.

Anonymous, "Forest Practice Handbook," West Coast Lumbermen's Association and Pacific Northwest Loggers Association, 1937.

Anonymous, "Michigan Log Marks," *Mem. Bul.*, 4, Mich. State Col. and Agr. Exp. Sta., 1942.

Anderson, J. V., "Application of Selective Logging to a Ponderosa Pine Operation in Western Montana," *State Univ. Mont. Studies*, No. 2, 1935.

Bentley, A. W., J. D. B. Harrison, and B. F. Avery, "Organization and Administration of Woods Operations," Woodlands Section, Canadian Pulp and Paper Association, 1938.

Brandstrom, Axel J. F., "Analysis of Logging Costs and Operating Methods in the Douglas Fir Region," Charles Lathrop Pack Forestry Foundation, 1933.

Brown, Nelson Courtlandt, "Logging—Principles and Practices," John Wiley & Sons, Inc., New York, 1934.

Brown, Nelson Courtlandt, "Logging—Transportation," John Wiley & Sons, Inc., New York, 1938.

Defebaugh, J. E., History of the lumber industry of America, *Amer. Lumberman*, 1906.

Garver, R. D., and Raphael Zon, "Selective Logging in the Northern Hardwoods of the Lake States," *U.S. Dept. Agr., Tech. Bul.*, 164, 1930.

Garver, R. D., and C. F. Korstian, "Selective Logging in the Loblolly Pine–Hardwood Forests of the Middle Atlantic Coastal Plain with Special Reference to Virginia," *Va. Forest Serv. Pub.*, 43, 1931.

Garver, R. D., and R. H. Miller, "Selective Logging in the Shortleaf and Loblolly Pine Forests of the Gulf States Region," *U.S. Dept. Agr., Tech. Bul.*, 375, 1933.

Hilton, C. Max, Rough pulpwood operating in northwest Maine, 1935–1940, *Maine Bul.* vol. XLV, No. 1, Univ. Maine Studies, Second Ser., No. 57, 1942.

Horn, Stanley F., "This Fascinating Lumber Business," Bobbs-Merrill Company, Indianapolis, 1943.

Hoyle, Raymond J., "Harvesting and Marketing Timber in New York," *Bul.* 1-*a*, vol. XI; N. Y. State Col. Forestry at Syracuse Univ., 1938.

Koroleff, Alexander M., and Ralph Clement Bryant, "The Transportation of Wood in Chutes," *Yale Univ. School Forestry Bul.*, 34, 1932.

Koroleff, A., "Pulpwood Cutting," Woodlands Section, Canadian Pulp and Paper Association, 1941.

Koroleff, A., J. F. Walker, and D. R. Stevens, "Pulpwood Hauling with Horse and Sleigh," Woodlands Section, Canadian Pulp and Paper Association, 1943.

Nix, L. A., "Woods Labour," Woodlands Section, Canadian Pulp and Paper Association, 1939.

Simmons, Fred C., "Northeastern Loggers' Handbook," U.S. Forest Service, Northeastern Forest Experiment Station, 1945.

Stovall, Charles, A manual for fallers, buckers, scalers, and bullbuckers, *Timberman*, Portland, Oreg., 1943.

Townsend, C. R., "The Motor Truck in Woods Operations," Woodlands Section, Canadian Pulp and Paper Association, 1937.

Townsend, C. R., "Mechanized Logging," Woodlands Section, Canadian Pulp and Paper Association, 1938.

Van Tassel, Alfred J., "Mechanization in the Lumber Industry," Report M-5, Work Projects Administration, National Research Project, 1940.

Worthington, Robert E., "Costs of Tractor Logging in Southern Pine," *U.S. Dept. Agr., Tech. Bul.*, 700, 1939.

INDEX

O

P

R

S